Texts in Applied Mathematics 40

Springer
New York
Berlin
Heidelberg
Hong Kong
London
Milan
Paris
Tokyo

Texts in Applied Mathematics

1. *Sirovich:* Introduction to Applied Mathematics.
2. *Wiggins:* Introduction to Applied Nonlinear Dynamical Systems and Chaos, 2nd ed.
3. *Hale/Koçak:* Dynamics and Bifurcations.
4. *Chorin/Marsden:* A Mathematical Introduction to Fluid Mechanics, 3rd ed.
5. *Hubbard/West:* Differential Equations: A Dynamical Systems Approach: Ordinary Differential Equations.
6. *Sontag:* Mathematical Control Theory: Deterministic Finite Dimensional Systems, 2nd ed.
7. *Perko:* Differential Equations and Dynamical Systems, 3rd ed.
8. *Seaborn:* Hypergeometric Functions and Their Applications.
9. *Pipkin:* A Course on Integral Equations.
10. *Hoppensteadt/Peskin:* Modeling and Simulation in Medicine and the Life Sciences, 2nd ed.
11. *Braun:* Differential Equations and Their Applications, 4th ed.
12. *Stoer/Bulirsch:* Introduction to Numerical Analysis, 3rd ed.
13. *Renardy/Rogers:* An Introduction to Partial Differential Equations.
14. *Banks:* Growth and Diffusion Phenomena: Mathematical Frameworks and Applications.
15. *Brenner/Scott:* The Mathematical Theory of Finite Element Methods, 2nd ed.
16. *Van de Velde:* Concurrent Scientific Computing.
17. *Marsden/Ratiu:* Introduction to Mechanics and Symmetry, 2nd ed.
18. *Hubbard/West:* Differential Equations: A Dynamical Systems Approach: Higher-Dimensional Systems.
19. *Kaplan/Glass:* Understanding Nonlinear Dynamics.
20. *Holmes:* Introduction to Perturbation Methods.
21. *Curtain/Zwart:* An Introduction to Infinite-Dimensional Linear Systems Theory.
22. *Thomas:* Numerical Partial Differential Equations: Finite Difference Methods.
23. *Taylor:* Partial Differential Equations: Basic Theory.
24. *Merkin:* Introduction to the Theory of Stability of Motion.
25. *Naber:* Topology, Geometry, and Gauge Fields: Foundations.
26. *Polderman/Willems:* Introduction to Mathematical Systems Theory: A Behavioral Approach.
27. *Reddy:* Introductory Functional Analysis with Applications to Boundary-Value Problems and Finite Elements.
28. *Gustafson/Wilcox:* Analytical and Computational Methods of Advanced Engineering Mathematics.
29. *Tveito/Winther:* Introduction to Partial Differential Equations: A Computational Approach.
30. *Gasquet/Witomski:* Fourier Analysis and Applications: Filtering, Numerical Computation, Wavelets.

(continued after index)

Fred Brauer Carlos Castillo-Chávez

Mathematical Models in Population Biology and Epidemiology

With 117 Illustrations

Springer

Fred Brauer
Department of Mathematics
University of British Columbia
121-1984 Mathematics Road
Vancouver, British Columbia V6T 1Z2
Canada
brauer@math.ubc.ca

Carlos Castillo-Chávez, Director
Mathematical and Theoretical Biology Institute
Professor of Biomathematics
Dept. of Theoretical and Applied Mechanics
317 Kimball Hall
and
Biometrics Department
431 Warren Hall
Cornell University
Ithaca, NY 14853
USA
cc32@cornell.edu

Series Editors

J.E. Marsden
Control and Dynamical Systems, 107–81
California Institute of Technology
Pasadena, CA 91125
USA

L. Sirovich
Division of Applied Mathematics
Brown University
Providence, RI 02912
USA

M. Golubitsky
Department of Mathematics
University of Houston
Houston, TX 77204-3476
USA

Mathematics Subject Classification (2000): 92-01, 92Bxx, 92D40

Library of Congress Cataloging-in-Publication Data
Brauer, Fred.
Mathematical models in population biology and epidemiology/Fred Brauer, Carlos Castillo-Chávez.
p. cm. — (Texts in applied mathematics; 40)
Includes bibliographical references (p.).
ISBN 978-1-4419-3182-5
1. Population biology—Mathematical models. 2. Epidemiology—Mathematical models.
I. Castillo-Chávez, Carlos. II. Title. III. Series.
QH352.B73 2000
577.8'8'015118—dc21 00-045033

Printed on acid-free paper.

Production managed by Frank McGuckin; manufacturing supervised by Jeffrey Taub.
Camera-ready copy prepared from the authors' LaTeX2e files using Springer's svsing2e.sty macro.

Printed in the United States of America.

9 8 7 6 5 4 3 2

Springer-Verlag New York Berlin Heidelberg
A member of BertelsmannSpringer Science+Business Media GmbH

Series Preface

Mathematics is playing an ever more important role in the physical and biological sciences, provoking a blurring of boundaries between scientific disciplines and a resurgence of interest in the modern as well as the classical techniques of applied mathematics. This renewal of interest, both in research and teaching, has led to the establishment of the series: *Texts in Applied Mathematics (TAM)*.

The development of new courses is a natural consequence of a high level of excitement on the research frontier as newer techniques, such as numerical and symbolic computer systems, dynamical systems, and chaos, mix with and reinforce the traditional methods of applied mathematics. Thus, the purpose of this textbook series is to meet the current and future needs of these advances and encourage the teaching of new courses.

TAM will publish textbooks suitable for use in advanced undergraduate and beginning graduate courses, and will complement the *Applied Mathematical Sciences (AMS)* series, which will focus on advanced textbooks and research level monographs.

Pasadena, California
Providence, Rhode Island
Houston, Texas

J.E. Marsden
L. Sirovich
M. Golubitsky

Preface

This book is intended to inspire students in the biological sciences to incorporate mathematics in their approach to science. We hope to show that mathematics has genuine uses in biology by describing some models in population biology and the mathematics that is useful in analyzing them, as well as some case studies representing actual, if somewhat idealized, situations. A secondary goal is to expose students of mathematics to the process of modeling in the natural and social sciences.

A realistic background in mathematics for studying this book is a year of calculus, some background in elementary differential equations and a little matrix theory. The mathematical treatment is based less on techniques for obtaining explicit solutions in "closed form," to which students in elementary mathematics courses may be accustomed than on approximate and qualitative methods. The emphasis is on describing the mathematical results to be used and showing how to apply them, rather than on detailed proofs of all results. References where proofs may be found are given. Our hope is that students in the biological sciences will cover enough mathematics in their first two years of University to make this book accessible in the third or fourth year.

For many problems, the use of a computer algebra system can give many insights into the behavior of a model, especially for generating graphic representations of solutions. Some of the exercises and projects in the book either require the use of a computer algebra system or are simplified considerably by one. At this writing, Mathematica and Maple are two widely used systems; students should become proficient in using at least one of them. In addition, the more specialized dynamical systems program XPP or its Windows version WinPP is very useful for studying dynamical systems and is especially valuable for differential-difference equations and equations with time lags. This program, created by Bard Ermentrout, may be downloaded from

http: //ftp.math.pitt.edu/pub/bardware/winpp.zip

The more elaborate version XPP may be run under Windows on an X-Windows server and may be downloaded from

http: //ftp.math.pitt.edu/pub/bardware/xpp4w32.zip

There are some topics that in earlier times would have been appendices in this book. These include some mathematical ideas such as Taylor approximation and the elements of linear algebra, and also some programs for solving problems with Maple, and WinPP. These topics, which we consider as the components of a virtual appendix and some detailed solutions to selected exercises will be found in the form of PDF files at a future web site.

Answers to selected exercises are still given in the book as an appendix. Genuine understanding of the material in the book requires working of exercises; this is not a spectator sport. Answers in the back of the book are to be used to *check* your work, not to lead you to the solution.

In addition to exercises, there are several more extended descriptions of models, which call for readers to fill in some gaps. These are designated as Projects, and they may be given as group assignments.

The book concentrates on population biology. One of the practical sides of population biology is resource management. Another aspect is the study of structured population models. Mathematical epidemiology is an example, with populations structured by disease status. The core of the book, which should be included in any beginning modeling course, is Chapters 1, 2, 4, and the first five sections of Chapter 5. These chapters cover elementary continuous and discrete models for single species populations and interacting populations. They include examples and exercises that may be too simplistic for more experienced students, who may progress through this material a little more rapidly than beginners. The first five sections of Chapter 7, on mathematical epidemiology, are also on a relatively elementary level and may be studied by students with relatively little background. Chapter 3, on continous models with delays, Chapter 6 on harvesting and its implications in resource management, and Chapter 8, on population models with age structure, as well as the later sections in chapters 5 and 7 are more demanding mathematically. This material should probably be reserved for students with more mathematical background and some experience in biology.

The bibliography includes not only the books and papers to which reference is made in the text but also related works which pursue further some of the topics in the book. The book is meant to be an introduction to the principles and practise of mathematical modeling in the biological sciences, one which will start students on a path; it is certainly not the last word on the subject.

Vancouver, B.C., Canada
Ithaca, N.Y., U.S.A.
December 2000

Fred Brauer
Carlos Castillo-Chavez

Acknowledgements

The book is based primarily on notes growing out of lectures given by Fred Brauer in a modeling course at the University of Wisconsin, Madison, beginning in 1981. It reflects the input and criticism, not always followed, of faculty, colleagues, friends and students from biology and mathematics, including those who took Brauer's course. Carlos Castillo-Chavez used the notes in his Mathematical Ecology course, a course for students in the natural and mathematical sciences that he has taught or co-taught for the last 10 years at Cornell University. Castillo-Chavez also used preliminary versions of this manuscript over the last four summers as part of a core intensive seven-week research undergraduate experience (REU) in mathematical and theoretical biology at Cornell's Mathematical and Theoretical Biology Institute (MTBI). MTBI faculty and students contributed to and influenced the content and philosophy of this book. Castillo-Chavez has used portions of this manuscript in the undergraduate differential equations course that he has taught over the last two years at Cornell University's Department of Theoretical and Applied Mechanics. Several of the projects included in this book resulted from undergraduate research carried out at MTBI's Cornell-SACNAS REU program. This research was co-directed by Castillo-Chavez and Abdul-Aziz Yakubu.

The contributions of Stephen P. Blythe, Carlos Castillo-Garsow, Carlos Hernández Suarez, Maia Martcheva, Ricardo Saenz, Baojun Song, Steve Tennenbaum, and Jorge Velasco-Hernández are too many to list. In particular, Section 4.9 and Section 8.4 are modified versions of some of the lectures that Maia Martcheva gave at MTBI and the index was created by Carlos Castillo-Garsow.

Héctor Miguel Cejudo Camacho devoted a considerable amount of time and energy to the drawing or redrawing of most of the figures in this book and he helped put the manuscript into Springer format. In the course of a sequence of revisions Ricardo Oliva facilitated the transfer of files between authors and his expertise with both text and figures helped us turn the various pieces of text and figures into what we hope is a coherent book.

We would also like to acknowledge some individuals who have taught us directly or indirectly, or who have shaped in many ways the material in this book. They are: Zvia Agur, Roy Anderson, Viggo Andreasen, Juan Aparicio, Sally Blower, Stavros Busenberg, Angel Capurro, Colin Clark, Charles Conley, Kenneth Cooke, Jim Cushing, Odo Diekmann, Klaus Dietz, Leon Glass, Zhilan Feng, Karl Hadeler, Tom Hallam, Hans Heesterbeek, Wenzhang Huang, Herbert Hethcote, John Jacquez, Simon Levin, Jia Li, Donald Ludwig, Leah Keshet, Denise Kirschner, Christopher Kribs, Tom Kurtz, Michael Mackey, Robert May, Charles McCulloch, Roger Nisbet, Joel Robbin, David Sánchez, Lisa Sattenspiel, Lee Segel, Carl Simon, Hal Smith, Avrum Soudack, Steve Strogatz, Horst Thieme, Pauline van den Driessche, Julio Villarreal, Paul Waltman, Steve Wirkus, Gail Wolkowicz, Abdul-Aziz Yakubu, and others.

We want to thank several anonymous reviewers for their input as well as those members of the editorial board of the Springer-Verlag series of Texts in Applied Mathematics who were involved in the oversight of this book. We thank the editorial staff of Springer-Verlag for their patience and understanding in the preparation of this book. In particular, we would like to recognize the efforts of Frank McGuckin, Springer-Verlag's Production Editor who kept this project alive despite our failures in meeting self-imposed deadlines. We also thank Kathy McKenzie for her editing services. Springer-Verlag Senior Editor, Achi Dosanjh encouraged and supported this project from the beginning. We are grateful for her patience and great sense of humor.

Finally, we would like to remind the reader that despite everybody's support in the production of this book, at the end, the responsibility for the final product is entirely ours.

Contents

Prologue: On Population Dynamics

As the world population exceeds the 6 billion mark the question "How many people can the earth support and under what conditions?" becomes at least as pressing as when Malthus (1798) posed it at the end of the 18th century in *An Essay on the Principle of Population.*

The likelihood of whether or not it is possible to support growing populations within existing economic systems and environments has been one of the main concerns of societies throughout history. *How Many People Can the Earth Support?* is a recent book by J.E. Cohen (1995) in which he tackles from historic and scientific perspectives possible responses to this question. Historical "solutions" to the question of overpopulation have had as their basis two underlying assumptions: First, that under constant positive per capita rates of population growth a population increases exponentially, that is, population "explosion" is observed; second, that resource limitations necessarily limit or control the magnitude of such an explosion. The uselfuness and validity of both assumptions are naturally limited as the environment, often called the "carrying capacity," does not remain fixed. Per capita rates of population growth are not fixed, but are functions of changing environments. A limiting factor in the development of a useful (in both practical and theoretical terms) theory of population dynamics lies in the inability of theoreticians to provide models and frameworks with environmental plasticity. The environmental landscape in which we live is dynamic and often experiences dramatic shifts due to technological innovations (such as birth control, disease, famine, and war) that periodically alter the bounds of what we think is possible. Cohen observes that population patterns and hence "growth" rates depend on our scale of observation in both time and space. By some scales, they are definitely not constant. For example, in the 14th century, Cohen notes that repeated waves of Black Death, a form of bubonic plague, together with wars, heavy taxes, insurrections and poor and sometimes malicious governments ... killed an estimated one third of the population living in India and Iceland, and that the population of Meso-America fell by perhaps 80 percent or 90 percent during the sixteenth century. [pp 38-41]. However, in spite of such sharp short-term decreases, world population size has actually grown steadily since prehistoric times, although not at a constant rate.

Local human populations have exhibited wide fluctuations throughout time and their growth may still be responding to environmental changes in modern times. Local variability in population growth rates is high in wars and during epidemics and famine. It may be affected dramatically by advances in housing, agricultural practices, health care, and so on. In retrospect, it is not surprising to see dramatic changes in the carrying capacity of the earth over time, because these changes are driven by strong environmental shifts and events that include: the effects of the global agricultural revolution observed from the 17th through the 19th centuries; the public health transformation experienced over the last five decades via the widespread use of antibiotics and the implementation of large-scale vaccination policies; and the fertility revolution of the last four decades due to the global availability of birth control measures (sometimes having a dramatic impact on per capita birth rates, as in China). Further, improvements in the economic state and declines in mortality from diseases in developing countries have often led to substantial declines in the birth rate. Since it is no longer necessary for a family to have as many children to assure the survival of enough children to care for their parents in old age, such improvements in the quality of life may lead to decreases in the rate of population growth. Hence, predicting how many individuals the earth can support becomes a rather complex problem with no simple answers particularly when different definitions of quality of life (e.g., Bangladesh as compared to Germany) are considered.

Questions and challenges raised by complex demographic processes may be addressed practically and conceptually through the use of mathematical models. Models may be particularly valuable when interactions with demographers, sociologists, economists, and health experts are at the heart of the model-building process. Simple models cannot by their own nature incorporate simultaneously many of the factors described above. However, they often provide useful insights, as will be shown in the following chapters, to help our understanding of complex processes. The usefulness of simple models to predict is limited and their use often may lead to misleading results in the hands of "black box" users. Simple population models such as the Malthus (exponential) and the Verhulst (logistic) models represent a natural starting point in the study of demographic processes. Their main role here is to help in our *understanding* of the dynamics of basic *idealized* demographic phenomena in the social and natural sciences.

If the population of individuals at time t is denoted by $x(t)$ then Malthus' law (1798) arises from the solution of the initial value problem

$$\frac{dx}{dt} = rx, \qquad x(0) = x_0,$$

where $r = b - \mu$ denotes the constant per capita growth rate of the population, that is, the average per person number of offspring b less the per person average number of deaths μ per unit of time, and $x_0 > 0$ denotes the initial population size. Since Δx denote the change in population from t to $t + \Delta$, the dynamics are approximated over a short time period by

$$\Delta x(t) \approx (\text{births in } (t, t+\Delta)) - (\text{deaths in } (t, t+\Delta))$$

or under our simplistic modeling assumptions by

$$\Delta x(t) \approx b\; x(t)\Delta - \mu\; x(t)\Delta = (b - \mu)x(t)\Delta = rx(t)\Delta.$$

In one of the most influential papers in history a variant of this model was introduced by Malthus in 1798. The assumption of a constant per capita growth rate leads to the solution $x(t) = x_0 e^{rt}$, which predicts population explosion if $r > 0$, extinction if $r < 0$, or no change if $r = 0$. This model may be useful in situations when the environment is not being taxed, the time scale of observation is small enough to make it acceptable to assume that r remain nearly constant, resources appear to be unlimited, and x_0 is small. This is a reasonable model in estimating the rate of growth of a parasite when first introduced in the blood stream of an individual (such as the malaria parasite), in the study of the rate of growth in the number of new cases of infection at the *beginning* of an epidemic, in the estimation of the rate of growth of a pest that has just invaded a field, in estimating the rate of decay of the effect of a drug (antibiotic) in the blood stream of an individual, or in estimating the rates of extinction of endangered species. The model may be inadequate when the number of generations gets large enough for other factors, such as density dependence to come into play. The assumption of a constant $r > 0$ implies that a generation not only replaces itself over its life span but also contributes to the growth of its population generation after generation, while the assumption of a constant $r < 0$ implies that generations do not contribute in a significant manner to the future of a population, that is, generations are not capable of replacing themselves. An alternative way of thinking about this demographic process is via the *basic reproductive number* or *ratio* R_0. This dimensionless quantity is used to denote the average number of offspring produced by a "typical" member of the population during its reproductive life when resources are unlimited which typically occurs when x_0 is small. Here $R_0 = b/\mu$, and if $R_0 > 1$, the population will grow, while if $R_0 < 1$ the population will eventually go extinct. The case $r = 0$ or $R_0 = 1$ represents stasis, that is, a situation where each individual on the average replaces itself before it dies and the population size on the average will not change. The case $r = 0$ represents a transition from $r < 0$ to $r > 0$ (or from $R_0 < 1$ to $R_0 > 1$), that is, from population decay to population explosion and vice versa. It is common to see that as a parameter–here the per capita growth rate–crosses a "tipping" or "threshold" value, the population dynamics change drastically from a situation in which we have population extinction to that where we observe population explosion (and vice versa).

The acknowledgement of the existence of finite resources defined by the carrying capacity of an ecosystem demands the introduction of models that cannot support exponential growth indefinitely. The simplest version is obtained when it is assumed that the per capita growth rate G depends on the size of the population. In mathematical terms, we have the model

$$\frac{dx}{dt} = xG(x), \qquad x(0) = x_0.$$

The most common example is provided by the logistic equation (introduced by Verhulst (1838, 1845) in which $G(x)$ is assumed to be a linear function often parametrized as

$$G(x) = r\left(1 - \frac{x}{K}\right).$$

The ideas embodied in the logistic model (exponential growth when x is small compared to K; little change when x is near K) led biologists to the formulation of a theory that characterizes environments in terms of those that "favor" r selection, the exponential component of growth, versus those that "favor" K selection, where K is a measure of its carrying capacity. Environments were classified as roughly belonging to two types: those that favored growth versus those that did not (because they were near their carrying capacity). The concept of r-K selection has been applied not only to populations of individuals but also to populations of species (environments in the r-phase can support additional species while environments in the K-phase cannot; see for example [May (1974)].

In order to give demographic meaning to the definition of $G(x)$, it is often convenient to redefine it as the sum of two functions,

$$G(x) = -\mu(x) + \beta(x)$$

where $\mu(x)$ and $\beta(x)$ denote the per capita mortality and fertility rates, respectively. A constant per capita mortality rate μ may be appropriate in situations where birth and other processes respond rapidly to changes in the population dynamics while it may not be appropriate when population death rates are affected by demographic factors such as population density.

Thus limited growth may be obtained when the per capita birth rate decreases as the population increases under the assumption of a constant per capita death rate. The simplest mathematical form for a decreasing per capita birth rate is the linear function

$$\beta(x) = a - bx,$$

which leads to a per capita growth rate

$$G(x) = \beta(x) - \mu = (a - bx) - \mu = (a - \mu) - bx,$$

or

$$G(x) = (a - \mu)\left(1 - \frac{b}{a - \mu}x\right),$$

which after renaming parameters

$$r = a - \mu, \qquad K = \frac{a - \mu}{b},$$

takes on the familiar logistic form

$$G(x) = rx\left(1 - \frac{x}{K}\right).$$

Malthus' model and the logistic model describe the dynamics of populations with overlapping generations, the kind that are best described using differential equations. However, the dynamics of some populations may not be appropriately described with differential equations. For example, salmon have an annual spawning season and births take place at essentially the same time each year. Systems with non-overlapping generations, like the salmon, are better described by discrete (difference equation) models, the topic of Chapter 2.

Impact on biological growth rates is not always instantaneous. In fact, it is often experienced after some delay (an egg stage duration, for example) and for these situations, the use of a time lag is often appropriate. The use of lags (or more generally, distributed delays) leads to the study of differential-difference equation models, the topic of Chapter 3.

The material in the first section will help the reader gain understanding of the dynamics of single species models. The material throughout the book should help the modeler and scientist understand the value of working with simple analyzable "realistic" models rather than with detailed "unsolvable" models.

Part I, therefore, focuses on the study of single species models including those commonly used to predict the growth of human and animal populations. Through a dissection of the underlying assumptions behind each model, it is possible to determine its usefulness as well as its limitations. Hence, single population models are the building blocks for additional detail such as population structure, the subject of Part III.

One-dimensional models assume implicitly that population growth is affected by intraspecies competition (competition such as for resources among members of the same species) in its various forms, including contest and scramble competition. Models with overlapping generations (here modeled via continuous time one-dimensional ordinary differential equations) exhibit simple dynamical behavior while models with discrete, nonoverlapping generations, may exhibit complex dynamics. Specific time predictions are nearly impossible even when plenty of data is available as is the case in the lake eutrophication example developed in Chapter 1. Nevertheless, one-dimensional models help clarify the world of possibilities once the emphasis has been shifted from a desire to fit specific (quantitative) dynamics to that of simply studying their qualitative behavior as *parameters are varied.* It should be clear that because most models of Part I are parametrized via the use of biologically derived constants that come from a multi-dimensional parameter space, we are in fact dealing with a world of possibilities with "simple" models. The spruce budworm project of Chapter 1 provides a powerful example of this view and, consequently, of the value of one-dimensional continuous time systems.

Rich and complex dynamics are obtained in Part I where the dynamics of discrete time models with nonoverlapping generations are discussed. Simple nonlinear models, like the discrete logistic equation, generate complex dynamics, including chaotic behavior. Hence, the view (dramatically set in a beautiful paper by May, 1976) that *complex dynamics are not necessarily the result of complex rules of interaction* makes a compelling and powerful argument on the search for "simple" explanations for the phenomena observed in biological systems.

Part I focusses on the description and analysis of simple models, including differential, difference, and delay equations, which have played a useful role in theoretical biology (see [May, (1974)]). Models of Part I do not incorporate age structure, nor do they take into account gender-related factors, such as mating, which are often central to the study of the life history of a population. In other words, single species models for homogeneously mixing populations can be used only to address limited (albeit often important) life history questions. Part III looks at the role of population structure on population dynamics in some depth.

Part II of this book looks at the dynamics that result from the interactions of populations that would *typically* behave as those of Part I when in isolation. Mechanisms that drive multispecies interactions include competition, mutualism, and predator-prey interactions. In Part II we often establish conditions for species coexistence (both species survive) or for competitive exclusion (one species survives but the other becomes extinct). The beginning of Chapter 4 is historical in nature as it focuses on the study of the Lotka-Volterra equations which provide the model prototype of predator-prey interactions, and the chemostat, a laboratory biological system used to cultivate bacteria and the subject of intense mathematical study [Smith & Waltman (1995)]. These classical models motivate the introduction of two-dimensional systems, where we reintroduce the ideas of stability and instability (including oscillations) in two-dimensional continuous time systems. Chapter 5 focuses on species competition, predator-prey systems, and mutualism, with a brief foray into the much more complicated situation where more than two species interact. Chapter 6 focuses on intervention and it intersects heavily with the field known as *bioeconomics* (see [Clark (1990)] and references therein).

Chapters 2 and 3 cover limited population structure. In Chapter 2 we look at populations with two age classes, while in Chapter 3 we consider age in various implicit forms. Individuals are allowed to remain in each stage a variable amount of time; in other words, their history plays a role. In this fashion, we are able to move away from models with no history (exponentially distributed waiting times) to models in which the time spent in a particular state is variable. In Part III, we focus on population structure. We introduce structure in two separate ways. We introduce general waiting times in the context of epidemiologically structured models. Chapter 7 classifies individuals according to their epidemiological status and it also incorporates additional heterogeneity through the introduction of variables that measure the variability in an individual's ability to fight infections. In Chapter 8, basic generalizations that incorporate age structure in the Malthus' model in discrete and continuous time are introduced. Chapter 8 focuses on the prototypes of age structured models: the (discrete) Leslie and the (continuous) McKendrick-Von Foerster model.

An additional objective of Part III is to provide an introduction to some of the modeling and mathematical challenges faced in the field of structured populations [Webb (1985); Metz & Diekmann (1986); Diekmann & Heesterbeek (2000)]. Some material is suggested for those individuals interested in applications to realistic biological models which have also motivated interesting mathematics. These generalizations are beyond the scope of this book, but we hope that this book provides a springboard to these topics.

Part I

Simple Single Species Models

1

Continuous Population Models

1.1 Exponential Growth

In this chapter we look at a population in which all individuals develop independently of one another. For this situation to occur these individuals must live in an unrestricted environment where no form of competition is possible. If the population is small then a stochastic model is more appropriate, as the likelihood that the population becomes extinct due to chance must be considered. However, a deterministic model may provide a useful way of gaining sufficient understanding about the dynamics of a population whenever the population is large enough. Furthermore, perturbations to populations at equilibrium often generate on short time scales independent individual responses, which are appropriately modeled by deterministic models. For example, the propagation of a disease in a large population via the introduction of a single infected individual leads to the generation of secondary cases. The environment is free of "intraspecific" competition, at least at the beginning of the outbreak, when a large population of susceptibles provides a virtually unlimited supply of hosts. In short, the spread of disease in a large population of susceptibles may be thought of as an invasion process via independent contacts with a few infectious individuals.

The population size of a single species at time t will be denoted by $x(t)$, where it is assumed that x is an everywhere differentiable, that is, a *smooth* function of t. Although this assumption is unrealistic in the sense that $x(t)$ is an integer-valued function and thus not continuous, for populations with a large number of members this assumption and the assumption of differentiability provide reasonable approximations. In many biological experiments the population biomass, which one might expect to be more nearly described by a smooth function than the population size, is often taken as the definition of $x(t)$.

The rate of change of population size can be computed if the births and deaths and the migration rates are known. A *closed* population has no migration either into or out of the population, that is, the population size changes only through births and deaths and the rate of change of population size is simply the birth rate minus the death rate. A formulation of a specific model requires explicit assumptions on the birth and death rates. Ideally, these assumptions are made with the goal of addressing *specific* biological questions such as under what conditions will intraspecific competition (competition for hosts or patches) lead to coexistence?

For microorganisms, which reproduce by splitting, it is reasonable to assume that the rate of birth of new organisms is proportional to the number of organisms present. In mathematical terms this assumption may be expressed by saying that if the population size at time t is x then, over a short time interval of duration h from time t to time $(t+h)$, the number of births is approximately bhx for some constant b, the *per capita birth rate.* Similarly, we may assume that the number of deaths over the same time interval is approximately μhx for some constant μ, the *per capita death rate.* Hence, the net change in population size from time t to time $(t+h)$, which is $x(t+h)-x(t)$, may be approximated by $[(bh-\mu h)]x(t)$. The duration h of the time interval must be short to ensure that the population size does not change very much and thus that the numbers of births and deaths are approximately proportional to $x(t)$. We obtain the approximate equality

$$x(t+h)-x(t) \approx (b-\mu)x(t)h. \tag{1.1}$$

(The symbol $\approx$ is used to denote approximate equality in a sense that must be specified.) Division by h gives

$$\frac{x(t+h)-x(t)}{h} \approx (b-\mu)x \tag{1.2}$$

and passage to the limit as $h \to 0$ gives

$$\frac{dx}{dt} = (b-\mu)x \tag{1.3}$$

under the assumption that the function $x(t)$ is differentiable. The approximate equality in (1.1) means that the difference between the two sides of (1.1) is so small that the result of dividing this difference by h gives a quantity that approaches zero as $h \to 0$.

If the net growth rate is naturally defined as

$$r \equiv b - \mu$$

then another way of looking at this model is to observe that if the population size at time t is $x(t)$, then in the next *tiny* time interval of length h the *net* increase in population size due to a single organism will be rh. Since all individuals are independent (no competition in an unrestricted environment) then the *net* increase in population due to all $x(t)$ organisms will be $rhx(t)$ and thus we arrive again at the differential equation

$$\frac{dx}{dt} = rx. \tag{1.4}$$

This differential equation has the infinite family of solutions given by the one-parameter family of functions $x(t) = ke^{rt}$; hence, this one parameter family gives a solution of (1.4) for every choice of the constant k. The most convenient way to impose a condition that will describe the population dynamics of a specific population is by specifying the initial population size at time $t = 0$ as

$$x(0) = x_0 \tag{1.5}$$

this choice selects *the* solution, $x(t) = x_0e^{rt}$. Condition (1.5) is called an *initial condition* and the problem consisting of the differential equation (1.4) together with the initial condition (1.5) is called an *initial value problem*.

As pointed out above, the above initial value problem has the unique solution

$$x(t) = x_0e^{rt},$$

where $r > 0$ (or equivalently $b > \mu$) implies that the population size will grow unbounded as $t \to \infty$, while $r < 0$ (or $b < \mu$) implies that the population size will approach zero as $t \to \infty$. An alternative interpretation can be reached using some of the ideas discussed in detail in the Appendix to this chapter. In the absence of births ($b = 0$) the population is deplenished by deaths at the rate μ and, consequently, the average life-span of a member of this population is $1/\mu$ (see Section 1.7). If $b > 0$ then the average number of offspring over the lifetime of an average individual under the Malthus' model would be b/μ. If this ratio (usually referred to as the basic reproductive number or ratio R_0) is greater than one then births exceed deaths and the average number of offspring per person over her lifetime is greater than one, that is, the population explodes; if this ratio is less than one, then deaths exceed births, the average number of offspring per person is less than one, and the population dies out.

The prediction that population size will grow exponentially under these conditions was first stated by Malthus (1798). Malthus predicted disaster as food supplies could not possibly be increased to keep pace with population growth at a constant positive per capita growth rate. Populations that grow exponentially at first are commonly observed in nature. However, their growth rates usually tend to decrease as population size increases. In fact, exponential growth or decay may be considered *typical* local behavior. In other words, populations dynamics can usually be approximated by this simple model only for short periods of time; that is, the dynamics of a population may be handled well locally with linear models. The assumption that the rate of growth of a population is proportional to its size (linear assumption) is usually unrealistic on longer time scales. The next section considers nonlinear assumptions on the rate of population growth rates, which lead to quite different qualitative predictions.

Exercises

We put an asterisk on exercises that may not be straightforward. This notation is used throughout the text.

In Exercises 1 through 10, assume that the rate of change of population size is proportional to population size.

1. Suppose the growth rate of a population is 0.7944 per member per day. Let the population have 2 members on day zero. Find the population size at the end of 5 days.

2. Suppose a population has 100 members at $t = 0$ and 150 members at the end of 100 days. Find the population at the end of 150 days.

3. Suppose the growth rate of a given population is 0.21 per member per day. If the population size on a particular day is 100, find the population size 7 days later.

4. Suppose a population has 39 members at $t = 8$, and 60 members at $t = 12$. What was the population size at $t = 0$?

5. Suppose a population has 24 members at $t = 5$, and 15 members at $t = 15$. What the populations size at $t = 0$?

6. The population of the earth was about 5×10^9 in 1986. Use an exponential growth model with the rate of population increase of 2 percent per year observed in 1986 to predict the population of the earth in the year 2000.

7. Bacteria are inoculated in a petri dish at density of 10/ml. The bacterial density doubles in 20 hours. Assume that this situation is described by the differential equation

$$\frac{d}{dt}x = Cx,$$

where x is the bacterial density and C is a constant.

   (a) Integrate this equation giving x as a function of time.

(b) Find the value of C.

(c) How long does it take for the density to increase to 8 times its original value? To 10 times?

8. Suppose that a population has a constant growth rate r per member per unit time and that the population size at time $t = t_0$ is x_0. Show that the population size at time t is $x_0 e^{r(t-t_0)}$.

9. Suppose that a population has a growth rate of r per member in unit time, with $r > 0$. Show that the time required for the population to double its initial size (called the *doubling time*) is $(\log 2)/r$. [*Note*: We will always use "log" to denote the natural logarithm.]

10. Suppose that a population has a growth rate of r per member per unit time with $r < 0$. Show that the time required for the population to decrease to half its initial size (called the *half life*) is $-(\log 2)/r$.

11. The method of least squares finds the "best" straight line $y = ax + b$ through a set of data points $(x_1, y_1), (x_2, y_2), \ldots, (x_n, y_n)$ by choosing a and b to minimize

$$\sum_{i=1}^{n} (y_i - (ax_i + b))^2,$$

the sum of the squares of the vertical distances between the data and the line. The solution is

$$a = \frac{\sum_{i=1}^{n} (x_i - \bar{x})(y_i - \bar{y})}{\sum_{i=1}^{n} (x_i - \bar{x})^2}, \quad b = \bar{y} - \hat{a}\bar{x},$$

where

$$\bar{x} = \frac{1}{n}\sum_{i=1}^{n} x_i, \quad \bar{y} = \frac{1}{n}\sum_{i=1}^{n} y_i.$$

Table 1.1 gives the census data of the United States from 1790 to 1990. Assume that these data fit an exponential growth model $x(t) = x_0 e^{rt}$, or $\log x(t) = \log x_0 + rt$. Use the method of least squares on these data to estimate r. *Hint*: Here $\log x(t)$ corresponds to y and t corresponds to x.

Year	Population
1790	3,900,000
1800	5,300,000
1810	7,200,000
1820	9,600,000
1830	12,900,000
1840	17,100,000
1850	23,100,000
1860	31,400,000
1870	38,600,000
1880	50,200,000
1890	62,900,000

Year	Population
1900	76,000,000
1910	92,000,000
1920	105,700,000
1930	122,800,000
1940	131,700,000
1950	150,700,000
1960	179,000,000
1970	205,000,000
1980	226,500,000
1990	248,700,000

TABLE 1.1. Census data of US from 1790 to 1990.

12*. Suppose that a population has a growth rate $r(t)$, which depends on time, so that population size $x(t)$ is governed by the initial value problem.

$$\frac{dx}{dt} = r(t)x, \quad x(0) = x_0.$$

Show that the population size at time t is given by

$$x(t) = x_0 e^{\int_0^t r(s)ds}.$$

1.2 The Logistic Population Model

As before, $x(t)$ denotes the size of a population at time t, and dx/dt, or $x'(t)$ the rate of change of population size. We shall continue to assume that the population growth rate depends only on the population's size. Such an assumption appears to be reasonable for simple organisms such as micro-organisms. For more complicated organisms like animals or humans this is obviously an over-simplification as it ignores intra-species competition for resources as well as other significant factors, including age structure (the mortality rate may depend on age rather than on population density, while the birth rate may depend on the adult population size rather than on total population size). Furthermore, the possibility that birth or death rates may be influenced by the size of populations that interact with the population under study must also be considered (competition, predation, mutualism). We shall consider the effects of some of these factors in later chapters.

Here we study models in which the growth rate depends only on population size, because, in spite of their shortcomings, these models do predict the qualitative behavior of many real populations. The *per capita growth rate*, or rate of growth per member, is given by $x'(t)/x(t)$, which we are assuming is a function of $x(t)$. In the last section it was assumed that the total growth rate was proportional to population size (a linear model) or, equivalently, we took a constant per capita growth rate. In this section total growth rates that decrease as population size increases are considered.

The simplest population model in which the per capita growth rate is a decreasing function of population size is $\lambda - ax$. This assumption leads to the *logistic* differential equation

$$x' = x(\lambda - ax),$$

first introduced by Verhulst (1838) and later studied further by R. Pearl and L. J. Reed (1920). This equation is commonly written in the form

$$x' = rx\Big(1 - \frac{x}{K}\Big), \tag{1.6}$$

with parameters $r = \lambda$, $K = \lambda/a$. The parameters r and K, assumed positive, may then be given biological significance. It is observed that $x' \approx rx$ when x is small, and that $x' = 0$ when x is near K. In other words, when x is small the population experiences exponential growth, while when x is near K the population hardly changes.

Separation of variables allows us to rewrite equation (1.6) as

$$\int \frac{dx}{x(K-x)} = \frac{r}{K}\int dt,$$

while using partial fractions, giving

$$\frac{1}{x(K-x)} = \frac{1}{K}\Big(\frac{1}{x} + \frac{1}{K-x},\Big)$$

allows us to integrate it:

$$\begin{aligned}\frac{r}{K}t + c &= \int \frac{dx}{x(K-x)} = \frac{1}{K}\Big(\int \frac{dx}{x} + \int \frac{dx}{K-x}\Big)\\ &= \frac{1}{K}\big(\log x - \log(K-x)\big),\end{aligned}$$

where c is the constant of integration.

If the population size at time $t = 0$ is x_0, substitution of the initial condition $x(0) = x_0$ gives

$$c = \frac{1}{K}\big(\log x_0 - \log(K - x_0)\big).$$

We now have

$$\frac{1}{K}(\log x - \log(K-x)) = \frac{r}{K}t + \frac{1}{K}(\log x_0 - \log(K-x_0))$$

$$\log\left(\frac{x}{K-x}\right) = rt + \log\frac{x_0}{K-x_0}$$

$$\log\frac{x(K-x_0)}{x_0(K-x)} = rt$$

$$\frac{x(K-x_0)}{x_0(K-x)} = e^{rt}.$$

Further algebraic simplification gives

$$x(K-x_0) = x_0(K-x)e^{rt} = Kx_0e^{rt} - xx_0e^{rt}$$

$$x\left(K - x_0 + x_0e^{rt}\right) = Kx_0e^{rt}$$

and finally

$$x(t) = \frac{Kx_0e^{rt}}{K - x_0 + x_0e^{rt}} = \frac{Kx_0}{x_0 + (K-x_0)e^{-rt}}. \tag{1.7}$$

The above solution is valid only if $0 < x_0 < K$ so that the logarithms obtained in the integration are defined. To obtain the solution without this restriction our integration should have given logarithms of absolute values. Nevertheless, the formula (1.7) for the solution of the logistic equation is valid for all x_0, as could be verified by a more careful analysis.

The expression (1.7) for the solution of the logistic initial value problem shows that the population size $x(t)$ approaches the limit K as $t \to \infty$ if $x_0 > 0$. The value K is called the *carrying capacity* of the population because it represents the population size that available resources can continue to support. The value r is called the *intrinsic growth rate* because it represents the per capita growth rate achieved if the population size were small enough to ensure negligible resource limitations. The logistic model predicts rapid initial growth for $0 < x_0 < K$, then a decrease in growth rate as time passes so that the size of the population approaches a limit. (Figure 1.1). This behavior is in agreement with the observed behavior of many populations, and for this reason, the logistic model is often used as a means of describing population size.

Example 1. Census data for the population of the United States in millions fit the function

$$x(t) = \frac{265}{1 + 69e^{-0.03t}}$$

reasonably well, with the year 1790 taken as $t = 0$. We may compare this with (1.7) in the form

$$x(t) = \frac{K}{1 + \left(\frac{K-x_0}{x_0}\right)e^{-rt}}$$

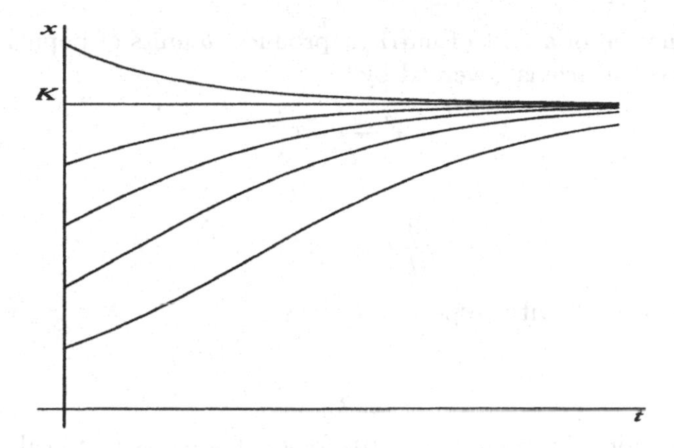

FIGURE 1.1. Solution of the logistic equation.

to see that this expression is a solution of the logistic model with $K = 265$, $r = 0.03$. If we use this logistic model to describe the population of the United States we would predict a carrying capacity of 265,000,000, and a 1990 census total of 226,300,000. (In fact, the population size found in the 1990 census was approximately 250,000,000.) The data from Table 1.1 plotted together with this solution are shown in Figure 1.2.

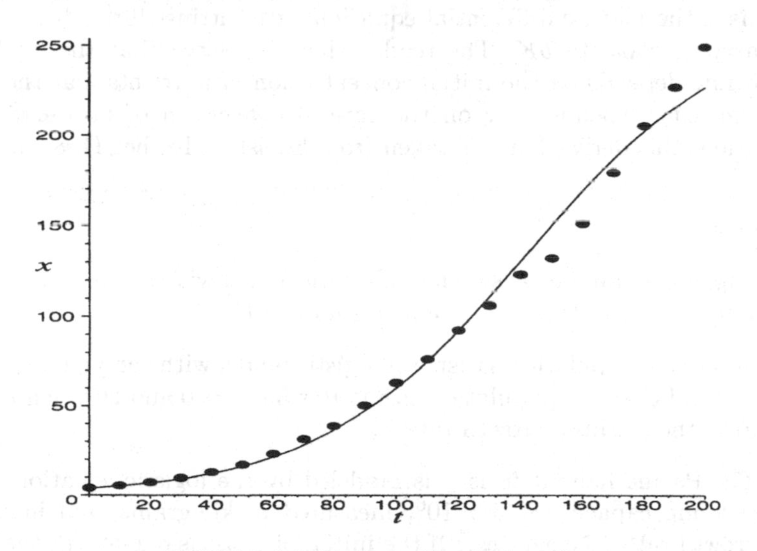

FIGURE 1.2. U. S. A. Population Size

It is possible to give a derivation of the logistic model based on a specific assumption about the resources on which a population depends. Let C denote the concentration of nutrients and assume that the per capita growth rate r is proportional to C, that is, $r = aC$ for some constant a. Assume also that we begin with a fixed concentration $C(0)$ of nutrients and that the consumption of a unit of nutrient produces b units of population size. Then population size is governed by

$$x' = aCx \tag{1.8}$$

with

$$\frac{1}{b}x' = -C'. \tag{1.9}$$

Integration of (1.9) with respect to t gives

$$C = -\frac{1}{b}x + K$$

and substitution of $x = x_0$, $C = C(0)$ for $t = 0$ enables us to calculate the constant of integration K as

$$K = C(0) + \frac{x_0}{b}.$$

Now (1.8) becomes

$$x' = ax\left(K - \frac{1}{b}x\right) = aKx\left(1 - \frac{x}{bK}\right)$$

and this is the logistic differential equation with intrinsic growth rate aK and carrying capacity bK. The reader should observe that the intrinsic growth rate depends on the initial concentration of nutrients, but the carrying capacity depends only on the rate of conversion of nutrients into population (this derivation was taken from Edelstein-Keshet (1988)).

Exercises

1. Suppose a population satisfies a logistic model with $r = 0.4$, $K = 100$, $x(0) = 5$. Find the population size for $t = 10$.

2. Suppose a population satisfies a logistic model with carrying capacity 100 and that the population size is 10 when $t = 0$ and 20 when $t = 1$. Find the intrinsic growth rate.

3. The Pacific halibut fishery is modeled by the logistic equation with carrying capacity 80.5×10^6, measured in kilograms, and intrinsic growth rate 0.71 per year. If the initial biomass is one-fourth the carrying capacity, find the biomass one year later and the time required for the biomass to grow to half the carrying capacity.

4. Use a logistic model with an assumed carrying capacity of 100×10^9, an observed population of 5×10^9 in 1986, and an observed rate of growth of 2 percent per year when population size is 5×10^9 to predict the population of the earth in the year 2000.

5*. Show that for a population which satisfies the logistic model, the maximum rate of growth of population size is $rK/4$, attained when population size is $K/2$.

6*. Show that for every choice of the constant c the function

$$x = \frac{K}{1 + ce^{-rt}}$$

is a solution of the logistic differential equation.

7*. Suppose a population satisfies a differential equation having the form of the logistic equation but with an intrinsic growth rate that depends on t:

$$x' = r(t)x\left(1 - \frac{x}{K}\right), \quad x(0) = x_0.$$

Show that the solution is

$$x(t) = \frac{Kx_0}{x_0 + (K - x_0)e^{-\int_0^t r(s)ds}}.$$

[Hint: Since there is an existence and uniqueness theorem which says that the initial value problem has exactly one solution, verification that the given function satisfies the differential equation and initial condition suffices to show that it is the solution. It is not necessary to "solve" the initial value problem and derive this solution.]

1.3 The Logistic Equation in Epidemiology

The logistic equation is most often associated with the study of the dynamics of a population under density-dependent birth or death rates. However, it also arises naturally in the study of epidemiological systems (as was first shown by Hethcote (1976)). A more thorough description of mathematical models in epidemiology may be found in Chapter 7, but some epidemiological examples will be introduced in earlier chapters. In this section a model for the transmission dynamics of gonorrhea in a homosexually active population is derived. $S(t)$ denotes all sexually active noninfected individuals and $I(t)$ denotes all sexually active infected (assumed infective) individuals with $N(t) = S(t) + I(t)$ being the total population size. It is assumed that new sexually active individuals arrive at the rate $\mu N(t)$ and that none of them is infected; that individuals also leave the sexually active class at the rate $\mu N(t)$ and that individuals leave the infective class by recovery, with return to the susceptible class, at the rate $\gamma I(t)$.

With these assumptions, we arrive at the following model:

$$\frac{dS}{dt} = \mu N(t) - B(S,I) - \mu S + \gamma I$$
$$\frac{dI}{dt} = B(S,I) - (\mu + \gamma)I,$$

where $B(S, I)$ denotes the incidence rate, that is, the number of *new* cases of infection per unit time. Obviously, $B(S, 0) = B(0, I) = B(0, 0) = 0$, that is, if there are no susceptibles or infected or both, then there should be no new cases of infection. Hence, it is plausible to assume

$$B(S, I) \propto SI.$$

If we let c (assumed to be constant) denote the average number of sexual partners per individual per unit time and ϕ (assumed constant) denote the average number of contacts per partner, then $\phi c S(t)$ denotes the total number of contacts per unit time of all susceptible individuals at time t. If all individuals mix at random (homogeneous mixing) then the number of sexual contacts per unit time between susceptibles would be

$$\phi c S(t) \frac{S(t)}{N(t)},$$

where $N(t) = S(t) + I(t)$, while the number of sexual contacts per unit time between susceptibles and infectives would then be

$$\phi c S(t) \frac{I(t)}{N(t)},$$

and, consequently,

$$B(S, I) \propto \phi c S(t) \frac{I(t)}{N(t)}.$$

In addition, we assume that only a fraction q $(0 \leq q \leq 1)$ of these contacts develop into new cases of infection. Defining the transmission rate as $\beta = q\phi c$, we arrive at the following expression for the incidence rate in a randomly mixing population:

$$B(S, I) = \beta S(t) \frac{I(t)}{N(t)}.$$

The model becomes

$$\frac{dS}{dt} = \mu N(t) - \beta S(t) \frac{I(t)}{N(t)} - \mu S + \gamma I \tag{1.10}$$
$$\frac{dI}{dt} = \beta S(t) \frac{I(t)}{N(t)} - (\mu + \gamma)I,$$

with $N = S + I$, and initial conditions $S(0) = S_0 > 0$, $I(0) = I_0 > 0$.

Since

$$\frac{d}{dt}(S+I) = \frac{d}{dt}N(t) = 0,$$

the birth rate equals the death rate and the total population size is constant. The substitution

$$S(t) = N - I(t),$$

where N is a constant, reduces the solution of the system (1.10) to the solution of the single differential equation

$$\frac{dI}{dt} = \beta(N - I(t))\frac{I}{N} - (\mu+\gamma)I(t) \tag{1.11}$$

or

$$\frac{dI}{dt} = \beta I\left(1 - \frac{I}{N}\right) - (\mu+\gamma)I(t)$$

or

$$\frac{dI}{dt} = (\beta - (\mu+\gamma))I\left(1 - \frac{I}{N(1 - \frac{\mu+\gamma}{\beta})}\right)$$

or

$$\frac{dI}{dt} = \beta\left(1 - \frac{1}{R_0}\right)I\left(1 - \frac{I}{K(1 - \frac{1}{R_0})}\right)$$

where $R_0 = \beta/(\mu+\gamma)$.

The definitions $r = \beta(1 - 1/R_0)$ and $K = N(1 - 1/R_0)$ reveal the logistic form

$$\frac{dI}{dt} = rI\left(1 - \frac{I}{K}\right).$$

Hence, if $R_0 > 1$ then $r > 0$, $K > 0$, and $I \to K$ (which is less than N). If $R_0 < 1$ then $r < 0$, $K < 0$, and $I \to 0$ (see Exercise 1 below). Of course, $K < 0$ makes no biological sense. Therefore we conclude that a positive (endemic) equilibrium $I_\infty > 0$ exists and it is approached by solutions if and only if $R_0 > 1$; otherwise the only biological equilibrium is $I_\infty = 0$, which is approached by solutions if $R_0 \leq 1$ but not if $R_0 > 1$.

Exercises

1. Show that if $r < 0$ and $K < 0$, every solution of the logistic equation with $x(0) \geq 0$ approaches zero as $t \to \infty$.

2*. Consider the system

$$\frac{dS}{dt} = \Lambda - \beta S(t)\frac{I(t)}{N(t)} - \mu S + \gamma I \tag{1.12}$$
$$\frac{dI}{dt} = \beta S(t)\frac{I(t)}{N(t)} - (\mu+\gamma)I,$$

where Λ denotes the total recruitment rate, assumed constant.

(a) Look at $dN/dt = d(S+I)/dt$ and solve the resulting differential equation for N, obtaining $N(t) = K + \big(K - N(0)\big)e^{-\mu t}$. This shows that the system is equivalent to the solution of the single *nonautonomous* differential equation

$$\frac{dI}{dt} = \beta(N(t) - I)\frac{I}{N(t)} - (\mu+\gamma)I, \tag{1.13}$$

where

$$N(t) = K + \big(K - N(0)\big)e^{-\mu t}, \tag{1.14}$$

with $K = \Lambda/\mu$.

(b) Show that $N(t) \to K$ as $t \to \infty$.

(c) Choose $K = 1,000$, $1/\mu = 10$ years, and two initial population sizes, $N(0) = 1,200$ and $N(0) = 700$. Using a differential equation solver find $I(10)$, $I(20)$, and $I(50)$ using values of the parameters that give $R_0 > 1$.

(d) If we look at the right hand side of equation (1.13) and let $t \to \infty$ and replace $S(t)$ by $K - I$, then we arrive formally at the following "asymptotic" differential equation

$$\frac{dI}{dt} = \beta(K - I)\frac{I}{K} - (\mu+\gamma)I(t), \tag{1.15}$$

where

$$K \equiv \frac{\Lambda}{\mu} = \lim_{t\to\infty} N(t).$$

Here, without justification, $S(t)$ has been replaced by $K - I$ and hence, Equation (1.15) and Equation (1.11) are not the same. However recent work [Castillo-Chavez and Thieme (1995)] has shown that these equations have the same qualitative dynamics. Compare the values found in (c) with those found using the limiting equation (1.15) numerically.

3*. Consider the following SIS epidemic model with variable population size (birth rate is different from death rate):

$$\frac{dS}{dt} = bN - \beta S\frac{I}{N} + \gamma I - \mu S$$
$$\frac{dI}{dt} = \beta S\frac{I}{N} - (\gamma + \mu)I.$$

(a) Check that

$$\frac{dN}{dt} = rN,$$

where $r = (b - \mu)$, and that, consequently, $N(t) = N(0)e^{rt}$.

(b) Rescale the above system by introducing the new variables

$$x(t) = \frac{S(t)}{N(t)} \text{ and } y(t) = \frac{I(t)}{N(t)}.$$

Note that

$$\frac{dx}{dt} = \frac{1}{N}\frac{dS}{dt} - \frac{S}{N^2}\frac{dN}{dt}$$

and that $x(t) + y(t) = 1$ for all t.

(c) Verify that

$$\frac{dy}{dt} = (\beta - (\gamma + b))y\left(1 - \frac{y}{1 - \frac{(\gamma+b)}{\beta}}\right).$$

(d) Let $R_0 = \beta/(\gamma + b)$. Can you interpret R_0? What is the qualitative behavior of the above equation?

(e) What is the meaning in terms of the original variables S and I as $y \to 0$ and $y \to y^* \in (0, 1)$.

1.4 Qualitative Analysis

When using the logistic model in practice one normally assumes that a population is indeed described by a logistic model and then attempts to choose the parameters r and K and the initial population size x_0 to give the best fit with experimental data. It is important to remember that the values of r, K, and x_0 are therefore subject to error. However, errors in r and x_0 do not affect our prediction of the ultimate population size K. The property that a small change in the initial size x_0 of a solution has only a small effect on the behavior of the solution as $t \to \infty$ is called *stability* of the solution. We will require stability of any solution to which we ascribe biological significance; if a small disturbance can cause a large change in the solution it is unreasonable to consider the solution meaningful.

It is also important to remember that the logistic model is an assumed form, not a consequence of a fundamental law. We will want to consider larger classes of models and examine properties that are valid for these larger classes rather than those properties that depend on the specifics of the logistic model. A property that holds for a large class of models is said to be *robust*, to indicate that it is more likely, in some sense, to have biological significance.

The information that we derived about the behavior of solutions of the logistic model was obtained from the explicit solution by separation of variables. If we wish to search for robust properties we must learn to deduce properties of solutions from the differential equation directly, without depending on analytic expressions for solutions.

The derivative of a solution $x(t)$ at a point $(t, x(t))$ is the right side of the logistic equation

$$\frac{dx}{dt} = rx\left(1 - \frac{x}{K}\right).$$

This is positive if $0 < x < K$, zero if $x = 0$ or $x = K$, and negative if $x < 0$ or $x > K$. Thus, a solution $x(t)$ is an increasing function of t when $0 < x(t) < K$ and a decreasing function of t when $x(t) > K$. (We ignore the case $x(t) < 0$ because it has no biological significance.) The constants $x \equiv 0$ and $x \equiv K$ are solutions. Differentiation of the logistic equation with respect to t gives

$$\begin{aligned}\frac{d^2x}{dt^2} &= \frac{d}{dx}\left(rx\left(1 - \frac{x}{K}\right)\right)\frac{dx}{dt} \\ &= r\left(1 - \frac{2x}{K}\right)\frac{dx}{dt} = r^2x\left(1 - \frac{2x}{K}\right)\left(1 - \frac{x}{K}\right).\end{aligned}$$

From this we deduce that $(d^2x)/(dt^2)$ changes sign as x crosses the horizontal line $x = K/2$ and thus, that a solution that crosses this line has an inflection point at the crossing. The solution curves must be as shown in Figure 1.3.

If $x(t) < K$ for some t then the graph of $x(t)$ cannot cross the line $x = K$ (to see this, we must use the *uniqueness of solutions*; if the graph did cross the line $x = K$ there would be two solutions passing through the point of crossing, and this is impossible) and is increasing for all t. Thus, $x(t)$ tends to a limit as $t \to \infty$, but the only possible limits are values of x for which the right side of the differential equation is zero, namely $x = 0$ or $x = K$. Since solutions near $x = 0$ tend away from $x = 0$, they cannot approach zero, but must tend to K as $t \to \infty$. A solution $x(t)$ that is above K decreases for all t and by a similar argument must tend to K as $t \to \infty$.

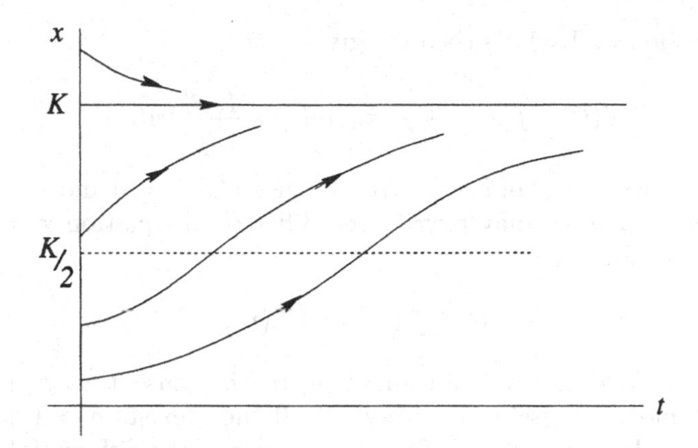

FIGURE 1.3. Solution curves of the logistic equation.

Thus, we see that every nonnegative solution except the constant solution $x \equiv 0$ tends to K as $t \to \infty$, and we have obtained this information *without explicitly solving the differential equation.* The method we have used may be adapted to more general first order differential equations.

We will consider *autonomous* first order differential equations, that is equations of the form

$$x' = f(x), \tag{1.16}$$

in which the right side does not contain the independent variable t. Sometimes we will write the equation in the form

$$x' = xr(x), \tag{1.17}$$

with $r(x)$ representing the per capita growth rate. We define an *equilibrium* of the differential equation (1.16) to be a value x_∞ such that $f(x_\infty) = 0$. An equilibrium corresponds to a constant solution $x(t) \equiv x_\infty$ of the differential equation.

If $x(t)$ is a solution of a differential equation $x' = f(x)$ that tends to a limit as $t \to \infty$ then it is not difficult to show that its limiting value must be an equilibrium. In fact, for a first order differential equation every solution must either tend to an equilibrium as $t \to \infty$ or be unbounded. However, not every equilibrium is a limit of non-constant solutions. For example, the only solution of the logistic equation that tends to zero as $t \to \infty$ is the identically zero solution.

In order to describe the behavior of solutions near an equilibrium we introduce the process of *linearization.* If x_∞ is an equilibrium of the differential equation $x' = f(x)$ so that $f(x_\infty) = 0$, we make the change of variable $u(t) = x(t) - x_\infty$, representing deviation of the solution from the equilibrium value. Substitution gives

$$u'(t) = f\big(x_\infty + u(t)\big),$$

and application of Taylor's theorem gives

$$u'(t) = f(x_\infty) + f'(x_\infty)u(t) + \frac{f'(c)}{2!}(u(t))^2$$

for some c between x_∞ and $x_\infty + u(t)$. We use $f(x_\infty) = 0$ and write $h(u) = (f'(c)/2!)u^2$. Then we may rewrite the differential equation $x' = f(x)$ in the equivalent form

$$u' = f'(x_\infty)u + h(u).$$

The function $h(u)$ is small for small $|u|$ in the sense that $h(u)/u \to 0$ as $u \to 0$; more precisely, for every $\epsilon > 0$ there exists $\delta > 0$ such that $|h(u)| < \epsilon|u|$ whenever $|u| < \delta$. The *linearization* of the differential equation at the equilibrium x_∞ is defined to be the linear homogeneous differential equation

$$v' = f'(x_\infty)v, \tag{1.18}$$

obtained by neglecting the higher order term $h(u)$ in $u' = f'(x_\infty)u + h(u)$. The importance of the linearization lies in the fact that the behavior of its solutions is easy to analyze, and this behavior also describes the behavior of solutions of the original equation (1.16) near the equilibrium.

Theorem 1.1. *If all solutions of the linearization* (1.18) *at an equilibrium* x_∞ *tend to zero as* $t \to \infty$ *then all solutions of* (1.16) *with* $x(0)$ *sufficiently close to* x_∞ *tend to the equilibrium* x_∞ *as* $t \to \infty$.

We have stated the theorem in a form that generalizes readily to results for systems of differential equations, as well as to results to be given later, when equations with delay are discussed. In the specific situation covered in Theorem 1.1 the condition that all solutions of the linearization tend to zero is $f'(x_\infty) < 0$. For an equilibrium x_∞ with $f'(x_\infty) < 0$ we must have $f(x) > 0$ for $x < x_\infty$ and $f(x) < 0$ for $x > x_\infty$ if x is sufficiently close to x_∞. Thus, the direction field is as shown in Figure 1.4.

A solution with $x(0) > x_\infty$ is monotone decreasing but bounded below by x_∞, and therefore tends to a limit as $t \to \infty$. As the only possible limits of solutions are equilibria, such a solution must tend to x_∞ (if there are no other equilibria between $x(0)$ and x_∞). By a similar argument a solution with $x(0) < x_\infty$ increases monotonically to x_∞ if there is no other equilibrium between $x(0)$ and x_∞. Thus, all solutions with $x(0)$ sufficiently close to x_∞ approach x_∞ as $t \to \infty$. Indeed, for first order differential equations we can be more precise: If x_∞ is an equilibrium with $f'(x_\infty) < 0$ then every solution whose initial value $x(0)$ is between x_∞ and the next equilibrium, in either direction, must tend to x_∞ as $t \to \infty$.

An equilibrium x_∞ is said to be *stable* if for every $\epsilon > 0$ there exists $\delta > 0$ such that $|x(0) - x_\infty| < \delta$ implies $|x(t) - x_\infty| < t$ for all $t > 0$. It is

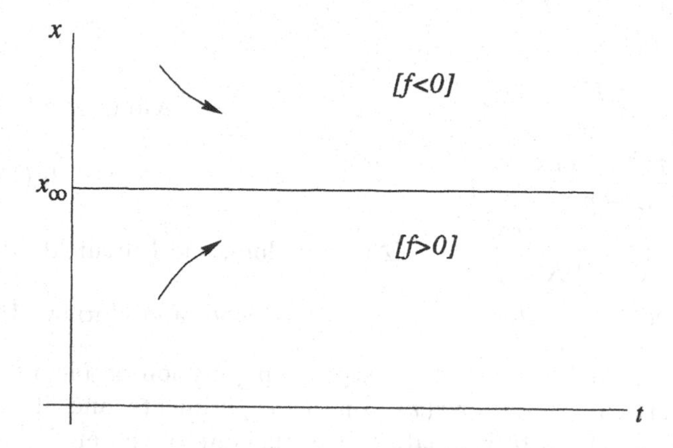

FIGURE 1.4. Direction fields of the autonomous equation.

implicit in this definition that the existence of the solution $x(t)$ is required for $0 \leq t < \infty$. An equilibrium x_∞ is said to be *asymptotically stable* if it is stable and if in addition $|x(0) - x_\infty| < \delta$ implies $\lim_{t \to \infty} x(t) = x_\infty$. Thus, stability means roughly that a small change in initial value produces only a small effect on the solution and this condition is a natural requirement for an equilibrium to be biologically meaningful. It is possible for systems to have equilibria for which all solutions starting near the equilibrium tend toward the equilibrium but only after traveling away from the equilibrium. Such an equilibrium would not be stable, but our definition of asymptotic stability requires stability in order to exclude this possibility. In biological applications we will ordinarily require asymptotic stability rather than stability, both because asymptotic stability can be determined from the linearization, while stability cannot, and because an asymptotically stable equilibrium is not disturbed greatly by a perturbation of the differential equation. In terms of asymptotic stability we may restate Theorem 1.1 and a corresponding instability result proved in the same way as follows:

Corollary 1.2. *An equilibrium x_∞ of (1.16) with $f'(x_\infty) < 0$ is asymptotically stable, while an equilibrium x_∞ with $f'(x_\infty) > 0$ is unstable.*

Actually, this result can be proved by an examination of the direction field without making use of the linearization. We introduce the linearization approach here because it is essential for the generalizations of Theorem 1.1 in later chapters.

We have already mentioned the logistic model, with $r(x) = r(1 - x/K)$, as an example. Other examples that have been used in population models include

$$r(x) = r \log \frac{K}{x} \qquad \text{[Gompertz (1825)]}$$

$$r(x) = \frac{r(K-x)}{K+ax} \qquad \text{[F. Smith (1963)]}$$

$$r(x) = r\left(1 - \left(\frac{x}{K}\right)^{\theta}\right) \qquad \text{[Ayala, Gilpin, and Ehrenfeld (1973)]}$$

$$r(x) = re^{1-x/K} - d \qquad \text{[Nisbet and Gurney (1982)]}$$

In using a model of this type to study a population problem one would assume a particular form for $r(x)$, conduct experiments, and fit the resulting data to this form to estimate the parameters of the model, and then compare other observations with the predictions of the model to judge its validity.

Every autonomous differential equation of the form $x' = f(x)$ or $x' = xr(x)$ has separable variables and thus can in principle be solved by integration. The reader should observe that for each of the above examples the necessary integration is sufficiently complicated to make a qualitative approach attractive.

If in the model

$$x' = xr(x) \tag{1.17}$$

the function $r(x)$ is nonnegative and decreasing for $0 \leq x \leq K$ then it is said to be a *compensation model.* If the per capita growth rate $r(x)$ is increasing for small x the model is said to be a *depensation model.* If the per capita growth rate is actually negative for small x, then the model is said to be a *critical depensation model.* While compensation models are the ones most commonly examined, both depensation and critical depensation models arise in fishery studies.

A compensation model is characterized by the conditions

$$r(x) \geq 0, \quad r'(x) \leq 0 \quad \text{for } 0 \leq x \leq K.$$

For a depensation model we assume

$$r(x) \geq 0, \; r''(x) \leq 0 \quad \text{for } 0 \leq x \leq K$$

$$r'(x) > 0 \text{ for } 0 < x < K^*$$

$$r'(x) < 0 \text{ for } K^* < x < K.$$

Thus, $r(x)$ achieves a maximum at K^*, and since

$$\lim_{x \to 0} r(x) = \lim_{x \to 0} f(x)/x = f'(0)$$

it follows that $f'(0) < r(K^*)$. Thus, the line joining the origin to the point $(K^*, f(K^*))$ on the *growth curve* $y = f(x)$, which has slope $r(K^*)$, lies above the tangent to the growth curve at the origin. Further, since

$$f'(x) = xr'(x) + r(x), \quad f''(x) = xr''(x) + 2r'(x),$$

we have $f''(0) = 2r'(0) \geq 0$ and

$$f''(K^*) = K^*r''(K^*) + 2r'(K^*) = K^*r''(K^*) < 0.$$

This shows that the growth curve has an inflection point to the left of K^*.

For a critical depensation model we assume

$$\begin{aligned} f(x) &< 0 \text{ for } 0 < x < K_0 \\ f(x) &\geq 0 \text{ for } K_0 \leq x \leq K. \end{aligned}$$

Under this assumption it is not difficult to show that the equation $x' = f(x)$ has three equilibria–an unstable equilibrium at K_0 and asymptotically stable equilibria at 0 and K. Then if the initial population size is below K_0, the population will die out. As we will see in the next section, in the case of critical depensation hunting may drive a population to extinction by bringing the population size below the critical level K_0, and this trend to extinction will not be reversed if hunting ceases. The extinction through hunting of the passenger pigeon in the 19th and early 20th centuries from an original population of 7 billion may have been an example of critical depensation. This property is sometimes called the *Allee effect* [Allee (1931)].

In the three cases of compensation, depensation, and critical depensation the growth curve has the different forms shown in Figure 1.5.

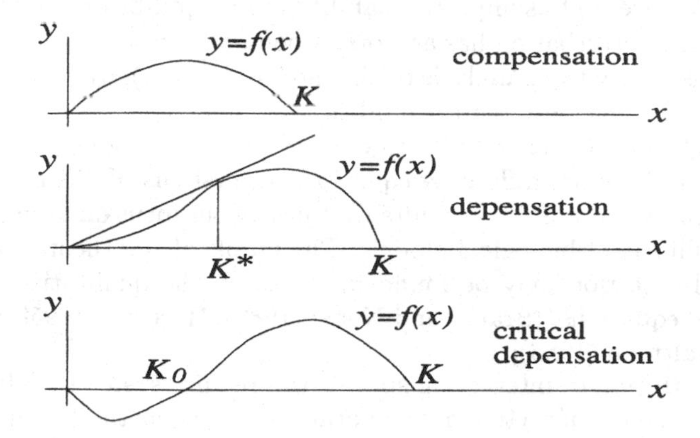

FIGURE 1.5. Growth curves for the cases of compensation, depensation and critical depensation.

In all of these models we have been assuming tacitly that the function $f(x)$ on the right side of the differential equation is exact. Any realistic study would recognize that the model can be at best an approximation, and that instead of $x' = f(x)$ we should really be studying a differential equation of the form

$$\frac{dy}{dt} = f(y) + h(y),$$

in which the term $h(y)$ represents the error made in the assumption of the specific form $f(y)$. It is in the nature of $h(y)$ that it cannot be known explicitly. Thus, instead of looking for explicit formulae for solutions of $y' = f(y) + h(y)$, we must be satisfied with qualitative information about the solutions for a given class of functions $h(y)$. It is possible to establish the following result, which justifies our interest in asymptotically stable equilibria of the model $x' = f(x)$.

Theorem 1.3. *Let x_∞ be an asymptotically stable equilibrium of $x' = f(x)$ with $f'(x_\infty) < 0$. Then*

(i) If $h(y)/(y - x_\infty) \to 0$ as $y \to x_\infty$, and if $|y(0) - x_\infty|$ is sufficiently small, the solution $y(t)$ of $y' = f(y) + h(y)$ tends to x_∞ as $t \to \infty$ i.e., x_∞ is an asymptotically stable equilibrium of $y' = f(y) + h(y)$.

(ii) If $|h(y)| \leq A$ for all y and A is sufficiently small, and if $|y(0) - x(0)|$ is sufficiently small, then $|y(t) - x(t)| \leq (KA)/||f'(x_\infty)|$ for some constant K i.e., solutions of $y' = f(y) + h(y)$ are close to x_∞ for all large t.

The essential content of Theorem 1.3 is that neglect of a perturbation $h(y)$ which tends to zero more rapidly than $(y - x_\infty)$ as $y \to x_\infty$ has no effect on the existence and asymptotic stability of an equilibrium, while neglect of a bounded perturbation has at worst a bounded effect on solutions. Thus conclusions drawn from analysis of the model $x' = f(x)$ are valid in a sense for a large class of more refined models.

Although we will not cite Theorem 1.3 or its analogues for difference equations, differential-difference equations, or systems of differential equations explicitly, these are the results that justify our focus on asymptotically stable equilibria of biological models. The proofs of the theoretical results cited in this section may be found in books on the qualitative theory of differential equations, [Brauer and Nohel (1989); Hurewicz (1958); Sánchez (1979); Waltman (1986)].

The description of interacting species will require systems of differential equations. The use of vector-matrix notation and the methods of linear algebra will make it possible to develop the theory of equilibria and asymptotic stability in a form analogous to what we have described here. The central result is the analogue of Theorem 1.1 in its form stated originally–that if all solutions of the linearization at an equilibrium tend to zero, then the equilibrium is asymptotically stable.

Exercises

In each of Exercises 1 through 5, find all equilibria and determine which are asymptotically stable.

1. $x' = rx\big(1 - x/K\big)$

2. $x' = rx \log \frac{K}{x}$

3. $x' = \frac{rx(K-x)}{K+ax}$

4. $x' = rx\left(1 - \left(\frac{x}{K}\right)^{\theta}\right) \quad (0 < \theta < 1)$

5. $x' = x\big(re^{1-x/K} - d\big)$

6. (a) A population is governed by the differential equation

$$x' = x(e^{3-x} - 1)$$

Find all equilibria and determine their stability.

(b) A fraction p $(0 < p < 1)$ of the population in part (a) is removed in unit time so that the population size is governed by the differential equation

$$x' = x(e^{3-x} - 1) - px$$

For what values of p is there an asymptotically stable positive equilibrium?

7. For which initial values $y(0)$ does the solution $y(t)$ of the differential equation

$$y' = y\big(e^{-y} - 2y\big)$$

approach zero as $t \to \infty$?

8. Let the concentration of a substance in a cell be C and the concentration outside the cell be Γ. Assume that the substance enters the cell by diffusion at a rate βC. This would give a differential equation

$$C' = \beta C(\Gamma - C).$$

Now include an absorption term to give a differential equation

$$C' = \beta C(1 - C) - \frac{C}{1 + C}$$

where β is a constant.

(a) Find the equilibria of this model and determine their stability.

(b) Compare the behavior of this model as $t \to \infty$ with the behavior of the model without absorption.

9. For the following equations compute the equilibrium points and analyze their local stability using linearization.

   (a) $p'(t) = \beta p(t)(1 - p(t)) - ep(t)$ where β and e are positive constants. Consider the cases $\beta/e > 1$ and $\beta/e < 1$. Sketch a graph of the solution.

   (b) $p'(t) = ap(t)e^{-kp(t)/p_0} - mp(t)$, where a, k, m, and p_0 are positive constants.

10. Discuss the model

$$x' = rx\left(1 - \frac{x}{K}\right)\left(\frac{x}{K_0} - 1\right),$$

where $0 < K_0 < K$. Find all limits of solutions with $x(0) > 0$ as $t \to \infty$ and find the set of initial values corresponding to each limit.

11*. (a) By sketching direction fields determine whether the equilibrium $x = 0$ of the differential equation $x' = x^3$ is asymptotically stable or unstable.

   (b) By sketching direction fields determine whether the equilibrium $x = 0$ of the differential equation $x' = -x^3$ is asymptotically stable or unstable.

   (c) Is an equilibrium x_∞ of a differential equation $x' = f(x)$ with $f(x_\infty) = 0$, $f'(x_\infty) = 0$, $f''(x_\infty) = 0$, $f'''(x_\infty) < 0$ asymptotically stable or unstable?

12*. Let $g(x)$ be a function such that $g(K) = 0$ and $g(x) > 0$ for $0 < x < K$ and suppose $0 < x_0 < K$.

   (a) Show that the function $x(t)$ defined implicitly by the relation.

$$\int_{x_0}^{x(t)} \frac{du}{ug(u)} = t \tag{1.19}$$

   is a solution of the initial value problem $x' = xg(x)$, $x(0) = x_0$. [*Remark:* This solution is obtained by separation of variables.]

   (b) Show that the integral on the left side of (1.19) is negative if $x(t) < x_0$ and positive if $x(t) > x_0$, and deduce that the solution of the initial value problem must satisfy $x(t) > x_0$ for $t > 0$.

   (c) Show that as $t \to \infty$ the integral on the left side of (1.19) becomes unbounded. Deduce that $g(x(t)) \to 0$, hence $x(t) \to K$.

13*. Populations of the North American spruce budworm have been modeled assuming that in the absence of predation the population would follow the logistic equation but that the population is subject to predation by birds. [Ludwig, Jones, and Holling (1978)]. The resulting model is

$$x' = xp(x) - xq(x)$$

with

$$p(x) = r\left(1 - \frac{x}{K}\right), \quad q(x) = \frac{cx}{x^2 + A^2}$$

(a) Show that for a differential equation $x' = xp(x) - xq(x)$ the equilibrium $x = 0$ is asymptotically stable if $p(0) < q(0)$ and that an equilibrium $x_\infty > 0$ is asymptotically stable if $p'(x_\infty) < q'(x_\infty)$.

(b) By sketching the graphs of $p(x)$ and $q(x)$ for different values of r show that the system has either one positive equilibrium that is asymptotically stable, or three positive equilibria of which two are asymptotically stable and one is unstable. It will be convenient to use a computer algebra system for making these sketches.

14*. For the differential equation

$$x' = f(x)$$

with $f(0) = 0$, $f(x) < 0$ $(0 < x < K_0)$, $f(K_0) = 0$, $f(x) > 0$ $(K_0 < x < K)$, $f(K) = 0$ (critical depensation), show that the equilibria at 0 and K are asymptotically stable and that the equilibrium at K_0 is unstable.

15. Imagine a small herd of cows in a field of modest size. The following example shows how the initial condition (the state of the field) might affect the final outcome. R.M. May (1974) developed a theoretical model to describe the dynamics of the amount of vegetation V:

$$\frac{dV}{dt} = G(V) - Hc(V), \tag{1.20}$$

where $G(V) = rV(1 - V/K)$ describes the growth of vegetation, r and K are positive constants; $c(V) = \beta V^2/(V_0^2 + V^2)$ is the consumption of vegetation per cow, β and V_0 are constants; H is the number of cows in the herd.

Choose $r = 1/3$, $K = 25$, $\beta = 0.1$, and $V_0 = 3$.

(1) Graph the functions $G(V)$ and $Hc(V)$ for different herd sizes: $H = 10, 20, 30$. What conclusions can you draw by examining these graphs?

(2) For the same parameter values as in part (a), graph dV/dt versus V. Use these graphs to determine all possible steady states of (1.20) and their stability for each value of H.

1.5 Harvesting in Population Models

We wish to study the effect on a population model of the removal of members of the population at a specified rate. If a population modeled by a differential equation

$$x' = f(x)$$

is subject to a harvest at a rate of $h(t)$ members per unit time for some given function $h(t)$ then the harvested population is modeled by the differential equation

$$x' = f(x) - h(t).$$

1.5.1 Constant Yield Harvesting

If the function $h(t)$ is a constant H, so that members are removed at the constant rate of H per unit time the model is

$$x' = f(x) - H.$$

This type of harvesting is called *constant rate* or *constant yield* harvesting. It arises when a quota is specified (for example, through permits as in deer hunting seasons in many states or by agreement as sometimes occurs in whaling).

If the population is governed by a logistic equation the model with harvesting is

$$\frac{dx}{dt} = rx\left(1 - \frac{x}{K}\right) - H \tag{1.21}$$

and equilibria of (1.21) may be found by solving the quadratic equation $rx(1 - x/K) - H = 0$, or $x^2 - Kx + KH/r = 0$. There are two equilibria,

$$x_L = \frac{K - \sqrt{K^2 - \frac{4HK}{r}}}{2}, \text{ and } x_U = \frac{K + \sqrt{K^2 - \frac{4HK}{r}}}{2},$$

provided $K^2 - 4HK/r \geq 0$, or $H \leq rK/4$. If $H > rK/4$ both roots are complex, $x'(t) < 0$ for all x, and every solution crashes, hitting zero in finite time. If a solution reaches zero in finite time, we consider the system to have collapsed. If $0 \le H < rK/4$, there are two equilibria: x_L, which increases from 0 to $K/2$ as H increases from 0 to $rK/4$, and x_U, which decreases from K to $K/2$ as H increases. The stability of an equilibrium x_∞ of $x' = F(x) - H$ requires $F'(x_\infty) < 0$, which for the logistic model means $x_\infty > K/2$. Thus, x_L is always unstable and x_U is always asymptotically stable. When H increases to the critical value $H_c = rK/4$ there is a discontinuity in the behavior of the system–the two equilibria coalesce and annihilate each other. For $H < H_c$ the population size tends to an equilibrium size that approaches $K/2$ as $H \to H_c$ (provided the initial population size is at least x_L), but for $H > H_c$ the population size reaches zero in finite time for all initial population sizes (Figure 1.6). Such a discontinuity is called a (mathematical) *catastrophe*; the biological implications are catastrophic to the species being modeled. [Brauer and Sánchez (1975)].

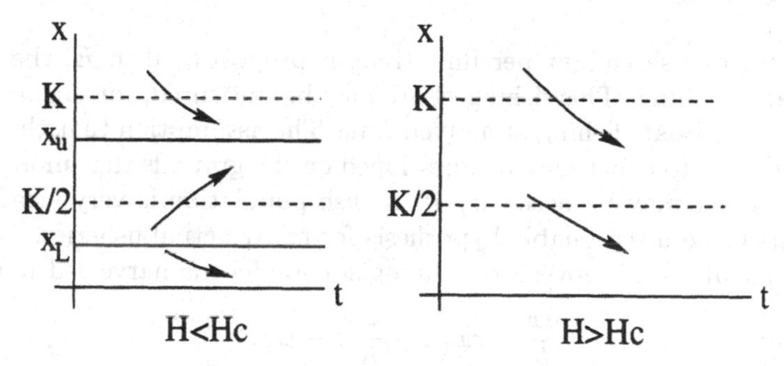

FIGURE 1.6. Behavior of solutions subject to constant yield harvesting

For a general model $x' = f(x) - H$ equilibria are found by solving $f(x) - H = 0$, that is, by finding values x_∞ of x for which the growth curve $y = f(x)$ and the harvest curve $y = H$ (a horizontal line) intersect. An equilibrium x_∞ is asymptotically stable if $\left(f(x) - H\right)'_{x=x_\infty} = f'(x_\infty) < 0$, that is, if at such an intersection the growth curve crosses the harvest curve from above to below as x increases. (Figure 1.7)

From Figure 1.7 it is clear that if $H > \max f(x)$ there is no equilibrium, and the critical harvest rate H_c at which two equilibria coalesce and disappear is $\max f(x)$.

1.5.2 *Constant Effort Harvesting*

If the function $h(t)$ is a linear function of population size $h(t) = Ex(t)$, the model is

$$x' = f(x) - Ex.$$

This type of harvesting is called *proportional* or *constant effort harvesting.* It arises in the modeling of fisheries, where it is often assumed that x,

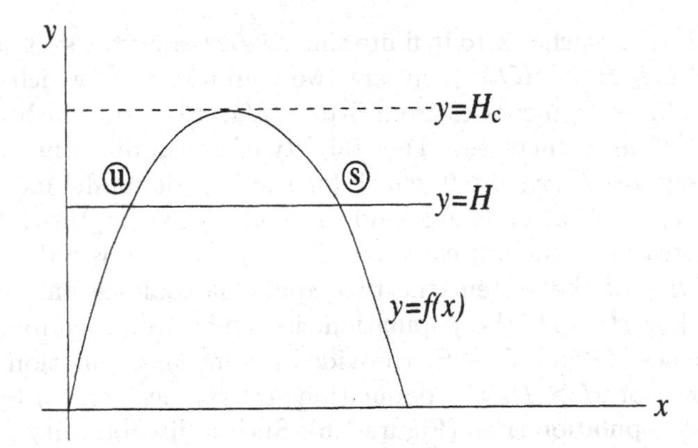

FIGURE 1.7. Intersections of the growth curve with the line of constant yield harvesting

the number of fish caught per unit time, is proportional to E, the effort expended in fishing. This fishing effort may be measured, for example, by the number of boats fishing at a given time. The assumption that the catch is proportional to effort may be questioned on the grounds that more effort per fish caught may be necessary if the fish population is very small, but it appears to be a reasonable hypothesis for many actual fisheries.

If the population is governed by a logistic model the harvested model is

$$\frac{dx}{dt} = rx\left(1 - \frac{x}{K}\right) - Ex,$$

and there are two equilibria, one at $x = 0$ and one obtained by solving $r(1 - x/K) - E = 0$, which we denote by $x_\infty(E) = K(r - E)/r$, provided $0 \leq E \leq r$. It is easy to verify that the equilibrium at $x = 0$ is unstable and the equilibrium at $x_\infty(E)$ is asymptotically stable for $0 \leq E \leq r$. As the effort increases from zero to r, the equilibrium decreases from K to zero. For a given effort E the yield is $Ex_\infty(E) = KE - KE^2/r$. This yield attains a maximum value of $rK/4$ for $E = r/2$, with $x_\infty(E) = K/2$; increasing the effort beyond $r/2$ is counter productive in that it decreases the yield.

For a general model $x' = f(x) - Ex$ the equilibria are found by solving $f(x) - Ex = 0$, that is by finding values $x_\infty(E)$ of x where the growth curve $y = f(x)$ and the harvest curve $y = Ex$ intersect. An equilibrium is asymptotically stable if

$$\left(f(x) - Ex\right)'_{x=x_\infty} = f'(x_\infty) - E < 0,$$

that is, if at such an intersection the growth curve crosses the harvest curve from above to below as x increases. If $f(0) = 0$ then $x = 0$ is an equilibrium which is unstable unless $x = 0$ is the only equilibrium.

For a given effort E the yield is $Y = Ex_\infty(E) = F(x_\infty)$, and the maximum yield is $\max F(x)$, obtained with $E = E_{\max}$ chosen so that the line $y = Ex$ passes through the maximum of $F(x)$.

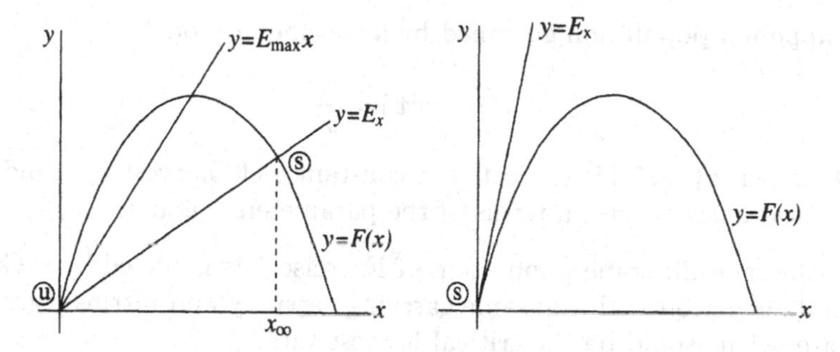

FIGURE 1.8. Intersections of the growth curve and the line of constant yeild effort harvesting.

The *yield effort curve* is the graph of yield against effort. In the case of compensation the yield increases as effort increases to a maximum, called the *maximum sustainable yield* (MSY), and then decreases continuously to zero, reaching zero when $E = f'(0)$. However, in the case of depensation there is a critical effort $E^* = r(K^*)$ where the yield drops to zero discontinuously (Figure 1.9). The same happens with critical depensation, but in critical depensation there is the additional property that if the effort is large enough the population size may fall below K_0 and then be drawn to the asymptotically stable equilibrium at zero.

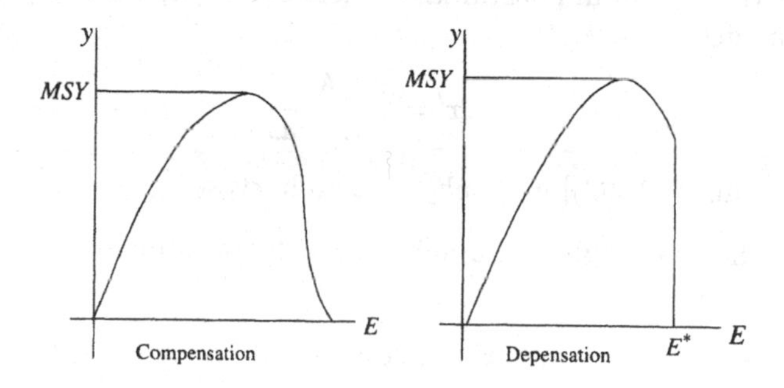

FIGURE 1.9. Yield effort curve.

Exercises

1. A population of sandhill cranes (*Grus canadiensis*) has been modeled by a logistic equation with carrying capacity of 194,600 members and intrinsic growth rate 0.0987 year^{-1}. Find the critical harvest rate for which constant yield harvesting will drive the population to extinction ,and find the equilibrium population size under constant yield harvesting of 3000 birds per year.

2. Suppose a population governed by a Gompertz model

$$x' = rx \log \frac{K}{x}$$

[Gompertz (1825)] is subjected to constant yield harvesting. Find the critical harvest rate in terms of the parameters r and K.

3. If the sandhill crane population of Exercise 1 is modeled by a Gompertz equation with the same carrying capacity and intrinsic growth rate, what would be the critical harvest rate?

4. Suppose a population governed by a logistic equation with carrying capacity K, intrinsic growth rate r, and initial size K is subjected to constant effort harvesting. By solving the initial value problem

$$x' = rx\left(1 - \frac{x}{K}\right) - Ex, \quad x(0) = K$$

analytically determine the population size $x(t)$ and verify that if $E \leq r$ then $\lim_{t\to\infty} x(t) = K(1 - E/r)$ while if $E > r$, $\lim_{t\to\infty} x(t) = 0$.

5. Find the maximum sustainable yield for a population governed by a Gompertz model and subjected to harvesting (either constant yield or constant effort).

6. Find the maximum sustainable yield for a population governed by the model

$$x' = \frac{rx(K - x)}{K + ax}$$

[F.E. Smith (1963)] and subjected to harvesting.

7. Find the maximum sustainable yield for a population governed by the model

$$x' = rxe^{(1-x/K)} - dx$$

[Nisbet and Gurney (1982)] and subjected to havesting.

8*. Show that the maximum sustainable yield under constant effort harvesting is equal to the maximum yield under constant yield harvesting.

1.6 Eutrophication of a Lake: A Case Study

A lake is a very complicated ecosystem. A full model for a lake would need to take into account its dimensions, depth, and temperature, the concentrations of a variety of organic and inorganic materials in the lake, the kinds and quantities of vegetation in the lake, and the variety of fish and other animal life in the lake. A lake may be oligotrophic, a state characterized by low inputs of nutrients and levels of plant production and relatively clear water. On the other hand, a lake may be eutrophic, a state characterized by high nutrient input and plant production, murky water, and toxicity. Avoidance or reversal of eutrophication has benefits, both aesthethic and commercial.

Ultimately, eutrophication is caused by excessive inputs of nutrients as a by-product of agriculture, forestry, or urban development. The primary cause of eutrophication is usually excessive inputs of phosphorus, mainly due to runoff from agricultural and urban lands. Phosphorus accumulates in sediment and recycles from sediment to water, and the phosphorus input from this recycling may exceed the phosphorus flow into the lake from outside. We shall describe a very simple model which displays many of the behaviors observed in real lakes. This model [Carpenter, Ludwig, and Brock (1999)] focuses on the phosphorus concentration in the lake and is given by the single differential equation

$$\frac{dp}{dt} = L - sp + r\frac{p^q}{m^q + p^q}. \tag{1.22}$$

Here p is the amount of phosphorus in the water. The rate of input of phosphorus from the watershed is L. The rate of loss of phosphorus from sedimentation, outflow, and absorption by consumers or plants is assumed to be proportional to the amount of phosphorus present and is given by sp. Study of limnological mechanisms suggest that the recycling rate is a sigmoid function

$$\frac{rp^q}{m^q + p^q}.$$

Here the exponent q ($q \geq 2$) describes the steepness of this function at its inflection point. The value of q may range from 20 for a shallow warm lake to 2 for a deep cold lake. The parameter r is the maximum recycling rate of phosphorus and m is the concentration of phosphorus at which recycling is half its maximum rate.

An equilibrium of the equation (1.22) is an intersection of the curve

$$Q = L + r\frac{p^q}{m^q + p^q}, \tag{1.23}$$

representing phosphorus input and the line

$$Q = sp, \tag{1.24}$$

representing phosphorus outflow. Because the line (1.24) starts at zero when $p = 0$ and is unbounded while the curve (1.23) has a nonnegative value L

when $p = 0$ and is bounded as $p \to \infty$, there is at least one equilibrium. There may be as many as three equilibria. A "typical" situation is that of a small value of L and an oligotrophic equilibrium; with a large value of L there is an eutrophic equilibrium; and with an intermediate value of L there are three equilibria: an oligotrophic equilibrium, a eutrophic equilibrium, and an intermediate unstable equilibrium which separates the domains of attraction of the other two equilibria.

The challenge of water quality management is to control a lake which is at a eutrophic equilibrium and move it to an oligotrophic equilibrium. Sometimes this may be accomplished by reducing the external phosphorus input. However, in some lakes reducing external input alone cannot reverse eutrophication due to the amount of recycling. In order to improve the water quality of such a lake additional methods of intervention to decrease recycling or increase sedimentation would be needed. Whether such additional methods are feasible depends on the properties of the lake.

We shall consider only that which can be accomplished by reducing phosphorus input. In terms of the model (1.22), we locate equilibria as intersections of the curve (1.23) and the line (1.24). Decreasing the external phosphorus input corresponds to moving the curve (1.23) down. In Figure 1.10 there is a eutrophic equilibrium for large L and an oligotrophic equilibrium for small L, while for intermediate values of L both equilibria are present.

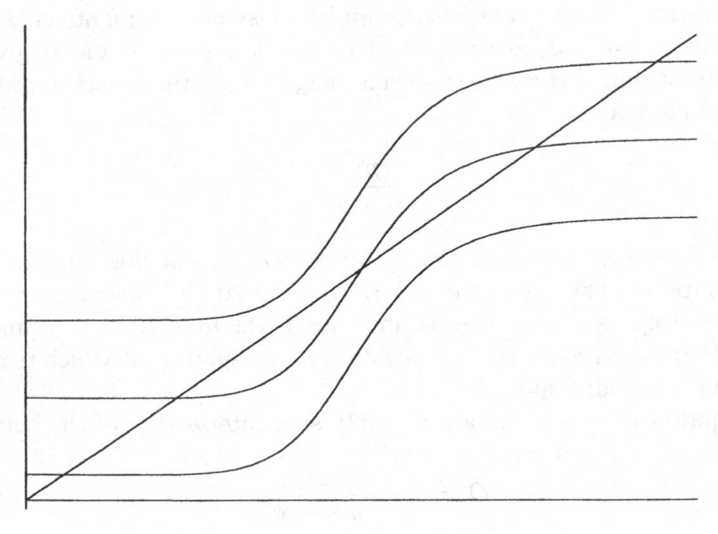

FIGURE 1.10.

In order to manage the lake to the oligotrophic equilibrium when both oligotrophic and eutrophic equilibria are present it is necessary to bring the phosphorus concentration below the unstable equilibrium. This may require additional intervention methods. In Figure 1.11 there is an input level for which the eutrophic and unstable equilibria coalesce and for which line and curve are tangent.

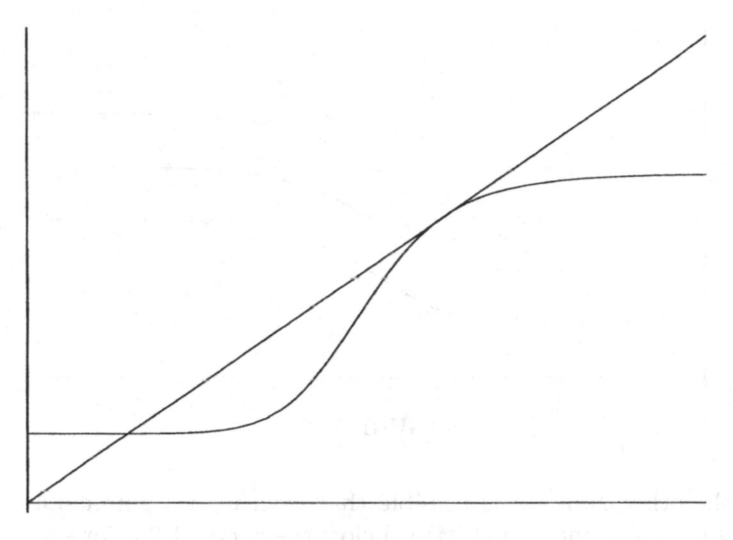

FIGURE 1.11.

When L is reduced below this critical level only the oligotrophic equilibrium remains. The equilibrium jumps to this level and the eutrophication of the lake is reversed. This equilibrium jump is known as *hysteresis*, and a lake displaying this property is said to be *hysteretic*.

The minimum phosphorus input to a lake is determined by factors such as soil chemistry, and this minimum input may not be low enough to move a hysteretic lake out of its eutrophic equilibrium. Even if a hysteretic lake could be moved to the oligotrophic equilibrium, changes in conditions may increase the minimum attainable phosphorus input and lead to eutrophication.

A hysteretic lake that is at its oligotropic equilibrium, if disturbed by a large input of phosphorus (even a single input and not a change in the normal input), may be moved to a state that is in the domain of attraction of the eutrophic equilibrium. Thus, it may switch rather rapidly from an oligotrophic state to a eutrophic state, and this switch may be quite difficult to reverse. Sudden shifts in the state of an ecosystem can lead to severe consequences. One much-studied example is the collapse of a fishery due to a shift that does not appear to influence the fishery.

If the slope of the line (1.24) is greater than the maximum slope of the curve (1.23) the system (1.22) has only an oligotrophic equilibrium (Figure 1.12). In this case the lake resists eutrophication, and a large input of phosphorus can be absorbed without significant harm. Such a lake is said to be *reversible*.

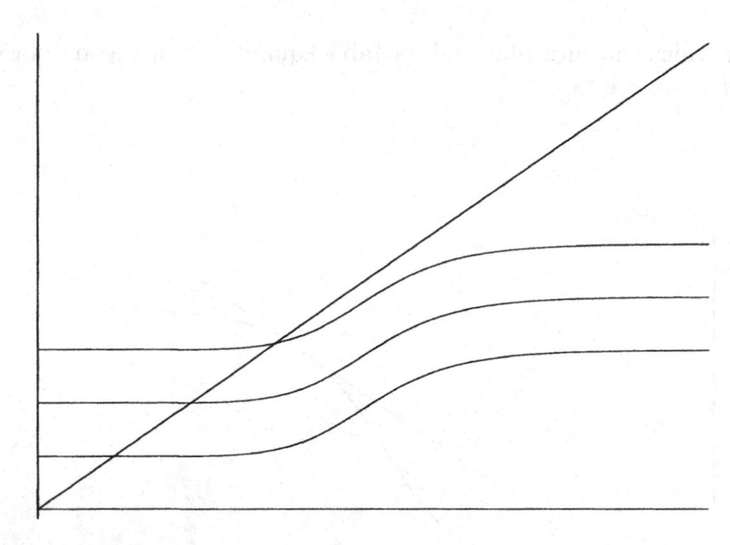

FIGURE 1.12.

On the other hand, it is possible that even for the minimum possible phosphorus input the line (1.24) is below the curve (1.23) for small p, and the system (1.22) has only a eutrophic equilibrium (Figure 1.13). Such a lake is said to be *irreversible*; it is not possible to bring it to an oligotrophic equilibrium by reducing the input of phosphorus.

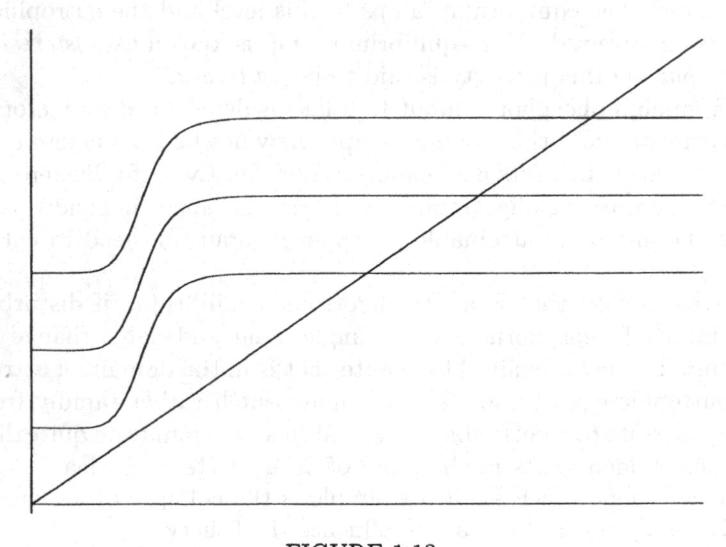

FIGURE 1.13.

Roughly speaking, a lake is reversible if the slope s of the line (1.24), representing the rate of loss of phosphorus, is sufficently large. If s is sufficiently small, a lake is irreversible, and intermediate values of s correspond to hysteretic lakes. The classification of a lake would depend both on s and on the minimum achievable input L.

One of the most studied lakes in the world is Lake Mendota, which abuts the campus of the University of Wisconsin-Madison. Detailed measurements of phosphorus input and phosphorus mass have been recorded there for more than 20 years, and estimates have been made of the parameters s, r, m, and q in equation (1.22), [Carpenter, Brock, and Ludwig(1999)], namely

$$\begin{aligned} s &= 0.817/\text{year}, \\ r &= 731,000 \text{ kg/year}, \\ m &= 116,000 \text{ kg}, \\ q &= 7.88 \end{aligned} \tag{1.25}$$

Unfortunately, there is a great deal of uncertainty in these estimates, with a 100-fold range for m and a 10-fold range for r. This means that an estimate of the state of the lake based on these parameters and the model (1.22) ranges from reversible to irreversible. Thus, even with this much experimental data there is much uncertainty about the lake's response to management.

Let us imagine that the parameter values given by (1.25) are correct and then use the model (1.22) to analyze the state of the lake. It is convenient to rescale the model by setting

$$p = mx$$

to give

$$m\frac{dy}{dt} = L - smx + r\frac{m^q x^q}{m^q + m^q x^q} = L - smx + r\frac{x^q}{1 + x^q}.$$

The equilibrium condition becomes

$$L - smx + r\frac{x^q}{1 + x^q} = 0.$$

We set

$$L = ra, \qquad s = \frac{rb}{m}$$

to make the equilibrium condition

$$a + \frac{x^q}{1 + x^q} = bx. \tag{1.26}$$

Thus, we attempt to determine the state of the lake by finding the intersections of the curve $y = a + x^q/(1 + x^q)$ and the line $y = bx$.

The model to which we have reduced (1.22) is

$$x' = \frac{r}{m}\left(a - bx + \frac{x^q}{1+x^q}\right). \tag{1.27}$$

The rate r/m affects the dynamics of the system, particularly the rate of approach to equilibrium but not the location or stability of equilibria. Equilibria depend on the three dimensionless parameters a, b, q. We may think of b and q as properties of the lake being studied (although some management methods may be able to change b) and a as a control parameter. We should note that uncertainty in the measurement of r and m translates into even greater uncertainty in the value of their ratio and thus into the value of b.

If we use the parameter values for Lake Mendota given by (1.25), we obtain $b = 0.130$. The minimum feasible phosphorus input rate is 3800 kg/year, corresponding to a value $a = 3800/731,000 = 0.0052$. With these values of a and b, and $q = 7.88$, we may plot the curve $y = a + x^q/(1+x^q)$ and the line $y = bx$ using a computer algebra system such as Maple or Mathematica to see that there are three equilibria. If is helpful to zoom in on the portion of the graphs for small x to identify the oligotrophic and unstable equilibria, as well as the overall structure (Figures 1.14, 1.15).

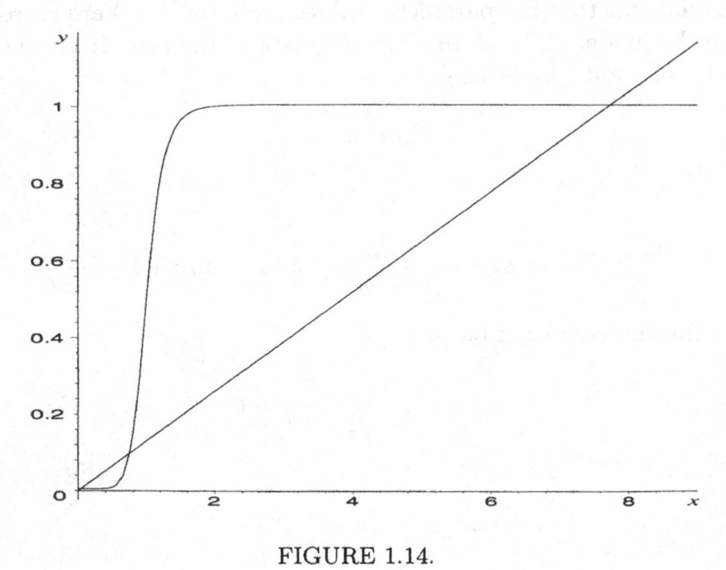

FIGURE 1.14.

We may estimate these equilibrium values as 0.04, 0.75, and 7.73. These correspond to phosphorus values of 4,640 kg (oligotrophic equilibrium), 87,000 kg (unstable equilibrium), and 897,000 kg (eutrophic equilibrium), respectively. The lake is currently at an equilibrium of 57,000 kg, which corresponds to $x = 0.49$. This is less than the unstable equilibrium of the system with minimum phosphorus input, and thus it is in the domain of

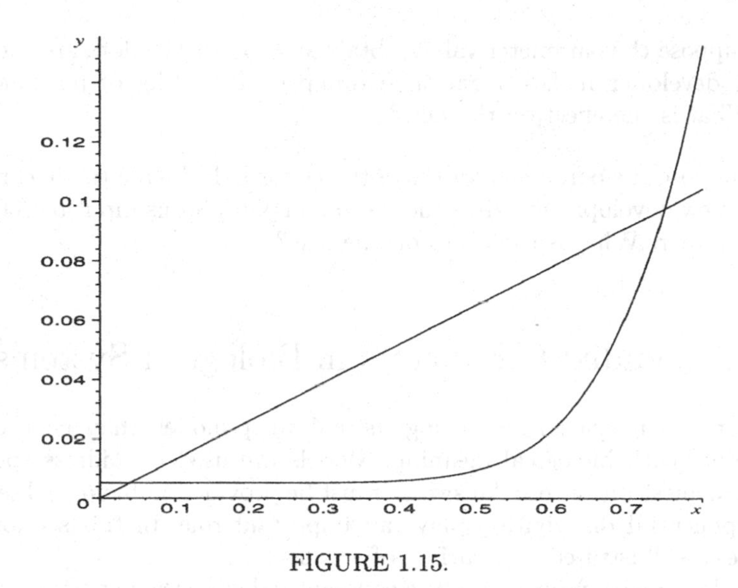

FIGURE 1.15.

atraction of the oligotrophic equilibrium. This indicates that the lake is hysteretic, but could be moved to an oligotrophic equilibrium. However, in the real world of substantial uncertanty about parameters values (not to mention model simplifications) it is not possible to draw an unequivocal conclusion.

Activities such as agriculture, forestry, and urban development that produce excessive nutrient inputs into a lake and lead to eutrophication have direct economic benefits. There is a tradeoff between these benefits and the cost from damage to the lake. The various interested groups, including farmers, foresters, developers, and environmental activists, are likely to have different estimates of the overall profits and losses of an activity. For each estimate one could formulate an economic optimization model, but there are also political questions involved in the decision of which model to use. As with any political question there are substantial opportunities for misunderstanding or misinterpretation of models and data.

Exercises

1. Suppose the values of s, r, and m obtained for Lake Mendota are correct, but a huge error has been made in the estimate of q, so that actually $q = 2$. What is the state of the lake and can it be brought to an oligotrophic equilibrium?

2. Suppose the values of s and q obtained for Lake Mendota are correct but the actual values of m and r are $m = 100,000$ kg and $r = 800,000$ kg/year. What is the state of the lake and can it be saved?

3. Suppose the parameter values obtained for Lake Mendota are correct. A developer makes a one-time dump of 40,000 kg of phosphorus. What is the effect on the lake?

4. Suppose the parameter values obtained for Lake Mendota are correct. A new development raises the minimum phosphorus input to 8000 kg per year. What is the effect on the lake?

1.7 Appendix: Parameters in Biological Systems

A challenge in mathematical biology is to develop models that incorporate parameters with biological meaning. Models are used to address specific biological questions whose "answers" must be provided in biological terms. The exponential distribution plays an important role. In this section its usefulness is illustrated in a variety of settings.

Consider a death process. Let the constant μ denote the per capita death rate of a population. Then its dynamics are modeled by the equation

$$\frac{dN(t)}{dt} = -\mu N(t), \qquad 0 \leq \mu < \infty, \qquad N(0) = N_0,$$

where $N(t)$ denotes the population size at time t. Hence,

$$\frac{N(t)}{N_0} = e^{-\mu t}, \qquad \text{for } t \geq 0,$$

that is, $e^{-\mu t}$ denotes the proportion of individuals who were alive at time $t = 0$ and who are still alive at time $t = t$ or, in probabilistic language, $e^{-\mu t}$ denotes the probability of a person being alive at time $t \geq 0$ given that he was alive at time $t = 0$.

Consequently,

$$F(t) = \begin{cases} 1 - e^{-\mu t} & \text{for } t \geq 0 \\ 0 & \text{for } t < 0 \end{cases}$$

gives the proportion of individuals who have died in $[0, t)$ or, in probabilistic lingo, $F(t)$ denotes the probability of dying in the time interval $[0, t)$. $F(t)$ is a *probability distribution*, that is, $F(t)$ satisfies the properties

(i) $F(t) \geq 0$,

(ii) $\lim_{t \to -\infty} F(t) = 0$,

(iii) $\lim_{t \to \infty} F(t) = 1$.

In fact $F(t)$ is the exponential cumulative probability distribution or exponential cumulative distribution function.

Probability distributions are often associated with random variables, that is, variables that take a value or a set of values with some probability. If we let X denote the time to death of an individual then, it is reasonable to assume that such a time is a random event and that X takes on the values $[0, \infty)$ with some probability. X is therefore a continuous random variable.

Modeling the time to death X with an exponential probability distribution is equivalent to the following probability statement:

$$\text{Prob}\,[X \leq t] \equiv F(t) = \begin{cases} 1 - e^{-\mu t} & \text{for } t \geq 0 \\ 0 & \text{otherwise.} \end{cases}$$

From the above expression we recover the probability density associated with $F(t)$ after we observe that

$$\text{Prob}\,[t < X \leq t + \Delta] \approx F(t + \Delta) - F(t)$$

or, approximately, when Δ is small, that

$$\text{Prob}\,[t < X \leq t + \Delta] \approx \Delta\left(\lim_{\Delta \to 0}\left(\frac{F(t+\Delta) - F(t)}{\Delta}\right)\right) = \Delta f(t),$$

where $f(t) = dF/dt$ is the probability density function of X. It satisfies the properties

(i) $f(t) \geq 0$,

(ii) $\displaystyle\int_{-\infty}^{\infty} f(t) = 1$,

(iii) $\displaystyle\text{Prob}\,[t < X \leq t + \Delta] = \int_{t}^{t+\Delta} f(e)de \approx f(t)\Delta.$

Therefore, when $F(t)$ denotes the exponential distribution,

$$f(t) = F'(t) = \begin{cases} \mu e^{-\mu t} & \text{for } t \geq 0 \\ 0 & \text{otherwise} \end{cases}$$

and

$$\text{Prob}\,[\text{dying in } (t, t + \Delta)] \approx \mu e^{-\mu t}\Delta.$$

Furthermore, the average time before death, or the life expectancy, is given by the mean or expected value of X, that is

$$E[X] \equiv \int_{-\infty}^{\infty} t f(t)dt$$

where

$$f(t) = \begin{cases} \mu e^{-\mu t} & \text{for } t \geq 0 \\ 0 & \text{otherwise.} \end{cases}$$

Using $E[X] = \int_0^\infty te^{-\mu t}dt$, after we integrate by parts we find that

$$E[X] = \frac{1}{\mu}.$$

We can do a little more and Bayes' theorem is useful in this respect. If A and B are two probabilistic events then

$$\text{Prob}\,[A|B] \cdot \text{Prob}\,[B] \equiv \text{Prob}\,\left[A \cap B\right]$$

or, in words, that *"The probability of the event A given that B has occurred times the probability of the event B equals the probability of the simultaneous occurrence of A and B"*.

If we let $B = \{X > t\}$, that is, the event that the time to death is greater than t; and $A = \{X \leq t + \Delta\}$, that is, the event time to death is less than or equal to $t + \Delta$, then the probability one dies before $t + \Delta$ *given* that one was alive at time t is computed as follows:

$$\text{Prob}\,[X \leq t + \Delta | X > t] \cdot \text{Prob}\,[X > t] = \text{Prob}\,[t < X \leq t + \Delta],$$

or

$$\text{Prob}\,[X \leq t + \Delta | X > t] = \frac{\text{Prob}\,[t < X \leq t + \Delta]}{\text{Prob}\,[X > t]},$$

or

$$\text{Prob}\,[X \leq t + \Delta | X > t] \approx \frac{f(t)\Delta}{1 - F(t)}.$$

In the case of the exponential distribution,

$$\text{Prob}\,[X \leq t + \Delta | X > t] \approx \frac{\mu e^{-\mu t}\Delta}{e^{-\mu t}} = \mu\Delta.$$

We note that $\text{Prob}\,[X \leq t + \Delta | X > t]$ is independent of time and approximately proportional to the length Δ of the time interval. This is why the constant of proportionality μ is referred to as the "probability of dying per unit time" $(0 \leq \mu < \infty)$.

Example 1. In our study of the logistic equation with harvesting we found that

$$\frac{dN}{dt} = rN\left(1 - \frac{N}{K}\right) - \alpha N.$$

Hence $1/\alpha$ is the average time before being caught, while, when N is small compared to K, we have that

$$\frac{dN}{dt} \approx rN$$

with

$$N(t) = N_0 e^{rt}, \qquad 0 \leq t < t_0,$$

with t_0 and N_0 sufficiently small. Hence, r denotes the "intrinsic" growth rate, that is, the per capita growth rate in the absence of (interference) competition.

If we define

$$R_0 = \frac{r}{\alpha} = \text{(intrinsic growth rate)} \times \text{(average time before death)},$$

then when $\alpha = r$,or $R_0 = 1$, we see that

$$\begin{aligned}
\frac{dN}{dt} &= -\frac{r}{K}N^2 \\
\frac{dN}{N^2} &= -\frac{r}{K}dt \\
\int_{N_0}^{N(t)} N^{-2} dN &= -\frac{r}{K}t \\
\frac{1}{N_0} - \frac{1}{N(t)} &= -\frac{r}{K}t \\
N(t) = \frac{1}{\frac{1}{N_0} + \frac{r}{K}t} &\to 0 \text{ as } t \to \infty.
\end{aligned}$$

While when $R_0 \neq 1$, we have that

$$\frac{dN}{dt} = r\left(1 - \frac{1}{R_0}\right)N\left(1 - \frac{N}{K(1 - \frac{1}{R_0})}\right)$$

and

$$N(t) \to K^* = K\left(1 - \frac{1}{R_0}\right) > 0$$

as $t \to \infty$ when $R_0 > 1$, while $N(t) \to 0$, the only biological equilibrium, as $t \to \infty$ when $R_0 \leq 1$.

Consequently, R_0 denotes the number of descendants of the initial small (compared to K) population N_0 of founders. If $R_0 > 1$ then $N(t)$ grows and establishes itself at the level $K^* = K^*(R_0)$, while if $R_0 \leq 1$, $N(t) \to 0$ and the population becomes extinct.

In the epidemiological (SIS) model for the transmission dynamics of gonorrhea in a homosexually active population (Section 1.3) we concluded that the dynamics were governed by the single equation

$$\frac{dI}{dt} = \beta I\left(1 - \frac{I}{K}\right) - \alpha I,$$

where $\beta = q\phi c$ denoted the transmission coefficient and α the treatment or recovery rate.

We observe that

$$R_0 = \frac{\beta}{\alpha} = \text{(effective contacts)} \times \text{(effective infective period)}.$$

Hence,

$$R_0 > 1 \Rightarrow I(t) \to K\left(1 - \frac{1}{R_0}\right) > 0 \text{ as } t \to \infty,$$

while

$$R_0 \leq 1 \Rightarrow I(t) \to 0 \text{ as } t \to \infty.$$

Hence, if the number of secondary infections (descendents) generated by the small (when compared to K) initial population of infectives is greater than one, then there is an epidemic: $I(t)$ grows initially, and the disease establishes itself; in fact, $I(t) \to K^* = K(1 - 1/R_0)$, as $t \to \infty$, while if $R_0 \leq 1$ then $I(t) \to 0$ as $t \to \infty$ and the disease dies out.

Exercises

1. Show that

$$f(t) = \begin{cases} \frac{1}{b-a} & \text{for } a \leq t \leq b, a < b \\ 0 & \text{otherwise} \end{cases}$$

is a probability density. Compute its mean value and its associated cumulative distribution.

2. Show that

$$f(t) = \begin{cases} 0.005e^{-0.005t} & \text{for } a \leq t \leq \infty \\ 0 & \text{otherwise} \end{cases}$$

is a probability density. Compute its mean value as well as Prob $[X < 5]$, Prob $[X \leq 5]$, Prob $[2 < X < 7]$, Prob $[X > 8]$, and compute its cumulative distribution. What is Prob $[X \leq 7.001|X > 7]$? What is Prob $[X \leq 8|X > 7]$?

3. Suppose that a disease has two stages, a low infective stage and a high infective stage. Suppose that we begin with $I_1(0) = I_0 > 0$ individuals at time $t = 0$ all in stage 1 and that their disease progression from low to high to no disease is modeled by

$$\frac{dI_1}{dt} = -\alpha_1 I_1$$
$$\frac{dI_2}{dt} = \alpha_1 I_1 - \alpha_2 I_2$$

where $1/\alpha_1$ denotes the average time in stage one and $1/\alpha_2$ denotes the average time in stage two (before total recovery). What is the average infective period? What is the probability density associated with the times as infective?

1.8 Project 1.1: The Spruce Budworm

The spruce budworm is an insect that lives in the spruce and fir forests of eastern Canada and northeastern US. Normally the spruce budworm exists in low numbers in these forests, kept in check by birds. However, every 40 years or so there is an outbreak of these pests and their numbers can defoliate and damage most of the fir trees in a forest in about 4 years. The trees (if they are not killed) can replace their foliage in about 7 to 10 years and their life span is about 100 to 150 years. The budworms, on the other hand, live for only a few months and can increase their numbers five fold in the course of a summer. Also since the birds do not feed exclusively on budworms, their numbers are for the most part independent of the budworm population [Yodzis (1989); Strogatz (1994)]. D. Ludwig, D.D. Jones, and C.S. Holling (1978) proposed the following model of the forest budworm system,

$$\frac{dN}{dt} = RN\left(1 - \frac{N}{K}\right) - BP\left(\frac{N^2}{A^2 + N^2}\right). \tag{1.28}$$

The first term on the right hand side is just logistic growth for the spruce budworm population. N is the density of the budworms. R is the maximum growth rate of the spruce budworm population. K is the carrying capacity of the forest (here we can assume it is directly proportional to the leaf biomass).

The second term is a functional response of the predator birds times the number of predators P, B is the maximum predation rate of an individual bird (on average), and A is the budworm population when the predation rate is at half the maximum.

Part 1

1. The forest biomass is directly proportional to K, which in this model is constant. Give a reasonable rationale for this, i.e., what assumptions do you think Ludwig et al. used to justify this.

2. The predator (bird) population is also taken as a constant. Explain.

Part 2

1. In order to more easily analyze the model we first non-dimensionalize it. We can write without loss of generality the following dynamically equivalent equation:

$$\frac{dx}{dt} = rx\left(1 - \frac{x}{k}\right) - \left(\frac{x^2}{1+x^2}\right) \tag{1.29}$$

Notice we now only have to deal with two parameters. Also, all the parameters and variables are dimensionless, which means we only have to know relative values of things rather than absolute values, which may depend on scale of measurement and other factors. What are the appropriate transformations of the variables N and t such that the model has only the two parameters r and k, and what are r and k in terms of the original parameters ($R, K, A, B,$ and P)?

2. Interpret in general terms (in the sense of "something relative to something else") the rescaled variables x and t and parameters r and k. [Hint: What has "disappeared" from the functional response?]

Part 3

1. What is the equilibrium that always exists?

2. Find the other equilibria graphically by plotting on the same graph

$$f(x) = \frac{x}{1+x^2} \quad \text{and} \quad g(x) = r\left(1 - \frac{x}{k}\right).$$

Show one example of each qualitatively different situation. Try $0 < x < 25$.

3. Show the stability of the equilibria found in Part (2) by plotting $\dot{x}/x^{3/4}$ versus $x/x^{3/4}$ and drawing (by hand is OK) arrows along the x-axis showing movement toward or away from the equilibrium points. (Dividing both the x-and the y-axis by $x^{3/4}$ will rescale so that the picture is clearer).

4. What is the minimum value of r and the maximum value of k for which it is *impossible* to have three nontrivial equilibria? You can find them analytically or graphically (to two decimal places).

Part 4

1. Below is a plot (Figure 1.16) of the nontrivial equilibria x^* versus. k. Draw a few vectors for Δx on both sides of the curve. This curve is called a *bifurcation curve.* A bifurcation is a point in parameter space where equilibria appear, disappear, or change stability, and the bifurcation curve indicates the parameter values for which a change may occur.

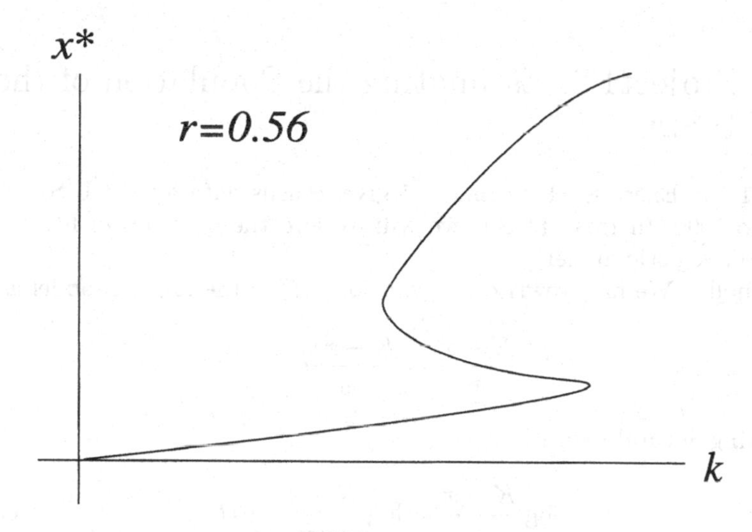

FIGURE 1.16.

2. Indicate where the curve represents the set of stable and unstable equilibria. (You should be able to do this using Part (1) without any calculations.

3. Assume the population tries to stay at equilibrium but that k (essentially the amount of leaf biomass) will drift up for the low stable x^* and will drift down for the high stable x^*. Show what happens to the population over time on the graph and explain.

Part 5

1. Compute and plot the bifurcation curves for this equation. This can be done most easily in parametric form with $r = r(x)$ and $k = k(x)$. You can see this easily by looking at your graph in Part 2. There are two conditions for the bifurcations,

   (a) x is at an equilibrium, i.e., $dx/d\tau = 0 \Rightarrow f(x) = g(x)$ and

   (b) $g(x)$ is tangent to $f(x)$ or $(d/dx)g(x) = (d/dx)f(x)$.

You should get two equations, one for $r(x)$ and the other for $k(x)$, which will define two separate curves in the (r, k)-plane (remember to restrict the range of r and k to positive values). Plot the curves in the (r, k)-plane and label the different regions with number of equilibria and whether they are "outbreak," " refuge," or "bistable."

2. The point where the two curves intersect and disappear is called the *cusp*. Calculate its coordinates.

1.9 Project1.2: Estimating the Population of the U.S.A.

Table 1.1 in Exercise 11, Section 1.1 gives census data for the U.S.A. from 1790 to 1990. In this project, we will explore the question of fitting this data to a logistic model.

Method I: We may rewrite the solution(1.7) of the logistic model as

$$\frac{K-x}{x} = \frac{K-x_0}{x_0}e^{-rt}$$

or, taking natural logarithms,

$$\log\frac{K-x}{x} = \log\frac{K-x_0}{x_0} - rt \qquad (1.30)$$

Question 1
Derive the relation (1.30) from (1.7)

Thus, if we plot $\log\frac{K-x}{x}$ against t we should obtain a straight line. However, there is a problem: We do not know the value of K. We may try to avoid this problem by estimating K by eye from the graph of the data points(t, x). If we obtain similar curves when we fit the data to a logistic curve for several different values of K, so that our results are not very sensitive to changes in K, then we may have some confidence in these results.

Question 2
For each of the values $K = 200, K = 250, K = 300$, use the data of Table 1.1 to plot $\log\frac{K-x}{x}$ as a function of t.

Question 3
For each of the values $K = 200, K = 250, K = 300$, use the method of least squares (Exercise 11,Section1.1) to estimate the slope $-r$ and the intercept $\log\frac{K-x_0}{x_0}$

Question 4
For each of the values $K = 200, K = 250, K = 300$, use the results obtained in Question 3 to give a function describing the population size, and use this function to predict the result of the Year 2000 census.

Method II: From the logistic model, we have

$$\frac{x'}{x} = r(1 - \frac{x}{K})$$

Thus if we plot $\frac{x'(t)}{x(t)}$ as a function of x, we should obtain a straight line with x−intercept K and slope $\frac{-r}{K}$. The problem here is that our data describes x, not x'. However, we can use the data to estimate x'. If x_i and x_{i+1} are two consecutive measurements taken with a time interval h, we may approximate x_i' by $\frac{x_{i+1}-x_i}{h}$. Actually, the approximation $\frac{x_{i+1}-x_{i-1}}{2h}$ is an approximation to x_i' which is significantly more accurate.

Question 5
Use the data of Table 1.1 to estimate x_i' and then plot $\frac{x_i'}{x_i}$ against x_i.

Question 6
Use the method of least squares to estimate r and K, and use your result to predict the result of the Year 2000 census.

In Question 6 you should find that the data is quite close to your straight line up to 1940 but not for 1950 and later.

Question 7
Using only the data from 1950 on, estimate x_i' and plot $\frac{x_i'}{x_i}$ against x_i. Then use the method of least squares to estimate r and K, and use your result to predict the result of the Year 2000 census.

Question 8
What reasons might there be for the apparent jump in carrying capacity between 1940 and 1950?

Over the past 150 years, immigration to the U. S. A. has varied considerably from year to year but has averaged more than 150,000 per year. The logistic model does not include any immigration.

Question 9
Suggest a modification of the logistic model which would include immigration. What would you expect to be the effect of this modification on the estimate of carrying capacity which would be obtained using the same data?

Question 10
Taking into account all that you have done in this project, what is your best guess for the result of the Year 2000 census?

2

Discrete Population Models

2.1 Introduction: Linear Models

In this chapter we shall consider populations with a fixed interval between generations or possibly a fixed interval between measurements. Thus, we shall describe population size by a *sequence* $\{x_n\}$, with x_0 denoting the initial population size, x_1 the population size at the next generation (at time t_1), x_2 the population size at the second generation (at time t_2), and so on. The underlying assumption will always be that population size at each stage is determined by the population sizes in past generations, but that intermediate population sizes between generations are not needed. Usually the time interval between generations is taken to be a constant.

For example, suppose the population changes only through births and deaths, so that $x_{n+1} - x_n$ is the number of births minus the number of deaths over the time interval from t_n to t_{n+1}. Suppose further that the birth and death rates are constants b and d, respectively (that is, if the population size is x then there are bx births and dx deaths in that generation). Then

$$x_{n+1} - x_n = (b - d)x_n,$$

or

$$x_{n+1} = x_n + (b - d)x_n = (1 + b - d)x_n.$$

We let $r = 1 + b - d$ and obtain the *linear homogeneous difference equation*

$$x_{n+1} = rx_n.$$

This together with the prescribed initial population size x_0 determines the population size in each generation. By a *solution* of the difference equation $x_{n+1} = rx_n$ with initial value x_0 we mean a sequence $\{x_n\}$ such that $x_{n+1} = rx_n$ for $n = 0, 1, 2, \cdots$, with x_0 as prescribed.

It is easy to solve the difference equation $x_{n+1} = rx_n$ algebraically. We begin by observing that $x_1 = rx_0$, $x_2 = rx_1 = r^2x_0$, $x_3 = rx_2 = r^3x_0$, and then we guess (and prove by induction) that the unique solution is $x_n = r^n x_0$ $(n = 0, 1, 2, \cdots)$. It follows that if $|r| < 1$ then $x_n \to 0$ as $n \to \infty$, while if $|r| > 1$ then x_n grows unbounded as $n \to \infty$. More precisely, if $0 \le r < 1$, x_n decreases monotonically to zero; if $-1 < r < 0$, x_n oscillates, alternating positive and negative values, but tends to zero; if $r > 1, x_n$ increases to $+\infty$; if $r < -1, x_n$ oscillates unboundedly. Negative values of x_n for this difference equation have no biological meaning, but we soon will consider difference equations in which the unknown is a deviation from equilibrium (which may be either positive or negative) rather than a population size. For this reason we have used the difference equation $x_{n+1} = rx_n$ as our first example, even though a more plausible model for a real population might be

$$x_{n+1} = \begin{cases} rx_n & \text{for } x_n > 0 \\ 0 & \text{for } x_n \le 0 \end{cases}$$

which says that the population becomes extinct once it becomes zero in any generation. This will occur if and only if $r \le 0$. The model $x_{n+1} = rx_n$ also arises under the assumption that all members of each generation die, but there is a constant birth rate b to form the next generation. In this case $d = 1$, so that $r = b$. We may form a different model by allowing migration and assuming a constant migration rate β per generation, with positive β denoting immigration and negative β denoting emigration. This leads to the linear nonhomogeneous difference equation

$$x_{n+1} = rx_n + \beta,$$

which may also be solved iteratively,

$$\begin{aligned} x_1 &= rx_0 + \beta \\ x_2 &= rx_1 + \beta = r(rx_0 + \beta) + \beta = r^2x_0 + r\beta + \beta \\ x_3 &= rx_2 + \beta = r(r^2x_0 + r\beta + \beta) + \beta = r^3x_0 + r^2\beta + r\beta + \beta \\ &\vdots \end{aligned}$$

Again we may guess, and then prove by induction, that

$$\begin{aligned} x_n &= r^n x_0 + \beta(r^{n-1} + r^{n-2} + \cdots + r + 1) \\ &= r^n x_0 + \frac{\beta(1 - r^n)}{1 - r} = \left(x_0 - \frac{\beta}{1 - r}\right) r^n + \frac{\beta}{1 - r}. \end{aligned}$$

If $r > 1$, then x_n grows unbounded for $\beta > -(r-1)x_0$ but x_n reaches zero if $\beta < -(r-1)x_0$; thus sufficiently large emigration will wipe out a population that would otherwise grow unbounded. If $0 < r < 1$, then x_n tends to the limit $\beta/(1-r) > 0$ for $\beta > 0$, while x_n reaches zero for $\beta < 0$. Thus, immigration may produce survival of a population that would otherwise become extinct.

The assumption of a constant growth rate independent of population size is unlikely to be reasonable for real populations except possibly while the population size is small enough not to be subject to the effects of overcrowding. Various nonlinear difference equation models have been proposed as more realistic. For example, the difference equations

$$x_{n+1} = \frac{rx_n}{x_n + A} \quad \text{[Verhulst (1845)]}$$

and

$$x_{n+1} = \frac{rx_n^2}{x_n^2 + A}$$

have been suggested as descriptions of populations that die out completely in each generations and have birth rates that saturate for large population sizes. The difference equations

$$x_{n+1} = x_n + rx_n\Big(1 - \frac{x_n}{K}\Big) \quad \text{and} \quad x_{n+1} = rx_n\Big(1 - \frac{x_n}{K}\Big),$$

both called the *logistic* difference equation, and essentially equivalent, describe populations with growth rates that decrease to zero as the population grows large. Neither should be taken too seriously for large population sizes as x_{n+1} becomes negative if x_n is too large. Another form, which could with some justification also be called the logistic equation is

$$x_{n+1} = x_n e^{r\left(1 - x_n/K\right)}.$$

Here the growth rate decreases to zero as $x_n \to \infty$, but x_{n+1} cannot become negative. Other difference equations, which have in fact been used as models to try to fit field data, are

$$x_{x+1} = rx_n(1 + \alpha x_n)^{-\beta} \quad \text{[Hassell (1975)]}$$

and

$$x_{n+1} = \begin{cases} r\epsilon^{\beta} x_n^{1-\beta} & \text{for } x_n > \epsilon \\ rx_n & \text{for } x_n < \epsilon. \end{cases}$$

It should be recognized that none of these models is derived from actual population growth laws. Rather, they are attempts to give quantitative expression to rough qualitative ideas about the biological laws governing the population. For this reason, we should be skeptical of the biological significance of any deduction from a specific model that holds only for that model. Our goal should be to formulate principles that are *robust*, that is, valid for a large class of models (ideally for all models that embody some set of qualitative hypotheses). In Section 2.5 we will describe some difference equation models which have been used to model fish populations and which are based on biological assumptions. Such models give some insight into the types of qualitative hypotheses which may be realistic.

Exercises

1. Find the solution of the difference equation $x_{n+1} = \frac{1}{2}x_n,\ x_0 = 2$.

2. Find the solution of the difference equation $x_{n+1} = \frac{1}{2}x_n + 1,\ x_0 = 2$.

3. Find by calculating recursively the solution of the second order difference equation $x_{n+2} = \frac{1}{2}x_n,\ x_0 = 1,\ x_1 = -1$.

4. Consider the second order difference equation
$$x_{n+2} - 3x_{n+1} + 2x_n = 0.$$
   (a) Show that the general solution to the equation is of form
$$x_n = A_1 + 2^n A_2,$$
   where A_1 and A_2 are any constants.
   (b) Suppose that x_0 and x_1 are given. Then A_1 and A_2 must satisfy the system of equations
$$A_1 + A_2 = x_0$$
$$A_1 + 2A_2 = x_1.$$
   (c) From the general solution, solve for the specific solution with initial conditions $x_0 = 10$ and $x_1 = 20$.

5. Find by calculating recursively the solution of the second order difference equation $x_{n+2} = rx_n$, $x_0 = 1$, $x_1 = -1$.

6. Find the general form of the solution of the difference equation
$$x_{n+1} = 1 - x_n$$
   for an arbitrary initial value $x_0 = a$.

7. Consider the model
$$x_{n+1} = rx_n\left(1 - \frac{x_n}{K}\right), \quad r > 0.$$

(a) Show that $x_{n+1} < 0$ if and only if $x_n > K$.

(b) Show that $x_{n+1} > K$ is possible with $0 < x_n < K$ only for $r > 4$.

(c) What conditions on x_0 are necessary and sufficient to guarantee $x_n > 0$ for $n = 1, 2, 3, \cdots$?

8. Find the general form of the solution of the difference equation

$$x_{n+1} = 1 - x_n$$

for an arbitrary initial value $x_0 = a$.

9*. The solution of the difference equation $x_{n+2} = x_n + x_{n+1}$, $x_0 = 0$, $x_1 = 1$ is called the *Fibonacci sequence* (originally formulated by Leonardo Fibonacci (1202) to describe the number of pairs of rabbits under the hypothesis that each pair of rabbits reproduces only at age one month and age two months and produces exactly one pair of offspring on each of these two occasions, with all rabbits living exactly two months).

(a) Calculate the first eight terms of the Fibonacci sequence.

(b) Suppose it can be shown that the ratio of successive terms x_{n+1}/x_n of the Fibonacci sequence tends to a limit τ as $n \to \infty$. Show that $\tau = \frac{1}{\tau} + 1$.

(c) Deduce that $\tau = \frac{1+\sqrt{5}}{2}$.

10*. For a general, not necessarily linear, first order difference equation

$$x_{n+1} = f(x_n),$$

show that if a solution $\{x_n\}$ approaches a limit x_∞ as $n \to \infty$, then the limit x_∞ must satisfy the equation

$$x_\infty = f(x_\infty).$$

2.2 Graphical Solution of Difference Equations

There is a way of solving difference equations graphically, called the *cobwebbing method*, which we illustrate for the simple linear homogeneous example $x_{n+1} = rx_n$. We begin by drawing the *reproduction curve* $y = rx$ in the (x, y)-plane. Then we mark x_0, go vertically to the reproduction curve and from there horizontally to the line $y = x$ at the point (x_1, x_1). Then we go vertically to the reproduction curve and from there horizontally to the line $y = x$ at the point (x_2, x_2), and so on. There are four separate cases: $r > 1$, $0 < r < 1$, $-1 < r < 0$, and $r < -1$, corresponding to different

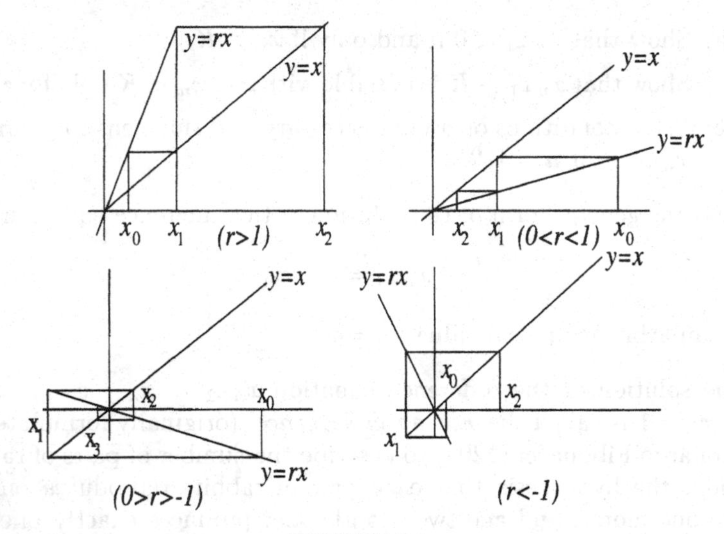

FIGURE 2.1.

relative positions of the reproduction curve $y = rx$ and the line $y = x$. In each case, the graphical solution illustrates the behavior already obtained analytically (Figure 2.1).

The cobwebbing method can be applied to any difference equation of the form $x_{n+1} = f(x_n)$ using the reproduction curve $y = f(x)$ and the line $y = x$; it gives information about the behavior of solutions. This is particularly useful for difference equations whose analytic solution is complicated. We give two more illustrative examples.

Example 1. (Verhulst equation) For the equation

$$x_{n+1} = \frac{rx_n}{x_n + A}$$

the reproduction curve is $y = rx/(x + A)$. Its slope is given by $dy/dx = rA/(x + A)^2$, which has the value r/A at $x = 0$. This means that we must distinguish the cases $r < A$, for which the line $y = x$ lies below the reproduction curve, and $r > A$, for which the line $y = x$ intersects the reproduction curve (Figure 2.2).

If $r > A$ every solution, regardless of the initial value x_0, tends to the limit $x_\infty = r - A$ where the line $y = x$ and the reproduction curve $y = rx/(x + A)$ intersect. If $r < A$, every solution tends to the limit zero.

Example 2. For the equation

$$x_{n+1} = \frac{rx_n^2}{x_n^2 + A}$$

the reproduction curve is $y = rx^2/(x^2 + A)$, which intersects the line $y = x$ at $x = 0$ and at $x = (r \pm \sqrt{r^2 - 4A})/2$. Thus for $r > 2\sqrt{A}$ there are three

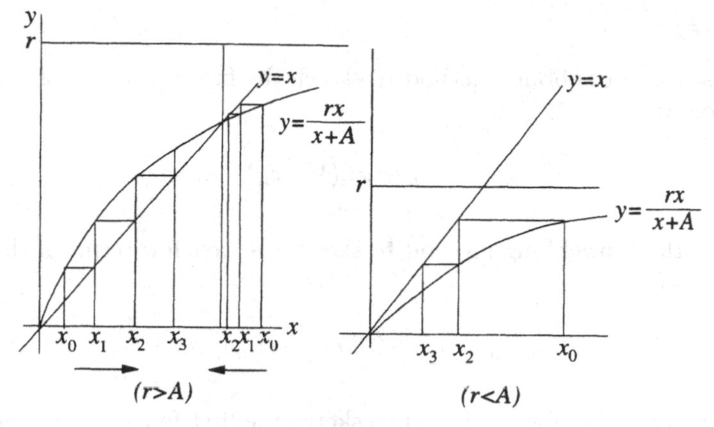

FIGURE 2.2.

real intersections, and for $r < 2\sqrt{A}$ the only real intersection is at $x = 0$ (Figure 2.3).

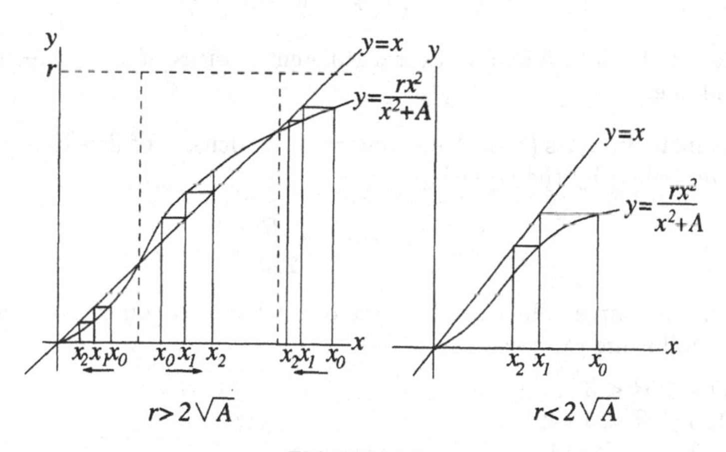

FIGURE 2.3.

If $r > 2\sqrt{A}$, all solutions with $x_0 < (r - \sqrt{r^2 - 4A})/2$ tend to zero and solutions with $x_0 > (r - \sqrt{r^2 - 4A})/2$ tend to the limit $x_\infty = (r + \sqrt{r^2 - 4A})/2$. If $r < 2\sqrt{A}$, all solutions tend to the limit zero. This model attempts to describe populations that collapse if their initial size is too small but survive if their initial size is large enough. This is analogous to the depensation model, or Allee effect, described for continuous population models in Section 1.4.

Exercises

1. Use the cobwebbing method to sketch the first few terms of the solution of

$$x_{n+1} = x_n + x_n(1 - x_n), \; x_0 = \frac{1}{2}.$$

2. Use the cobwebbing method to sketch the first few terms of the solution of

$$x_{n+1} = x_n + 2.2x_n(1 - x_n), \; x_0 = \frac{1}{2}$$

3. Use the cobwebbing method to sketch the first few terms of the solution of

$$x_{n+1} = x_n + 3x_n(1 - x_n), \; x_0 = \frac{1}{2}$$

4. Consider the difference equation

$$x_{n+1} = \begin{cases} x_n^{1/2} & \text{for } x_n > \frac{1}{4} \\ 2x_n & \text{for } x_n < \frac{1}{4}. \end{cases}$$

Sketch the solutions for several different choices of x_0 between zero and one.

5. [Kaplan & Glass (1995)] Assume that the density of flies in a swamp is described by the equation

$$x_{n+1} = Rx_n - \frac{R}{2000}x_n^2.$$

Consider three values of R, where one value of R comes from each of the following ranges:

(1) $1 \leq R < 3$,
(2) $3 \leq R < 3.449$,
(3) $3.57 \leq R < 4$.

For each value of R graph x_{n+1} as a function of x_n. Using the cobweb method follow x_n for several generations. Describe the qualitative behavior found for $R = 2$.

2.3 Equilibrium Analysis

In the examples of the preceeding section we observed a tendency for solutions to approach a limit that is the x-coordinate of an intersection of the reproduction curve and the line $y = x$. Such a value of x is a constant solution of the difference equation. This motivates the following definition of *equilibrium* of a difference equation:

$$x_{n+1} = f(x_n). \tag{2.1}$$

Definition. An *equilibrium* of a difference equation (2.1) is a value x_∞ such that $x_\infty = f(x_\infty)$, so that $x_n = x_\infty$ $(n = 0, 1, 2, \cdots)$ is a constant solution of the difference equation.

In order to describe the behavior of solutions near an equilibrium, we introduce the process of *linearization* just as we did in Section 1.4 for first order differential equations. If x_∞ is an equilibrium of the difference equation $x_{n+1} = f(x_n)$, so that $x_\infty = f(x_\infty)$, we make the change of variable $u_n = x_n - x_\infty$ $(n = 0, 1, 2, \cdots)$. Thus u_n represents deviation from the equilibrium value. Substitution gives

$$x_\infty + u_{n+1} = f(x_\infty + u_n),$$

and application of Taylor's theorem gives

$$x_\infty + u_{n+1} = f(x_\infty + u_n) = f(x_\infty) + f'(x_\infty)u_n + \frac{f'(c)}{2!}u_n^2$$

for some c between x_∞ and $x_\infty + u_n$. We write $h(u_n) = f'(c)u_n^2/2!$ and use the relation $x_\infty = f(x_\infty)$ to form the difference equation equivalent to the original difference equation (2.1),

$$u_{n+1} = f'(x_\infty)u_n + h(u_n). \tag{2.2}$$

The function $h(u)$ is small for u small in the sense that $|h(u)/u| \to 0$ as $|u| \to 0$; more precisely, for every $\epsilon > 0$ there exists $\delta > 0$ such that $|h(u)| < \epsilon|u|$ whenever $|u| < \delta$. The *linearization* of the difference equation $x_{n+1} = f(x_n)$ at the equilibrium x_∞ is defined to be the linear homogeneous difference equation

$$v_{n+1} = f'(x_\infty)v_n, \tag{2.3}$$

obtained by neglecting the higher order term $h(u_n)$ in (2.2). The importance of the linearization lies in the fact that the behavior of its solutions describes the behavior of solutions of the original equation (2.1) near the equilibrium. The behavior of solutions of the linearization has been described completely in Section 2.1. The following result explains the significance of the linearization at an equilibrium.

Theorem 2.1. *If all solutions of the linearization* (2.3) *at an equilibrium* x_∞ *tend to zero as* $n \to \infty$, *then all solutions of* (2.1) *with* x_0 *sufficiently close to* x_∞ *tend to the equilibrium* x_∞ *as* $n \to \infty$.

Proof. For convenience we write $\rho = |f'(x_\infty)|$. The assumption that all solutions of the linearization tend to zero is equivalent to $\rho < 1$. Now choose $\epsilon > 0$ so that $\rho + \epsilon < 1$. The difference equation $x_{n+1} = f(x_n)$ is equivalent to $u_{n+1} = f'(x_\infty)u_n + h(u_n)$. Then

$$\begin{aligned}|u_{n+1}| &\leq |f'(x_\infty)||u_n| + |h(u_n)| \\ &< \rho|u_n| + \epsilon|u_n|\end{aligned}$$

provided $|u_n| < \delta$, where δ is determined by the condition that $|h(u)| < \epsilon|u|$ for $|u| < \delta$. Thus, $|u_{n+1}| \leq (\rho + \epsilon)|u_n|$ provided $|u_n| < \delta$. If $|u_0| < \delta$, it is easy to show by induction that $|u_{n+1}| < \delta$ for $n = 0, 1, 2, \cdots$. This establishes $|u_{n+1}| \leq (\rho + \epsilon)|u_n|$ for $n = 0, 1, 2, \cdots$. Now it is easy to show, again by induction, that

$$|u_n| \leq (\rho + \epsilon)^n |u_0|, \; n = 0, 1, 2, \cdots.$$

Since $\rho + \epsilon < 1$, it follows that $u_n \to 0$, and thus that $x_n \to x_\infty$ as $n \to \infty$. □

In Section 2.1 we observed that if $|f'(x_\infty)| < 1$ then the solutions of $v_{n+1} = f'(x_\infty)v_n$ all tend to zero, and further that this approach is monotone if $0 < f'(x_\infty) < 1$ and oscillatory if $-1 < f'(x_\infty) < 0$. It is possible to refine Theorem 2.1 to show that the approach to an equilibrium x_∞ of $x_{n+1} = f(x_n)$ is monotone if $0 < f'(x_\infty) < 1$ and oscillatory if $-1 < f'(x_\infty) < 0$. That this is true is suggested by the cobwebbing method (Figure 2.4).

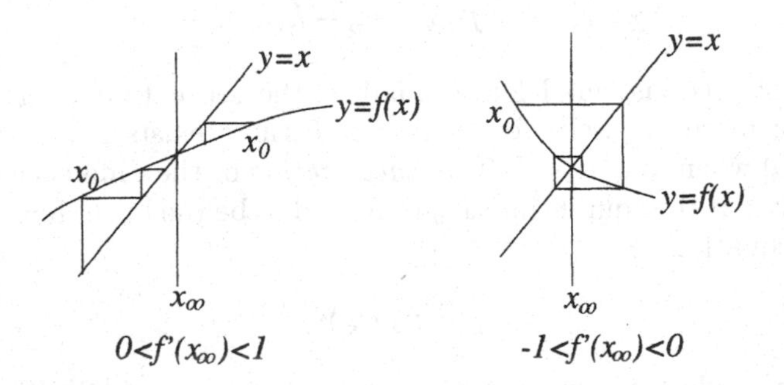

FIGURE 2.4.

The content of Theorem 2.1 is that an equilibrium x_∞ with $|f'(x_\infty)| < 1$ has the property that every solution with x_0 close enough to x_∞ remains close to x_∞ and tends to x_∞ as $n \to \infty$. This property is called *asymptotic stability* of the equilibrium x_∞. The condition $f'(x_\infty) < 1$ means that the curve $y = f(x)$ crosses the line $y = x$ from above to below as x increases, while the condition $f'(x_\infty) > -1$ means that the curve $y = f(x)$ cannot be too steep at the crossing. If $|f'(x_\infty)| > 1$, it is not difficult to show that except for the constant solution $x_n = x_\infty$ $(n = 0, 1, 2, \cdots)$, solutions cannot remain close to x_∞. This property is called *instability* of the equilibrium x_∞. An unstable equilibrium has no biological significance since any deviation, however small, is enough to force solutions away.

We emphasize that Theorem 2.1 applies to solutions whose initial value x_0 is close enough to the equilibrium x_∞. This is because the nonlinear term $h(u_n)$ in the difference equation $u_{n+1} = f'(x_\infty)u_n + h(u_n)$ is small enough to have an almost negligible effect on the solution only near the equilibrium x_∞. Theorem 2.1 gives no explicit method of computing how close to x_∞ is close enough for the solution with a given initial value to tend to x_∞. Often this can be seen in practice by using the cobwebbing method of constructing solutions graphically, as we have shown in Section 2.2. Proofs of the theorems in this section may be found in such books as [Elaydi (1996)] or [Sandefur(1990)].

Example 1. For the *logistic* difference equation

$$x_{n+1} = x_n + rx_n\left(1 - \frac{x_n}{K}\right)$$

with $f(x) = (1+r)x - rx^2/K$ and $f'(x) = (1+r) - 2rx/K$, it is easy to find equilibria by solving the quadratic equation $x = x + rx(1 - x/K)$ and obtaining the roots $x = 0$ and $x = K$. Since $f'(0) = 1 + r$, the equilibrium $x = 0$ is asymptotically stable if $-1 < 1+r < 1$, or $-2 < r < 0$. Since $r > 0$ in applications, this means that the equilibrium $x = 0$ is unstable. Since $f'(K) = 1 - r$, the equilibrium $x = K$ is asymptotically stable if $0 < r < 2$. It is not difficult to show that for $0 < r < 2$, every solution tends to the equilibrium K. If $r > 2$, the equilibrium $x = K$ is unstable and there is no asymptotically stable equilibrium to which solutions can tend. In the following section, we shall explore the behavior of solutions if $r > 2$ in more detail.

The logistic difference equation is sometimes presented in the form

$$x_{n+1} = rx_n\left(1 - \frac{x_n}{K}\right).$$

The study of the equation in this form is quite similar to the previous discussion; there is an equilibrium at $x = 0$ which is asymptotically stable if $r < 1$, in which case every solution tends to zero, and an equilibrium at $x = K(1 - 1/r)$ which is asymptotically stable if $1 < r < 3$, in which case every solution tends to $K(1 - 1/r)$, and if $r > 3$ there is no asymptotically stable equilibrium.

Example 2. For the Verhulst equation

$$x_{n+1} = \frac{rx_n}{x_n + A},$$

we have $f(x) = rx/(x + A)$; $f'(x) = rA/(x + A)^2$. The solution of $x = rx/(x + A)$ gives two roots, $x = 0$ and $x = r - A$. Thus, if $r < A$ the only equilibrium corresponding to a nonnegative population size is $x = 0$. Since $f'(0) = r/A < 1$, this equilibrium is asymptotically stable and every solution tends to zero. If $r > A$ there are two equilibria, $x = 0$, and $x = x_\infty = r - A$. Since $f'(0) = r/A > 1$, the equilibrium at $x = 0$ is unstable. Since $f'(x_\infty) = A/r < 1$, the equilibrium x_∞ is asymptotically stable. We have seen in Section 2.2 (graphically) that in fact every solution approaches x_∞, that is, that the equilibrium x_∞ is globally asymptotically stable.

Exercises

In Exercises 1 through 6 find each equilibrium of the given difference equation and determine whether it is asymptotically stable or unstable.

1. $x_{n+1} = \frac{rx_n^2}{x_n^2+A}$ (r and A are nonnegative)

2. $x_{n+1} = x_n e^{r\left(1-x_n/K\right)}$

3. $x_{n+1} = rx_n(1+\alpha x_n)^{-\beta}$

4. $x_{n+1} = \begin{cases} x_n^{1/2} & \text{for } x_n > \frac{1}{4} \\ 2x_n & \text{for } x_n < \frac{1}{4} \end{cases}$

5. $x_{n+1} = \frac{2x_n}{1+x_n}$

6. $x_{n+1} = x_n \log x_n$

7. (a) A population is governed by the difference equation

$$x_{n+1} = x_n e^{3-x_n}$$

Show that all equilibria are unstable.

(b) The population of part (a) is to be stabilized by removing a fraction p $(0 < p < 1)$ of the population in each time period after all births and deaths have taken place, to give the model

$$x_{n+1} = (1-p)x_n e^{3-(1-p)x_n}$$

For what values of p does the population have an asymptotically stable positive equilibrium?

8. (a) In the Fibonacci equation (see Exercise 9, Section 2.1) $x_{n+2} = x_n + x_{n+1}$, make the change of variable $u_n = x_{n+1}/x_n$ and obtain the transformed difference equation $u_{n+1} = 1 + 1/u_n$.

(b) Find all equilibria of the transformed difference equation of part (a) and determine which are asymptotically stable.

9. (a) Find the nonnegative equilibria of a population governed by

$$x_{n+1} = \frac{3x_n^2}{x_n^2 + 2}$$

and check for stability.

(b) Suppose a fraction a is removed from the population in each generation, so that the model becomes

$$x_{n+1} = \frac{3x_n^2}{x_n^2 + 2} - ax_n.$$

For what values of a is there a stable equilibrium only at $x = 0$?

10. [Kaplan & Glass (1995)] The following equation plays a role in the analysis of nonlinear models of gene and neural networks:

$$x_{n+1} = \frac{\alpha x_n}{1 + \beta x_n},$$

where α and β are positive numbers and $x_n \geq 0$.

(a) Algebraically determine the fixed points. For each fixed point give the range of α and β for which it exists, indicate whether the fixed point is stable or unstable, and state whether the dynamics in the neighborhood of the fixed point are monotonic or oscillatory. For parts (b) and (c) assume $\alpha = \beta = 1$.

(b) Sketch the graph of x_{n+1} as a function of x_n. Graphically iterate the equation starting from the initial condition $x_0 = 10$. What happens as the number of iterates approaches ∞?

(c) Algebraically determine x_{n+2} as a function of x_n, and x_{n+3} as a function of x_n. Based on these computations what is the algebraic expression for x_{n+k} as a function of x_n? What is the behavior of x_{n+k} as $k \to \infty$? This should agree with what you found in part (b).

11. Consider the following pair of difference equations:

$$x_{n+1} = f(n, x_n)$$
$$y_{n+1} = g(n, y_n)$$

where f and g are nonnegative functions defined on $[0, \infty)$. Assume that $f(n, x_n) \leq g(n, x_n)$ for each nonnegative integer n and each nonnegative x_n and $g(n, y(n))$ is nondecreasing with respect to the second argument y_n. Prove that if $\{x_n\}_{n \geq 0}$ is a solution of the first equation and $\{y_n\}_{n \geq 0}$ is a solution of the second equation with $x_0 \leq y_0$, then $x_n \leq y_n$ for all $n = 0, 1, 2, 3, \cdots$.

12. Consider the single species discrete-time population model

$$x_{n+1} = x_n e^{\beta \frac{1-x_n}{1+x_n}}$$

where $x_n \geq 0$ is the nonnegative population density in generation n, and the positive constant $\beta > 4$. Let $\{x_1, x_2\}$ be a 2-cycle of this model, where $x_1 > 0$, $x_2 > 0$ and $x_1 \neq x_2$. Decide if the 2-cycle $\{x_1, x_2\}$ is asymptotically stable. Explain.

2.4 Period-Doubling and Chaotic Behavior

For the logistic difference equation

$$x_{n+1} = x_n + rx_n\left(1 - \frac{x_n}{K}\right),$$

we have seen that the equilibrium $x_\infty = K$ is asymptotically stable for $0 < r < 2$. How do solutions behave if $r > 2$? We may think of r as a parameter that may be varied, and as r passes through the value 2 there must be a fundamental change in the behavior of solutions. While there is an equilibrium of K for all r, every solution tends to this equilibrium if $0 < r < 2$, but no solution other than the constant solution $x_n = K$ $(n = 0, 1, 2, \cdots)$ tends to this equilibrium if $r > 2$. What happens when r increases past 2 is that a solution of period 2 appears. By this we mean that there are two values, x^+ and x^-, with $f(x^+) = x^-$, $f(x^-) = x^+$ so that the alternating sequence $x^+, x^-, x^+, \cdots$ is a solution of the difference equation.

To establish the existence of this periodic solution, we take

$$f(x) = x + rx\left(1 - \frac{x}{K}\right) = (1+r)x - \frac{r}{K}x^2$$

and define

$$\begin{aligned}
f_2(x) = f(f(x)) &= (1+r)f(x) - \frac{r}{K}\left(f(x)\right)^2 \\
&= (1+r)^2 x - \frac{r(1+r)}{K}x^2 - \frac{r}{K}\left((1+r)x - \frac{r}{K}x^2\right)^2 \\
&= (1+r)^2 x - \frac{r(1+r)(2+r)}{K}x^2 + \frac{2r^2}{K^2}(1+r)x^3 - \frac{r^3}{K^3}x^4.
\end{aligned}$$

We now look for equilibria of the second order difference equation

$$x_{n+2} = f_2(x_n).$$

Such equilibria give solutions of period 2 for the original difference equation $x_{n+1} = f(x_n)$. These equilibria are solutions of the fourth degree polynomial equation

$$x = (1+r)^2 x - \frac{r(r+1)(r+2)}{K}x^2 + \frac{2r^2(1+r)}{K^2}x^3 - \frac{r^3}{K^3}x^4,$$

giving

$$x\left(r^3\left(\frac{x}{K}\right)^3 - 2r^2(1+r)\left(\frac{x}{K}\right)^2 + r(r+1)(r+2)\left(\frac{x}{K}\right) - r(r+2)\right) = 0$$

or

$$x\left(\left(\frac{x}{K}\right) - 1\right)\left(r^2\left(\frac{x}{K}\right)^2 - r(r+2)\left(\frac{x}{K}\right) + (r+2)\right) = 0.$$

There are four roots, namely $x = 0, x = K$, and the roots x_+, x_- of the quadratic equation $r^2(x/K)^2 - r(r+2)(x/K) + (r+2) = 0$. Thus

$$x_+ = \frac{(r+2)+\sqrt{r^2-4}}{2r}K, \qquad x_- = \frac{(r+2)-\sqrt{r^2-4}}{2r}K,$$

and these roots are real if $r \geq 2$. We also have

$$\begin{aligned}
f(x_+) &= (1+r)x_+ - \frac{r}{K}x_+^2 \\
&= (r+1)\frac{r+2}{2r}K + (r+1)\frac{\sqrt{r^2-4}}{2r}K \\
&\quad - \frac{r}{K}\frac{K^2}{4r^2}\left((r+2)^2 + r^2 - r + 2(r+2)\sqrt{r^2-4}\right) \\
\frac{2r}{K}f(x_+) &= (r+1)(r+2) + (r+1)\sqrt{r^2-4} \\
&\quad - \frac{1}{2}\left(r^2 + 4r + 4 + r^2 - 4 + 2(r+2)\sqrt{r^2-4}\right) \\
&= (r+2) - \sqrt{r^2-4} = \frac{2r}{K}x_-.
\end{aligned}$$

Thus $f(x_+) = x_-$, and since $f_2(x_+) = f\big(f(x_+)\big) = x_+$, we have $f(x_-) = x_+$. We have now shown that if $r > 2$ there is a periodic solution of period 2 of $x_{n+1} = f(x_n)$ given by $x_n = x_+$ (if n is odd), $x_n = x_-$ (if n is even).

In order to test the stability of this periodic solution, we must compute $f_2'(x_+)$, which may be done by starting with

$$\begin{aligned}
f_2(x) - x &= -rx\left(\left(\frac{x}{K}\right) - 1\right)\left(r^2\left(\frac{x}{K}\right)^2 - r(r+2)\left(\frac{x}{K}\right) + (r+2)\right) \\
&= r\left(x - \frac{x^2}{K}\right)\left(r^2\left(\frac{x}{K}\right)^2 - r(r+2)\left(\frac{x}{K}\right) + (r+2)\right).
\end{aligned}$$

Differentiation (using the product rule) gives

$$\begin{aligned}
f_2'(x) - 1 &= r\left(1 - \frac{2x}{K}\right)\left(r^2\left(\frac{x}{K}\right)^2 - r(r+2)\left(\frac{x}{K}\right) + (r+2)\right) \\
&\quad + r\left(x - \frac{x^2}{K}\right)\left(2r\frac{x}{K^2} - \frac{r(r+2)}{K}\right).
\end{aligned}$$

Since $r^2(x_+/K)^2 - r(r+2)(x_+/K)r(r+2) = 0$, we have

$$\begin{aligned}
f_2'(x_+) - 1 &= r\left(x_+ - \frac{x_+^2}{K}\right)\left(2r\frac{x_+}{K^2} - \frac{-r(r+2)}{K}\right) \\
&= \frac{r(r+2) + r\sqrt{r^2-4}}{2}\left(1 - \frac{(r+2)+\sqrt{r^2-4}}{2r}\right)\sqrt{r^2-4} \\
&= \frac{1}{4}\left((r+2)+\sqrt{r^2-4}\right)\left((r-2)-\sqrt{r^2-4}\right)\sqrt{r^2-4} \\
&= 4 - r^2.
\end{aligned}$$

We now have $f_2'(x_+) = 5 - r^2$. If we accept the theorem that a constant solution $x_n = \bar{x}$ $(n = 1, 2, \cdots)$ of the second order difference equation $x_{n+2} = f_2(x_n)$ is asymptotically stable if $|f_2'(\bar{x})| < 1$, a theorem analogous to the one established in Section 2.3 for first order difference equations (which will be described further in Exercises 2 and 3 below), then we see that this periodic solution is asymptotically stable if $-1 < 5 - r^2 < 1$, or $2 < r < \sqrt{6} = 2.449$. Thus, if $2 < r < 2.449$, there is a solution of period 2 to which every solution of $x_{n+1} = f(x_n)$ tends.

For $r > \sqrt{6}$, the solution of period 2 is unstable, but it can be shown that a solution of period 4 appears and that this solution is asymptotically stable if $\sqrt{6} < r < 2.544$. When it becomes unstable, a solution of period 8 appears, which is asymptotically stable for $2.544 < r < 2.564$. This *period-doubling* phenomenon continues until $r = 2.570$ when periodic solutions whose periods are not powers of 2 begin to appear, but these solutions are unstable. In addition, for many values of $r > 2.570$ solutions are *aperiodic*, that is, they never settle down to either an equilibrium or a periodic orbit [Strogatz (1994)]. It is possible to show analytically that a solution of period 3 appears when $r = \sqrt{8} = 2.828$ [Saha and Strogatz (1995)]. For $r > \sqrt{8}$ there is a periodic solution of period k for every integer k, but different initial values give different solutions. There are also solutions whose behavior is apparently random; such solutions are called *chaotic* (see Figure 2.5, a bifurcation diagram generated by a program in (the virtual) Appendix C). The existence of a solution of period 3 implies chaotic behavior [Li and Yorke (1975)].

These facts, whose proofs require a close examination of the properties of continuous functions and fixed points of iterates of continuous functions, are not restricted to the logistic difference equation. It is a remarkably robust fact that for every difference equation $x_{n+1} = rf(x_n)$ with $f(x)$ a function increasing to a unique maximum and then decreasing, the period-doubling phenomenon and the onset of chaos occur. In fact, if r_n is the value of r for which the asymptotically stable solution of period 2^n appears, then

$$\lim_{n\to\infty} \frac{r_{n+1} - r_n}{r_{n+2} - r_{n+1}} = 4.6692016\ldots,$$

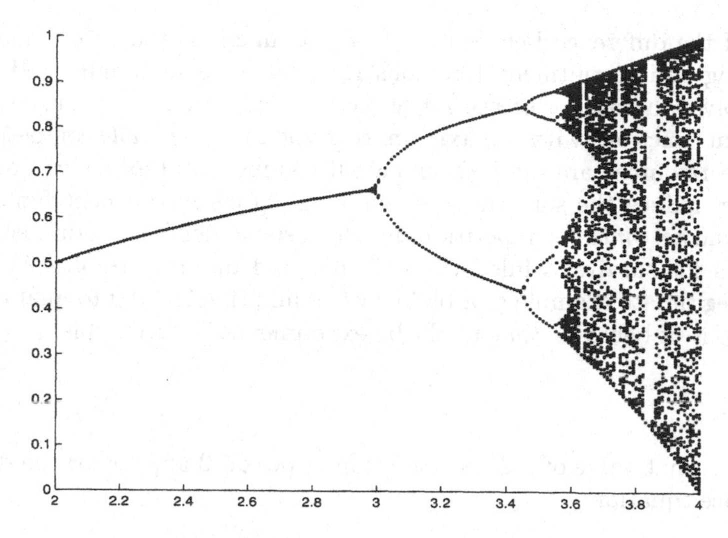

FIGURE 2.5. Bifurcation diagram.

the *Feigenbaum constant*. For the logistic equation, $r_1 = 2.000$, $r_2 = 2.449$, $r_3 = 2.544$, and $(r_2 - r_1)/(r_3 - r_2) = 4.73$; usually the limiting value is approached very rapidly. This says that the period-doubling values of r occur closer and closer together [Feigenbaum (1980)].

From a biological point of view, these results are also remarkable. One interpretation is that even very simple models can produce apparently unpredictable behavior and this suggests the possibility that the governing laws may be relatively simple and therefore discoverable [May (1976)]. There do appear to be experimental observations supporting the possibility of chaotic behavior [Gurney, Blythe, and Nisbet (1980)]. On the other hand, the fact that such simple models lead to unpredictable results suggests that experimental results and observations may not be repeatable. This suggests that one should focus on the range of values of r in which the behavior is predictable and in the chaotic ranges look for properties of solutions, such as upper and lower bounds, which are verifiable.

For models of the form $x_{n+1} = rf(x_n)$ with $f(x)$ a bounded monotone increasing functions, such as the Verhulst equation

$$x_{n+1} = \frac{rx_n}{x_n + A},$$

it is easy to verify that, since $rf(x)$ is bounded while x is not, there is a largest equilibrium x_∞ at which $y = rf(x)$ crosses the line $y = x$ from above to below. This implies $0 < rf'(x_\infty) < 1$, and shows that the equilibrium x_∞ is asymptotically stable; in fact solutions approach x_∞ monotonically. Thus, there is no possibility of period-doubling or chaotic behavior, or even of stable oscillations. This fact is also robust in that it is valid for all bounded increasing recruitment functions $f(x)$. The biological significance of the difference between recruitment functions that are monotone increasing and recruitment functions that rise to a maximum and then fall involves the nature of the intraspecies competition for resources. Recruitment functions with a maximum correspond to *scramble* competition, in which resources are divided among all members and excessive population sizes reduce the survival rate, while monotone recruitment functions correspond to *contest* competition, in which some members obtain enough resources for survival while others do not and die as a result. We now have a legitimate example of a biological assumption leading to qualitative predictions of behavior that might be experimentally observable.

Exercises

1. For what value of r does a solution of period 2 appear for the difference equation

$$x_{n+1} = r x_n e^{1-x_n}?$$

2. Let $\{x_+, x_-\}$ be a solution of period 2 of the difference equation

$$x_{n+1} = f(x_n).$$

Show that both x_+ and x_1 are equilibria of the second order difference equation.

$$x_{n+2} = f\big(f(x_n)\big).$$

3. Define a new index $k = n/2$ for n even and the iterated function

$$f_2(x) = f\big(f(x)\big).$$

Show that x_+ and x_- from Exercise 2 are equilibria of the first order difference equation

$$x_{k+1} = f_2(x_k).$$

[*Remark:* Exercise 3 together with Theorem 2.1 of Section 2.3 shows that an equilibrium x^* of the second order equation $x_{n+2} = f_2(x_n)$ is asymptotically stable if $|f_2'(x^*)| < 1$. Exercise 8 below gives another stability criterion for the asymptotic stability of a solution of period 2 of the difference equation $x_{n+1} = f(x_n)$.]

4. [Kaplan & Glass(1995)] Consider an ecological system described by the finite diference equation

$$x_{n+1} = C x_t^2 (2 - x_n), \qquad \text{for } 0 \le x_n \le 2,$$

where x_n is the population density in year n and C is a positive constant that we assume is equal to 25/16.

(a) Sketch the graph of the right hand side of this equation. Indicate the maxima, minima, and inflection points.

(b) Determine the fixed points of this system.

(c) Determine the stability at each fixed point and describe the dynamics in the neighborhood of the fixed points.

(d) In a brief sentence or two describe the expected dynamics starting from initial values of $x_0 = 1/3$ and also $x_0 = 1$ in the limit as $n \to \infty$. In particular, comment on the possibility that the population may go to extinction or to chaotic dynamics in the limit $n \to \infty$.

5. [Kaplan and Glass(1995)] The following finite difference equation has been considered as a mathematical model for a periodically stimulated biological oscillator [Bélair and Glass (1983)].

$$\phi_{n+1} = \begin{cases} 6\phi_n - 12\phi_n^2 & \text{for } 0 \le \phi_n < 0.5 \\ 12\phi_n^2 - 18\phi_n + 7 & \text{for } 0.5 \le \phi_n \le 1 \end{cases}$$

(a) Sketch ϕ_{n+1} as a function of ϕ_n for $0 \le \phi_n \le 1$. Be sure to show all maxima and minima and compute the values of ϕ_{n+1} at these extreme points.

(b) Compute all fixed points. What are the qualitative dynamics in the neighborhood of each fixed point?

(c) If you have done part (a) right, you should be able to find a cycle of period 2. What is this cycle? Show it on your sketch.

6. For the logistic difference equation $x_{n+1} = x_n + r(1 - x_n/K)$ with $r > 2$, show that

$$0 < x_- < x_\infty < x_+ < K.$$

7*. (a) Let $\{x_+, x_-\}$ be a solution of period 2 of the difference equation $x_{n+1} = f(x_n)$. Use the chain rule of calculus to show that if $f_2(x) = f(f(x))$, then

$$f_2'(x_+) = \frac{d}{dx} f(f(x))\bigg|_{x=x_+} = f'(x_-)f'(x+).$$

(b) Deduce from part (a) that the solution of period 2 is asymptotically stable if

$$|f'(x_-)| \cdot |f'(x_+)| < 1.$$

8. [Kaplan and Glass (1995)] The finite difference equation

$$x_{n+1} = 0.5 + \alpha \sin(2\pi x_n), \quad 0 \le x < 1,$$

where $0 \le \alpha < 0.5$, has been used as a mathematical model for periodic stimulation of biological oscillators .

   (a) There is one steady state. Determine this steady state and its stability as a function of α.
   (b) Sketch x_{n+1} as a function of x_n for $\alpha = 0.25$. Be sure to indicate all maxima, minima, and inflection points.
   (c) For $\alpha = 0.25$ there is a stable period 2 orbit. What is it?

9*. For what value of r does a solution of period 2 appear for the difference equation

$$x_{n+1} = r x_n e^{1-x_n}?$$

HINT: Let $f(x)$ be the right hand side function, i.e., $f(x) = rxe^{1-x}$. Find the condition for r under which $f(f(x))$ has fixed points. If you find that the condition is $r > e$, you will find the solution.

10. The population of a species is described by the finite difference equation

$$x_{n+1} = a x_n \exp(-x_n) \ \text{ for } x_n \ge 0,$$

where a is a positive constant.

   (a) Determine the fixed points.
   (b) Evaluate the stability of the fixed points.
   (c) For what value of a is there a period-doubling bifurcation (using the conclusion of the previous exercise).
   (d) For what values of a will the population go extinct starting from any initial condition?

11. The objective of this problem is to get you to read and think about some of the work on difference-equation models in population biology. Read [May (1976)]. Write a summary that deals with critical ideas, methods, and presentation in the following article. The questions you might wish to answer are:

   (a) What is the main focus of this article? Is a particular question being addressed?
   (b) Do the mathematical models help illuminate the topics? If so, in what ways?
   (c) Are there alternative methods or approaches that might have been suitable for answering the questions the author addressed?

2.5 Discrete Time Metered Models

In many populations there is a recruitment cycle in which the population size at each stage is a function of the population size at the previous stage, but the form of this function is determined by a continuous birth and death process. In this case the population size is given by a difference equation

$$x_{n+1} = f(x_n)$$

describing what may be called the long-term dynamics of the model. The function $f(x)$ is constructed from assumptions on births and deaths occurring continuously in the intervals between stages and incorporates the short-term dynamics of the model. Such models are called *metered models*. As difference equations they may be analyzed by the methods of this chapter. What is new in this section is the use of models of the type considered in Chapter 1 to establish specific forms for the reproductive curve $y = f(x)$. In many fish populations there is an annual birth process, with the number of births depending on the adult population size at the time, followed by a continuous death rate until the next birth cycle at which time the survivors make up the adult population. Such populations lend themselves naturally to metered models.

To describe the general form of a metered model, we let x_n be the size of the adult population at the nth stage. Suppose this parent stock gives rise to B_n young and the survivors of this class at time T (often one year for fish populations) are the x_{n+1} adults at the next stage. More generally, we may assume that there are R_n surviving recruits of whom H_n are harvested with the remainder $R_n - H_n$ forming the adult population x_{n+1} at the next stage. This parent stock x_{n+1} is often called the *escapement* by fishery biologists. This description assumes that none of the adults' x_n parents survive to the next stage, but it it not difficult to relax this restriction. It also assumes that harvesting occurs just before the reproductive stage.

We shall assume constant fertility, that is, that the number of births B_n is proportional to the number of adults x_n, that is,

$$B_n = \alpha x_n.$$

We also assume that between birth times there is a per capita death rate that is a function of the number of survivors from the B_n newborn members. This means that if there are $z(t)$ survivors at time then there is a function $\phi(z)$ (the per capita death rate) such that

$$\frac{dz}{dt} = -z\phi(z).$$

Then the recruitment R_n is the value for $t = T$ of the solution of the initial value problem

$$\frac{dz}{dt} = -z\phi(z), \quad z(0) = B_n = \alpha x_n. \tag{2.4}$$

Formally, we can solve by separation of variables, obtaining

$$\int_{\alpha x_n}^{R_n} \frac{dz}{z\phi(z)} = -T.$$

The function f in the metered model $x_{n+1} = f(x_n)$ is given implicitly by the relation

$$\int_{\alpha x_n}^{f(x_n)} \frac{dz}{z\phi(z)} = -T.$$

Under harvesting, the model is

$$x_{n+1} = R_n - H_n = f(x_n) - H_n,$$

with the same function f.

Example 1. (The Beverton and Holt stock recruitment model) In some bottom-feeding fish populations, including the North Atlantic plaice and haddock studied by Beverton and Holt (1957), recruitment appears to be essentially unaffected by fishing, and this is true over a wide range of fishing effort. These species have very high fertility rates and very low survivorship to adulthood. The Beverton and Holt model assumes a linear per capita mortality rate, so that the differential equation describing survivorship has the form

$$\frac{dz}{dt} = -z(\mu_1 + \mu_2 z),$$

with μ_1 and μ_2 positive constants. Explicit solution of the initial value problem (2.4) leads to a recruitment function of the form

$$R_n = \frac{ax_n}{1 + bx_n},$$

where a and b are positive constants related to μ_1 and μ_2. In fact, the same form is valid if μ_1 and μ_2 are arbitrary nonnegative functions of t. This leads to the Beverton and Holt metered model.

$$x_{n+1} = \frac{ax_n}{1 + bx_n}.$$

The reader should observe that this is equivalent to the Verhulst model

$$x_{n+1} = \frac{rx_n}{x_n + A}$$

described earlier with $a = r/A$, $b = 1/A$. As we have seen for the Verhulst equation, there is an asymptotically stable positive equilibrium only if $r > A$, or equivalently if $a > 1$.

Example 2. (The Ricker stock recruitment model) It was observed by Ricker (1954, 1958) that some species of fish, including salmon, habitually cannibalize their eggs and young. The Ricker model assumes a per capita death rate proportional to the initial size of the young population. Then the survivorship differential equation has the form

$$\frac{dz}{dt} = -zB_n = -\alpha x_n z, \quad z(0) = \alpha x_n.$$

This has the solution

$$z = \alpha x_n e^{-\alpha x_n t}$$

and therefore

$$R_n = \alpha x_n e^{-\alpha T x_n}$$

which we write

$$R_n = \alpha x_n e^{-\beta x_n}$$

by letting $\beta = \alpha T$. This leads to the Ricker metered model

$$x_{n+1} = \alpha x_n e^{-\beta x_n}.$$

Exercises

1. (a) Show that the Ricker model $x_{n+1} = \alpha x_n e^{-\beta x_n}$ has an equilibrium $x = 0$ and a positive equilibrium $x_\infty = \log \alpha / \beta$ if $\alpha > 1$.
   (b) Determine the range of values of the parameter α for which each of these equilibria is asymptotically stable.
2. In the Beverton and Holt model

$$x_{n+1} = \frac{a x_n}{1 + b x_n},$$

   determine the constants a and b in terms of α and T if the survivorship differential equation is

$$\frac{dz}{dt} = -dz^2.$$

   (Or $\mu_1 = 0, \mu_2 = d$.)
3. Analyze the behavior of the continuous analogue of the metered Ricker model,

$$\frac{dx}{dt} = \alpha x e^{-\beta x} - x,$$

   and compare with the behavior of the metered model.

4. Analyze the behavior of the continuous analogue of the metered Beverton and Holt model
$$\frac{dx}{dt} = \frac{ax}{1+bx} - x.$$

2.6 A Two-Age Group Model and Delayed Recruitment

Suppose we are interested in studying a population which in the nth generation contains x_n immature members and y_n adult members, with a birth rate depending on the number of adult members and a transition rate from immature to adult members depending on the number of immature members. If the birth rate is α and the rate of transition is β, we are led to a system of two difference equations,

$$\begin{aligned} x_{n+1} &= \alpha y_n \\ y_{n+1} &= \beta x_n, \end{aligned}$$

assuming no survival of adult members to the next generation. Graphical methods of solving this system are cumbersome, but the method of analytic solution is easy. Iteration gives $x_1 = \alpha y_0$, $y_1 = \beta x_0$; $x_2 = \alpha y_1 = \alpha\beta x_0$, $y_2 = \beta x_1 = \alpha\beta y_0$, $x_3 = \alpha y_2 = \alpha^2\beta y_0$, $y_3 = \beta x_2 = \alpha\beta^2 x_0$. The pattern becomes apparent if we introduce vector-matrix notation. Define the two-dimensional column vector

$$\vec{z}_n = \begin{pmatrix} x_n \\ y_n \end{pmatrix}$$

and the *reproduction matrix*

$$A = \begin{pmatrix} 0 & \alpha \\ \beta & 0 \end{pmatrix}.$$

Then the system can be written

$$\vec{z}_{n+1} = A\vec{z}_n,$$

and now iterative solution gives

$$\vec{z}_n = A^n \vec{z}_0,$$

where A^n is the nth power of the matrix A. More generally, we could assume a nonlinear birth function $B(y)$ and a nonlinear mortality function $D(y)$, that is, a nonlinear system

$$\begin{aligned} x_{n+1} &= B(y_n) \\ y_{n+1} &= \alpha x_n - D(y_n). \end{aligned}$$

An equilibrium of this system is a solution (x_∞, y_∞) of the system $x_\infty = B(y_\infty)$, $y_\infty = \alpha x_\infty - D(y_\infty)$. We may linearize about the equilibrium and examine the asymptotic stability of the equilibrium by studying the linearized system

$$\begin{aligned} u_{n+1} &= B'(y_\infty)v_n \\ v_{n+1} &= \alpha u_n - D'(y_\infty)v_n, \end{aligned}$$

with coefficient matrix

$$A = \begin{pmatrix} 0 & B(y_\infty) \\ \alpha & -D'(y_\infty) \end{pmatrix}.$$

Such a study requires the machinery of linear algebra, which we shall not undertake here. Models with a larger number of age groups are also natural and their study leads to systems of difference equations each of whose dimension is equal to the number of age groups. Again, the use of linear algebra is essential. In order to study a two age group model such as

$$x_{n+1} = B(y_n), \quad y_{n+1} = \alpha x_n - D(y_n)$$

without being forced to use linear algebra we may eliminate by substituting $B(y_{n-1})$ for x_n in the second equation. We then obtain a single *second order* difference equation

$$y_{n+1} = \alpha B(y_{n-1}) - D(y_n),$$

using the relation $x_{n+1} = B(y_n)$ to determine x_n once this second order equation has been solved. An equilibrium of this second order equation is a value y_∞ such that

$$y_\infty = \alpha B(y_\infty) - D(y_\infty).$$

The linearization at the equilibrium is the second order linear homogeneous difference equation

$$u_{n+1} = \alpha B'(y_\infty)u_{n-1} - D'(y_\infty)u_n.$$

In order to study the stability of an equilibrium of a difference equation of order higher than one, we first state the following linearization theorem without proof.

Theorem 2.2. *If x_∞ is an equilibrium of the difference equation*

$$x_{n+k} = f(x_{n+k-1}, x_{n+k-2}, \ldots, x_{n+1}, x_n)$$

of order k, so that

$$x_\infty = f(x_\infty, x_\infty, \ldots, x_\infty),$$

the equilibrium is asymptotically stable if all solutions of the linearization at the equilibrium

$$u_{n+k} = \sum_{j=1}^{k} a_j u_{n+k-j}$$

(with $a_j = f_j(x_\infty, x_\infty, \ldots, x_\infty)$ and f_j denoting the partial derivative with respect to the jth variable) tend to zero.

In order to determine whether all solutions of a linear difference equation tend to zero, we look for solutions of the form $x_n = \lambda^n x_0$ and obtain a *characteristic equation*–a polynomial equation of degree k–for λ. For the difference equation $u_{n+k} = \sum_{j=1}^{k} a_j u_{n+k-j}$ this characteristic equation is $\lambda^{n+k} = \sum_{j=1}^{k} a_j \lambda^{n+k-j}$, or

$$\lambda^k - \sum_{j=1}^{k} a_j \lambda^{k-j} = 0.$$

If the roots of this characteristic equation, say $\lambda_1, \lambda_2, \ldots, \lambda_k$, are distinct, then every solution of the difference equation $u_{n+k} = \sum_{j=1}^{k} a_j u_{n+k-j}$ is a linear combination of $\lambda_1^n, \lambda_2^n, \ldots, \lambda_k^n$. If the characteristic equation has multiple roots, then there are also terms $\lambda_j^n \log \lambda_j$, but in any case the condition that all solutions of a linear homogeneous difference equation tend to zero is that all roots λ_j of the characteristic equation satisfy $|\lambda_j| < 1$.

Combination of this information about the solutions of linear difference equations with Theorem 2.2 gives the following extension of Theorem 2.1 of Section 2.3.

Theorem 2.3. *Let x_∞ be an equilibrium of the difference equation of order k*

$$x_{n+k} = f(x_{n+k-1}, x_{n+k-2}, \ldots, x_{n+1}, x_n).$$

If all roots of the characteristic equation

$$\lambda^k - \sum_{j=1}^{k} f_j(x_\infty, x_\infty, \ldots, x_\infty)\lambda^{k-j} = 0$$

of the linearization at this equilibrium satisfy $|\lambda| < 1$ then the equilibrium x_∞ is asymptotically stable.

For a first order difference equation $x_{n+1} = f(x_n)$ the characteristic equation is $\lambda - f'(x_\infty) = 0$, and thus the condition for asymptotic stability is $|f'(x_\infty) < 1$, as given in Theorem 2.1 of Section 2.3. For the equilibrium x_+ of the second order difference equation

$$x_{n+2} = f_2(x_n)$$

considered in Section 2.4 the characteristic equation is $\lambda^2 - f_2'(x_+) = 0$, with roots $\lambda = \pm\sqrt{|f_2'(x_+)|}$ or $\lambda = \pm i\sqrt{|f_2'(x_+)|}$, depending on whether $f_2'(x_+) > 0$ or $f_2'(x_+) < 0$. In either case the condition for asymptotic stability is $|f_2'(x_+)| < 1$, a fact used without proof in Section 2.4.

The results developed in Theorem 2.3 would enable us to study the delayed recruitment model $y_{n+1} = \alpha B(y_{n-1}) - D(y_n)$ formulated at the beginning of this section. However, we shall instead consider the model

$$x_{n+1} = ax_n + F(x_{n-\tau}),$$

which is often used to study whale populations. Here x_n represents the adult breeding population, a $(0 \le a \le 1)$ the survival coefficient, and $F(x_{n-\tau})$ the recruitment to the adult stage with a delay of τ years. Equilibrium population size is obtained by solving

$$x_\infty = ax_\infty + F(x_\infty),$$

or $F(x_\infty) = (1-a)x_\infty = Mx_\infty$, where $M = 1 - a$ is the annual mortality rate. More generally, we could consider a model of the form $x_{n+1} = G(x_n) + F(x_{n-\tau})$ with equilibrium population size determined from $x_\infty = G(x\infty) + F(x_\infty)$. To study the stability of equilibrium, we linearize about the equilibrium by setting $x_n = u_n + x_\infty$ and neglecting higher order terms, obtaining

$$u_{n+1} = au_n + F'(x_\infty)u_{n-\tau}.$$

We let $b = F'(x_\infty)$ to write this in the form

$$u_{n+1} = au_n + bu_{n-\tau}.$$

The characteristic equation is

$$\lambda^{\tau+1} - a\lambda^\tau - b = 0,$$

and asymptotic stability of equilibrium requires $|\lambda| < 1$ for all roots of this equation.

If $\tau = 0$, the characteristic equation is $\lambda - a - b = 0$ and the stability condition is $|a+b| < 1$, or $-1 - a < b < 1 - a$

If $\tau = 1$, the characteristic equation is $\lambda^2 - a\lambda - b = 0$ and has roots

$$\lambda = a \pm \frac{\sqrt{a^2+4b}}{2}.$$

If $a^2 + 4b \ge 0$ these roots are real and the condition $|\lambda| < 1$ is equivalent to $a + \sqrt{a^2+4b} < 2$ and $a - \sqrt{a^2+4b} > -2$. These conditions give $\sqrt{a^2+4b} < 2 - a$ and $\sqrt{a^2+4b} < 2 + a$. Since $2 - a \le 2 + a$ we have the single condition $\sqrt{a^2+4b} < 2 - a$, or $b < 1 - a$. If $a^2 + 4b < 0$, the roots are complex and

$|\lambda|^2 = a^2/4 + (-a^2 - 4b)4 = -b$. Since $b < 0$, we must have $-1 < b < 0$. Combining the cases $a^2 + 4b < 0$ and $a^2 + 4b \geq 0$, we see that for $\tau = 1$ the equilibrium x_∞ is asymptotically stable if $-1 < b < 1 - a$.

For values of $\tau > 1$, the stability condition is more difficult to analyze, but it is possible to establish the following result [Levin and May (1976)].

Theorem 2.4. *There is a function $z_\tau(a) \leq -1 + a$ with $z_\tau(a) \nearrow -1 + a$ as $\tau \to \infty$ such that the equilibrium x_∞ is asymptotically stable if*

$$z_\tau(a) < b < 1 - a.$$

We have shown that $z_0(a) = -1-a$, $z_1(a) = -1$. Since $z_\tau(a) < -1+a$ for all τ, the equilibrium is certainly asymptotically stable if $-1+a < b < 1-a$, or $|b| < 1 - a$.

The population of the antarctic fin whale has been studied using this model with $F(x) = rx(1 - x/K)$, $r = 0.12$, $a = 0.96$, $k = 600,000$, $\beta = 5$. The equilibrium population size is given by $rx_\infty(1 - x_\infty/K) = (1 - a)x_\infty$, or $x_\infty = K(1-(1-a)/r)$. If we use $M = 1-a$, we have $x_\infty = K(1-M/r)$. Since $F'(x) = 4 - 2rx/K$, $F'(x_\infty) = 2M - r$. The equilibrium is certainly asymptotically stable if $|2M - r| < 1 - a = M$, or $M < r < 3M$. With $K = 600,000$, $M = 0.04$, $r = 0.12$, this condition is not satisfied, since $r = 3M$. However, since $z_\tau(a)$ is actually less than $-1 + a$, the stability condition is satisfied.

Discrete single-species models do not involve merely first order difference equations. As we have seen, age-class models lead to systems of difference equations and delayed recruitment models lead to higher order difference equations. For a unified treatment, we would have to show how to write a difference equation of order k as a system of k first order difference equations, and then use vector-matrix notation and methods of linear algebra to develop the theory of equilibria and asymptotic stability.

Exercises

1. Convert the system of difference equations

$$x_{n+1} = 2y_n, \quad y_{n+1} = 3x_n$$

   to a second order difference equation and find the first three terms of the solution with $x_0 = y_0 = 1$.

2. Solve the second order difference equation

$$x_{n+2} - x_n = 0$$

   with $x_0 = 1, \quad x_1 = -1$.

3. Solve the second order difference equation

$$x_{n+2} + x_n = 0$$

with $x_0 = 1, \quad x_1 = -1$.

The Jury criterion states that the eigenvalues of a 2×2 matrix M have magnitude less than one if and only if $\left|\text{tr}\,(M)\right| < \det(M)+1 < 2$. Use it in problems 4 and 5.

4. Assume that the population (P) of a parasite and its host population (H) is modeled by the difference equations

$$P_{t+1} = \alpha H_t(1 - e^{-aP_t})$$
$$H_{t+1} = \alpha H_t e^{-aP_t},$$

where α, a are positive.

   (a) Calculate the equilibrium population sizes and show that they are positive only if $\alpha > 1$.
   (b) Use the Jury criterion to show that if $\alpha > 1$ then the equilibrium is unstable. *Hint*: The following relation holds:

$$\frac{\alpha}{\alpha - 1} \ln \alpha > 1 \text{ for all } \alpha > 1.$$

5*. Determine all equilibria and the stability of each equilibrium for the system

$$x_{n+1} = ax_n e^{-by_n}$$
$$y_{n+1} = cx_n(1 - e^{-by_n}).$$

(This system is known as the Nicholson and Bailey model (1935) for a host-parasite system; x_n denotes the number of hosts and y_n the number of parasites.)

5. One of the common discrete time models for the growth of a single species is the Pielou logistic equation

$$x_{n+1} = \frac{\alpha x_n}{1 + \beta x_n},$$

where $x_n \geq 0$ is the size of the population at generation n, $\alpha > 1$, and $\beta > 0$. If we assume that there is a delay of time period 1 in the response of growth rate per individual to density change, we obtain the delay difference equation model

$$x_{n+1} = \frac{\alpha x_n}{1 + \beta x_{n-1}}.$$

Determine the stability of all the nonnegative fixed points of this equation.

2.7 Systems of Two Difference Equations

In Section 2.6, we examined a system of two difference equations by reducing it to a single second order difference equation. In Section 2.8 we shall examine a system that cannot be reduced to a single equation of higher order. In this section, we shall outline the main results of the analysis of stability of equilibrium of a system of two first order difference equations.

We begin with a system of two difference equations,

$$\begin{aligned} x_{n+1} &= f(x_n, y_n) \\ y_{n+1} &= g(x_n, y_n). \end{aligned} \tag{2.5}$$

An equilibrium of the system (2.5) is a solution (x_∞, y_∞) of the system

$$f(x,y) = 0, \quad g(x,y) = 0.$$

Generally, $f(x,y) = 0$ and $g(x,y) = 0$ are represented by curves in the (x,y)-plane and an equilibrium is an intersection of the two curves. If (x_∞, y_∞) is an equilibrium of (2.5), then the system (2.5) has a constant solution $x_n = x_\infty$, $y_n = y_\infty$ $(n = 1, 2, \ldots$.

The description of the behavior of solutions near an equilibrium parallels the description given in Section 2.3 for a single first order difference equation. If (x_∞, y_∞) is an equilibrium of the system (2.5) we make the change of variables $u_n = x_n - x_\infty$, $v_n = y_n - y_\infty$ $(n = 0, 1, 2, \ldots)$ so that (u, v) represents deviation from the equilibrium. We then have the system

$$\begin{aligned} u_{n+1} &= f(x_\infty + u_n, y_\infty + v_n) - x_\infty \\ &= f(x_\infty + u_n, y_\infty + v_n) - f(x_\infty, y_\infty) \\ v_{n+1} &= g(x_\infty + u_n, y_\infty + v_n) - y_\infty \\ &= g(x_\infty + u_n, y_\infty + v_n) - f(x_\infty, y_\infty). \end{aligned} \tag{2.6}$$

If we use Taylor's theorem to approximate the functions $f(x_\infty + u_n, y_\infty + v_n)$ and $g(x_\infty + u_n, y_\infty + v_n)$ by their linear terms and neglect the remainder terms, we obtain a linear system

$$\begin{aligned} u_{n+1} &= f_x(x_\infty, y_\infty)u_n + f_y(x_\infty, y_\infty)v_n \\ v_{n+1} &= g_x(x_\infty, y_\infty)u_n + g_y(x_\infty, y_\infty)v_n. \end{aligned} \tag{2.7}$$

called *the linearization of the* system (2.5) *at the equilibrium* (x_∞, y_∞), which approximates the system 2.5 near the equilibrium. The analogue of Theorem 2.1, Section 2.3, which explains the significance of the linearization at an equilibrium, is valid.

Theorem 2.5. *If all solutions of the linearization* (2.7) *of the system* (2.5) *at an equilibrium* (x_∞, y_∞) *tend to zero as* $n \to \infty$, *then all solutions of* (2.7) *with* x_0 *and* y_0 *sufficiently close to* x_∞ *and* y_∞ *respectively tend to the equilibrium* (x_∞, y_∞) *as* $n \to \infty$.

The proof is more complicated than that given in Section 2.3 for $n = 1$, and we shall omit it.

The next problem is to determine conditions under which all solutions of the linear system (2.7) approach zero. The idea behind the solution of this problem, although there are some technical complications, is to look for solutions of the form $u_n = u_0\lambda^n$, $v_n = v_0\lambda^n$ and then determine conditions under which all values of λ for which this is possible satisfy $|\lambda| < 1$. (Recall that if $|\lambda| < 1$, then $\lambda^n \to 0$ as $n \to \infty$.) The basic fact is that all solutions of the linear system (2.7) approach zero if all roots of the characteristic equation

$$\lambda^2 - \operatorname{tr} A(x_\infty, y_\infty)\lambda + \det A(x_\infty, y_\infty) = 0$$

satisfy $|\lambda| < 1$. Here, $\operatorname{tr} A$ and $\det A$ are the trace and determinant of the 2×2 matrix

$$A(x_\infty, y_\infty) = \begin{pmatrix} f_x(x_\infty, y_\infty) & f_y(x_\infty, y_\infty) \\ g_x(x_\infty, y_\infty) & g_y(x_\infty, y_\infty) \end{pmatrix}.$$

This characteristic equation may also be written as a determinant, namely as

$$\det\Big(A(x_\infty, y_\infty) - \lambda I\Big) = 0, \tag{2.8}$$

where $I = \begin{pmatrix} 1 & 0 \\ 0 & 1 \end{pmatrix}$, the identity matrix. It arises from the condition that

$$A(x_\infty, y_\infty)\begin{pmatrix} u \\ v \end{pmatrix} = \lambda\begin{pmatrix} u \\ v \end{pmatrix}$$

has a non-trivial solution for the vector $\begin{pmatrix} u \\ v \end{pmatrix}$. In this vector-matrix form, the stability result generalizes to systems of arbitrary order.

Theorem 2.6. *If all roots of the characteristic equation (2.8) at an equilibrium satisfy $|\lambda| < 1$, then all solutions of the system (2.5) with initial values sufficiently close to an equilibrium approach the equilibrium.*

A proof of this result may be found in books that explore the theory of difference equations, such as Elaydi (1996) or Sandefur (1990).

The characteristic equation for a system of k difference equations at an equilibrium is a polynomial equation of degree k. Conditions are known under which all roots of a polynomial equation have absolute value less than 1. These conditions were originally derived to analyze some economic models [Samuelson (1941)]. For a quadratic equation

$$f(\lambda) = \lambda^2 + a_1\lambda + a_2 = 0 \tag{2.9}$$

both roots satisfy $|\lambda| < 1$ if and only if

$$1 + a_1 + a_2 > 0, \quad 1 - a_1 + a_2 > 0, \quad 1 - a_2 > 0.$$

These three conditions can be combined and written as

$$0 < |a_1| < a_2 + 1 < 2, \tag{2.10}$$

which is the *Jury criterion* (Exercise 3, Section 2.6).

To establish the Jury criterion, we begin by noting that $f(\lambda) \to +\infty$ as $\lambda \to \infty$ and $\lambda \to -\infty$. If $f(-1) < 0$ there is a root less than -1 and if $f(1) < 0$ there is root greater than 1. Further, the product of the roots of (2.9) is a_2; thus we must have $|a_2| < 1$, $f(-1) > 0$, $f(1) > 0$ in order to have all roots of (2.9) satisfy $|\lambda| < 1$. We may rewrite these conditions as $-1 < a_2 < 1$ or $0 < a_2 + 1 < 2$, $f(-1) = 1 - a_1 + a_2 > 0$, $f(1) = 1 + a_1 + a_2 > 0$. The conditions $f(-1) > 0$ and $f(1) > 0$ may be combined to give

$$-(1 + a_2) < a_1 < 1 + a_2,$$

or $|a_1| < 1 + a_2$. Thus, in order to have the roots of (2.9) satisfy $|\lambda| < 1$, the conditions in (2.10) must be satisfied.

To prove that the conditions in (2.10) imply that the roots of (2.9) satisfy $|\lambda| < 1$, we consider first the case where the roots of (2.9) are complex conjugate. In this case, both roots have the same absolute value, and $|a_2| < 1$ implies that this absolute value is less than 1. If the roots of (2.9) are real and $f(-1) > 0$, $f(1) > 0$, then either both roots are less than -1 (contradicted by $|a_2| < 1$), or both roots are greater than 1 (contradicted by $|a_2| < 1$), or both roots are between -1 and 1. Thus the conditions in (2.10) imply that both roots satisfy $|\lambda| < 1$, and the Jury criterion is established.

In the next section, we will examine a system of three first order difference equations. It can be shown that the conditions under which the roots of a cubic equation

$$\lambda^3 + a_1\lambda^2 + a_2\lambda + a_3 = 0$$

satisfy $|\lambda| < 1$ are

$$\begin{aligned} 1 + a_1 + a_2 + a_3 &> 0, & 1 - a_1 + a_2 - a_3 &> 0 \\ 3 + a_1 - a_2 - 3a_3 &> 0, & 1 + a_1a_3 - a_2 - a_3^2 &> 0 \end{aligned} \tag{2.11}$$

[Samuelson (1941)]. We will make use of this result.

Exercises

1. Find all equilibria of the system

$$\begin{aligned} x_{n+1} &= B(y_n) \\ y_{n+1} &= \alpha x_n - D(y_n) \end{aligned}$$

treated as a single second order difference equation in Section 2.6, and establish conditions for their stability.

2. For the delayed recruitment model.

$$x_{n+1} = ax_n + F(x_{n-r}),$$

with $0 \leq a \leq 1$, considered in Section 2.6, the characteristic equation at an equilibrium x_∞, that is, a solution of $F(x_\infty) = (1-a)x_\infty$, is

$$\lambda^{r+1} - a\lambda^r - F'(x_\infty) = 0.$$

Determine the conditions on a and $F'(x_\infty)$ for stability of equilibrium if $r = 2$ and write them in the form

$$z_r(a) < F'(x_\infty) < 1 - a,$$

i.e., determine the function $z_r(a)$.

3. Consider the two-dimensional system

$$x_{n+1} = \frac{\alpha y_n}{1 + (x_n)^2}$$
$$y_{n+1} = \frac{\beta x_n}{1 + (y_n)^2},$$

where α and β are positive constants. If $\alpha^2 < 1$ and $\beta^2 < 1$, prove that the origin $(0,0)$ is globally asymptotically stable.

4. Consider the single species, age structured population model

$$x_{n+1} = y_n \exp(r - ax_n - y_n)$$
$$y_{n+1} = x_n,$$

where $x_n \geq 0$, $y_n \geq 0$, and the constants a, r are positive. Show that all the solutions are bounded. Interpret your result.

2.8 Oscillation in Flour Beetle Populations: A Case Study

Some recent experimental studies of flour beetles *(Tribolium Castaneum)* have indicated a possibility of behavior in the laboratory that appears to be chaotic [R.F. Costantino, R.A. Desharnais, J.M. Cushing, B. Dennis,

(1997), (1995)]. We shall describe and attempt to analyze a model for such behavior, taking note of the properties of the life cycle of the flour beetle.

The life cycle consists of larval and pupal stages, each lasting approximately two weeks, followed by an adult stage. Both larvae and adults are cannibalistic, consuming eggs and thus reducing larval recruitment. In addition, there is adult cannibalism of pupae. We take two weeks as the unit of time and formulate a discrete model describing the larval population L, pupal population P, and adult population A at two-week intervals.

If there were no cannibalism, we could begin with a linear model

$$\begin{aligned} L_{n+1} &= bA_n \\ P_{n+1} &= (1-\mu_L)L_n \\ A_{n+1} &= (1-\mu_P)P_n + (1-\mu_A)A_n, \end{aligned}$$

where b is the larval recruitment rate per adult in unit time, and μ_L, μ_P, μ_A are the death rates in the respective stages. However, in practice $\mu_P = 0$ as there is no pupal mortality except for cannibalism. We assume that cannibalistic acts occur randomly as the organisms move through the container of flour that forms their environment. This suggests a metered model with cannibalism rates proportional to the original size of the group being cannibalized, as in the Ricker fish model (Section 2.5). We are led to a model

$$\begin{aligned} L_{n+1} &= bA_n e^{-c_{EA}A_n} e^{-c_{EL}L_n} \\ P_{n+1} &= (1-\mu_L)L_n \\ A_{n+1} &= P_n e^{-c_{PA}A_n} + (1-\mu_A)A_n \end{aligned} \tag{2.12}$$

with "cannibalism coefficients" c_{EA}, c_{EL}, c_{PA}. The fractions $e^{-c_{EA}A_n}$ and $e^{-c_{EL}L_n}$ are the probabilities that an egg is not eaten in the presence of A_n adults and L_n larvae through the larval stage. The fraction $e^{-c_{PA}A_n}$ is the survival probability of a pupa through the pupal stage in the presence of A_n adults.

Equilibria of our basic model (2.12) are solutions (L, P, A) of the system of equations

$$\begin{aligned} Le^{c_{EL}L} &= bAe^{-c_{EA}A} \\ P &= (1-\mu_L)L \\ \mu_A e^{c_{PA}A} &= P. \end{aligned} \tag{2.13}$$

This system has a solution $(0,0,0)$ corresponding to extinction and also has a solution with $L > 0$, $P > 0$, $A > 0$ corresponding to survival for some sets of parameter values. We are unable to find this survival equilibrium analytically, but we may solve numerically for a given choice of parameters.

We may rewrite the equilibrium conditions by eliminating P as

$$(1-\mu_L)L = \mu_A A e^{c_{PA}A}$$
$$Le^{c_{EL}L} = bAe^{-c_{EA}A}.$$

Division of the second equation by the first gives

$$e^{c_{EL}L} = \frac{b(1-\mu_L)}{\mu_A} e^{-(c_{EA}+c_{PA})A}. \tag{2.14}$$

The left side of (2.14) increases with L and is greater than one for all positive L, while the right side of (2.14) decreases with A and is between $b(1-\mu_L/)\mu_A$ and zero. Thus, if the quantity θ, defined by

$$\theta = \frac{b(1-\mu_L)}{\mu_A}, \tag{2.15}$$

is less than one there cannot be a solution of (2.14) and thus there cannot be a survival equilibrium. On the other hand, if θ is greater then one, the equation (2.14) represents a straight line from $(0, \log\theta/c_{EL})$ to $(\log\theta/(c_{EA}+c_{PA}), 0)$ in the (A, L)-plane. An equilibrium is an intersection of this line with the curve $(1-\mu_L)L = \mu_A e^{c_{PA}A}$, which starts from the origin and grows as A increases. Thus, if $\theta > 1$ there is always a survival equilibrium.

Some experiments have been carried out with flour beetle populations and fit to the model (2.12) with the parameter values $b = 7$, $c_{EA} = c_{EL} = 0.01$, $c_{PA} = 0.005$, $\mu_L = 0.2$, $\mu_A = 0.01$. [Costantino, Desharnais, Cushing, and Dennis (1997), (1995)] Since experimental data is inevitably noisy it is not possible to determine parameters exactly, but it is possible to obtain a confidence interval for the parameters. We take these values as a baseline and then compare the model with the experiment when some of the parameters are manipulated. For example, we may remove (harvest) some adults at each census and thus set μ_A arbitrarily. In real life outside the laboratory, adult mortality may be changed by spraying with a pesticide. It is also possible to manipulate the cannibalism coefficient c_{PA} by changing the supply of food; increasing the food supply reduces the rate of cannibalism of pupae by adults.

With the parameter values given above, we find $\theta = 560$, and there is a survival equilibrium $L = 36$, $P = 29$, $A = 398$, as well as the extinction equilibrium $L = 0$, $P = 0$, $A = 0$. In order to determine the stability of these equilibria we must compute the matrix of partial derivatives at an equilibrium and form the characteristic equation as in Section 2.7.

At an equilibrium (L, P, A) this matrix is

$$\begin{pmatrix} -c_{EL}bAe^{-c_{EA}A}e^{-c_{EL}L} & 0 & be^{-c_{EL}L}e^{-c_{EA}A}(1-c_{EA}A) \\ 1-\mu_L & 0 & 0 \\ 0 & e^{-c_{PA}A} & 1-\mu_A - c_{PA}Pe^{-c_{PA}A} \end{pmatrix} \tag{2.16}$$

At the extinction equilibrium $(0,0,0)$, it reduces to

$$\begin{pmatrix} 0 & 0 & b \\ 1-\mu_L & 0 & 0 \\ 0 & 1 & 1-\mu_A \end{pmatrix}.$$

The characteristic equation at $(0,0,0)$ is (after some manipulation of signs)

$$\det \begin{pmatrix} -\lambda & 0 & b \\ 1-\mu_L & -\lambda & 0 \\ 0 & 1 & 1-\mu_A-\lambda \end{pmatrix} = \lambda^2(\lambda-(1-\mu_A)) - b(1-\mu_L) = 0,$$

or $\lambda^3-(1-\mu_A)\lambda^2-b(1-\mu_L)=0$. Thus, it has the form $\lambda^3+a_1\lambda^2+a_2\lambda+a_3 = 0$ with $a_1 = -(1-\mu_A)$, $a_2 = 0$, $a_3 = -b(1-\mu_L)$. The conditions for asymptotic stability ((2.11), Section 2.7) are

$$\begin{aligned} &1+a_1+a_2+a_3>0, \qquad && 1-a_1+a_2-a_3>0, \\ &3+a_1-a_2-3a_3>0, && 1+a_1a_3-a_2-a_3^2>0, \end{aligned} \tag{2.17}$$

and these become

$$\begin{aligned} \mu_A - b(1-\mu_L) &> 0 \\ 2-\mu_A+b(1-\mu_L) &> 0 \\ 3-(1-\mu_A)+3b(1-\mu_L) &> 0 \\ 1+b(1-\mu_L)(1-\mu_A)-b^2(1-\mu_L)^2 &> 0. \end{aligned}$$

Because $0 \le \mu_A \le 1$, $0 \le \mu_L \le 1$, the second and third of these conditions are satisfied automatically. The first condition is satisfied if and only if $\mu_A > b(1-\mu_L)$, which is equivalent to $\theta < 1$.

The last condition,

$$\left(b(1-\mu_L)\right)^2 - b(1-\mu_L)(1-\mu_A) < 1,$$

is satisfied as well since

$$\left(b(1-\mu_L)\right)^2 - b(1-\mu_L)(1-\mu_A) < \mu_A^2 - b(1-\mu_L)(1-\mu_A) < \mu_A^2 < 1.$$

Thus, the extinction equilibrium is asymptotically stable if and only if $\theta < 1$, that is, if and only if the extinction equilibrium is the only equilibrium.

At a survival equilibrium we may use the equilibrium conditions (2.13) to simplify the coefficient matrix (2.16) to

$$\begin{pmatrix} -c_{EL}L & 0 & \frac{L}{A}-c_{EA}L \\ 1-\mu_L & 0 & 0 \\ 0 & e^{c_{PA}A} & 1-\mu_A-\mu_A c_{PA}A \end{pmatrix}.$$

In this case, the characteristic equation is

$$\lambda(\lambda + c_{EL}L)\left(\lambda - (1 - \mu_A - \mu_A c_{PA} A)\right) - \left(\frac{L}{A} - c_{EA}L\right)(1-\mu_L)e^{-c_{PA}A} = 0,$$

or

$$\lambda^3 + \left(c_{EL}L + \mu_A c_{PA}A - (1-\mu_A)\right)\lambda^2 - c_{EL}L(1-\mu_A)\lambda - \left(\frac{L}{A} - c_{EA}L\right)(1-\mu_L)e^{-c_{PA}A} = 0,$$

that is, a cubic equation $\lambda^3 + a_1\lambda^2 + a_2\lambda + a_3 = 0$ with coefficients

$$\begin{aligned} a_1 &= c_{EL} + \mu_A A c_{PA} - (1-\mu_A), \\ a_2 &= -c_{EL}L(1-\mu_A), \\ a_3 &= -\left(\frac{L}{A} - c_{EA}L\right)(1-\mu_L)e^{-c_{PA}A}. \end{aligned} \tag{2.18}$$

We are unable to analyze the stability of the survival equilibrium in general, but for a particular choice of parameters b, c_{EA}, c_{EL}, c_{PA}, μ_L, μ_A we can calculate the survival equilibrium (L, P, A) numerically and then use the values given by (2.18) to check the stability condition (2.17).

With the baseline parameters $b = 7$, $c_{EA} = c_{EL} = 0.01$, $c_{PA} = 0.005$, $\mu_L = 0.2$, $\mu_A = 0.01$, the survival equilibrium is $(36, 29, 398)$, and we find from (2.18) that $a_1 = -0.61$, $a_2 = -0.36$, $a_3 = 0.43$. The stability conditions (2.17) are satisfied, and thus the survival equilibrium is asymptotically stable. This agrees with experimental observations. However, this does not validate the model since the parameter values were chosen to fit the experimental data.

To obtain some validation of the basic model, we must manipulate some of the parameter values and see if experimental observations still agree with model predictions. Thus, we set $\mu_A = 0.96$, $c_{PA} = 0.5$ by harvesting adults and reducing the flour supply. With these parameter values, the model predicts a survival equilibrium $(12, 10, 3)$, and (2.18) gives $a_1 = 1.52$, $a_2 = -0.005$, $a_3 = -0.69$ in the cubic characteristic equation. Now the stability condition $1 + a_1a_3 - a_2 - a_3^2 > 0$ is violated, and our model predicts instability of the survival equilibrium. A more detailed study of the model indicates that with $\mu_A = 0.96$ the dynamics are very sensitive to changes in the cannibalism rate c_{PA}. For $c_{PA} = 0.5$ there is a solution of period 3 and a chaotic attractor, while for $c_{PA} = 0.55$ there are two attractors and a solution of period 8. Experimental observations indicate chaotic behavior, but it is not possible to be specific about the nature of the dynamics. Nevertheless, this does indicate some validity for the model and supplies what appears to be genuinely chaotic behavior in the laboratory.

Another way in which it is possible to perturb the model is to introduce periodic forcing by varying the volume of flour. Experiments indicate that cannibalism rates are inversely proportional to flour volume. Thus we may assume

$$c_{EL} = \frac{k_{EL}}{V}, \quad c_{EA} = \frac{k_{EA}}{V}, \quad c_{PA} = \frac{k_{PA}}{V},$$

where V is the volume of flour. We make flow volume oscillate with period 2 and amplitude αV_0 about a mean V_0, so that $V_n = V_0\big(1+\alpha(-1)^n\big)$. Then the cannibalism coefficients at stage n are

$$c_{EL} = \frac{k_{EL}}{V_0\big(1+\alpha(-1)^n\big)}, \quad c_{EA} = \frac{k_{EA}}{V_0\big(1+\alpha(-1)^n\big)}, \quad c_{PA} = \frac{k_{PA}}{V_0\big(1+\alpha(-1)^n\big)}.$$

If we let c_{EL}, c_{EA}, c_{PA} devote the cannibalism coefficients in the average flour volume, $c_{EL} = k_{EL}/V_0$, $c_{EA} = k_{EA}/V_0$, $c_{PA} = k_{PA}/V_0$, we obtain the periodic model

$$\begin{aligned} L_{n+1} &= bA_n \exp\left(-\frac{c_{EL}L_n + c_{EA}A_n}{1+\alpha(-1)^n}\right) \\ P_{n+1} &= (1-\mu_L)L_n \\ A_{n+1} &= P_n \exp\left(-\frac{c_{PA}A_n}{1+\alpha(-1)^n}\right) + (1-\mu_A)A_n. \end{aligned} \tag{2.19}$$

This model has an extinction equilibrium $(0,0,0)$ which may be shown to be asymptotically stable if $\theta < 1$. If $\theta > 1$, there is a solution of period 2 which is asymptotically stable if θ is close to 1 but for larger values of θ the dynamics may be chaotic. In addition, population sizes are considerably larger than in the unforced case, and this is borne out by experiment.

There are two important lessons that may be drawn from this model. The first is that trying to control a pest population by removing adults may have unintended consequences such as large fluctuations in the pest population size. The second is that periodic variation in the environment may produce substantial increases in population size.

The analysis of the model (2.12) becomes considerably simpler if we neglect cannibalism of eggs by larvae. Mathematically, this means taking $c_{EL} = 0$. The recruitment of larvae at equilibrium is then changed from $bA_n e^{-c_{EA}A_n} e^{-c_{EL}L_n}$ to $bA_n e^{-c_{EA}A_n}$. In order to make the parameter values correspond, we should replace b by $be^{-c_{EL}L^*}$, where c_{EL} is the original cannibalism coefficient and L^* is the equilibrium larval population. With our baseline parameters this would mean replacing $b = 7$ by $b^* = 7e^{-0.36} = 4.88$. We will not carry out the analysis of this reduced model

$$\begin{aligned} L_{n+1} &= b^* A_n e^{c_{EA} a_n} \\ P_{n+1} &= (1 - \mu_L) L_n \\ A_{n+1} &= P_n e^{-c_{PA} A_n} + (1 - \mu_A) A_n \end{aligned} \tag{2.20}$$

but will indicate it in a sequence of exercises.

Exercises

1. (a) Show that the survival equilibrium of (2.20) has

   $$A = \frac{1}{c_{EA} + c_{PA}} \log \theta,$$

   and once A has been calculated,

   $$L = b^* A e^{-c_{EA} A}, \quad P = b^*(1 - \mu_L) A e^{-c_{EA} A}.$$

   (b) Calculate the equilibrium population sizes for the parameter values $b^* = 4.88$, $c_{EA} = 0.01$, $\mu_L = 0.2$, and (*i*) $\mu_A = 0.81$, $c_{PA} = 0.005$, (*ii*) $\mu_A = 0.96$, $c_{PA} = 0.5$.

2. Show that the extinction equilibrium of the model (2.20) is asymptotically stable if and only if $\theta < 1$.

3. Show that at a survival equilibrium of (2.20) the characteristic equation is a cubic polynomial with

   $$\begin{aligned} a_1 &= \mu_A c_{PA} A - (1 - \mu_A) \\ a_2 &= 0 \\ a_3 &= -\left(\frac{L}{A} - c_{EA} L \right) (1 - \mu_L) e^{-c_{PA} A}. \end{aligned}$$

4. Show that with parameter values $b^* = 4.88$, $c_{EA} = 0.01$, $\mu_L = 0.2$, the survival equilibrium of (2.20) is asymptotically stable if $\mu_A = 0.01$, $c_{PA} = 0.005$, and unstable if $\mu_A = 0.96$, $c_{PA} = 0.5$.

5. Run simulations to compare the behaviors of the models (2.12) and (2.20) with the two sets of parameter values used in this section and a variety of initial values.

6. Show that it is possible to eliminate L and P from the model (2.20) and obtain a single third order difference equation,

   $$A_{n+3} = b^*(1 - \mu_L) A_n e^{-c_{EA} A_n} e^{-c_{PA} A_{n+2}} + (1 - \mu_A) A_{n+2}.$$

2.9 Project 2.1: A Discrete *SIS* Epidemic Model

In this project we outline of analysis of an *SIS* (Suceptible Infected Suceptible) discrete epidemic model in a population with variable size. The SIS model is given by the system

$$\begin{aligned} S_{n+1} &= f(T_n) + S_n\pi(n,n+1)h(I_n) + I_n\pi(n,n+1)[1-\xi(n,n+1)] \\ I_{n+1} &= S_n\pi(n,n+1)[1-h(I_n)] + I_n\pi(n,n+1)\xi(n,n+1)\zeta(n,n+1), \end{aligned}$$

with

$$T_n = S_n + I_n = f(T_n) + T_n\pi(n,n+1) + I_n\pi(n,n+1)[\zeta(n,n+1)-1],$$

where $\pi(n,n+1)$, $\xi(n,n+1)$, $\zeta(n,n+1)$ are assumed to be constants with α, μ, σ positive constants, that is,

$$1-\pi(n,n+1) = 1-e^{-\mu}$$

is the probability of death due to natural causes,

$$1-\xi(n,n+1) = 1-e^{-\sigma}$$

is the probability of recovering,

$$1-\zeta(n,n+1) = 1-e^{-\rho}$$

is the probability of death due to infection,

$$h(I_n) = e^{-\alpha I_n}$$

is the probability of not becoming infected, $f(T_n)$ is birth or immigration rate (2 cases). In this project we take $\rho = 0$, that is, the disease is not fatal.

The model assumes that the time step is one generation; from generation n to $n+1$, infections occur before deaths; there are no infected offspring, that is, all newborns or recruits enter into the susceptible class; in the case of nonconstant recruitment, if there are no people then there are no births, that is, $f(0) = 0$; if there are too many people, then there are not enough resources to sustain further reproduction, that is, $\lim_{T_n \to \infty} f(T_n) = 0$; the probability of not becoming infected when there are no people is one, that is $h(0) = 1$; the probability of not becoming infected as the number of infected increases is a strictly decreasing function, $h'(I_n) < 0$; as the number of infected people increases, the probability of not becoming infected gives to zero, that is, $\lim_{I_n \to \infty} h(I_n) = 0$.

Case A. Assume a constant recruitment rate, that is, let

$$f(T_n) = \Lambda > 0,$$

where Λ is a constant (immigration rate).

1. Show that the model becomes:

$$S_{n+1} = \Lambda + S_n\ e^{-\mu} e^{-\alpha I_n} + I_n e^{-\mu}[1 - e^{-\sigma}]$$

$$I_{n+1} = S_n e^{-\mu}[1 - e^{-\alpha I_n}] + I_n e^{-\mu} e^{-\sigma}, \qquad (2.21)$$

   with

$$T_{n+1} = \Lambda + T_n e^{-\mu}.$$

2. Show that

$$T_n = e^{-\mu n}\left(T_0 - \frac{\Lambda}{1 - e^{-\mu}}\right) + \frac{\Lambda}{1 - e^{-\mu}}$$

   and that

$$\lim_{n\to\infty} T_n = \frac{\Lambda}{1 - e^{-\mu}} \equiv T_\infty.$$

3. Set $T_0 = T_\infty$ (initial population size). This simply means that the population starts at its asymptotic limit, that is, the population is at a demographic equilibrium. Now substitute S_n by $S_n = T_\infty - I_n$ into (2.21) and show that

$$I_{n+1} = (T_\infty - I_n)e^{-\mu}[1 - e^{-\alpha I_n}] + I_n e^{-(\mu+\sigma)}. \qquad (2.22)$$

4. Show that $I^* = 0$ is a fixed point of (2.22) and also show that the basic reproductive number is

$$R_0 = \frac{\alpha T_\infty e^{-\mu}}{1 - e^{-(\mu+\sigma)}}.$$

5. Explain the biological meaning of R_0.
6. Show that if $R_0 < 1$ then $I^* = 0$ is a global attractor of $u(I_n)$.
7. Show that if $R_0 > 1$ then the disease-free equilibrium is unstable.
8. Show that there exists a unique fixed point $I^* > 0$ of $v(I_n)$ for $R_0 > 1$.

Case B. Assume a nonconstant recruitment rate of Ricker type, that is, let

$$f(T_n) = \beta T_n e^{-\gamma T_n}.$$

Then

$$\begin{aligned} S_{n+1} &= \beta T_n e^{-\gamma T_n} + S_n e^{-\alpha I_n} e^{-\mu} + e^{-\mu}[1 - e^{-\sigma}] I_n \\ I_{n+1} &= S_n e^{-\mu}[1 - e^{-\alpha I_n}] + I_n e^{-\mu} e^{-\sigma} \\ T_{n+1} &= S_{n+1} + I_{n+1} = \beta T_n e^{-\gamma T_n} + T_n e^{-\mu} \end{aligned} \qquad (2.23)$$

where β = maximal birth rate/person/generation
Let $R_d = \beta/(1 - e^{-\mu})$,

1. Show that if $R_d < 1$ then there is no positive fixed point.

2. Show that if $R_d > 1$ then there exist two fixed points, $T'_\infty = 0$ and $T^2_\infty > 0$.

3. Show that if $1 < R_d < e^{2/(1-e^{-\mu})}$ then T^2_∞ is locally stable.

4. What is the biological interpretation of R_d? Assume that (2.23) can be "reduced" into a single "equivalent" limiting equation

$$I_{n+1} = (T_\infty - I_n)e^{-\mu}(1 - e^{-\alpha I_n}) + I_n e^{-\mu} e^{-\sigma} \tag{2.24}$$

when $1 < R_d < e^{2/(1-e^{-\mu})}$ and where $T^2_\infty = \frac{1}{\gamma}\ln(R_d)$.

5. Show that the basic reproductive number is

$$R_0 = \frac{\alpha e^{-\mu}\frac{1}{\gamma}\ln R_d}{1 - e^{-(\mu+\sigma)}}.$$

6. Show that T^2_∞ is a global attractor if $R_0 < 1$. Show that if $R_0 > 1$ then the endemic equilibrium of (2.24) is a global attractor.

7. Simulate the full system (2.23) in the region $1 < R_d < e^{2/(1-e^{-\mu})}$, where T_∞ is a fixed point, and in the regions $R_d > e^{2/(1-e^{-\mu})}$, where period-doubling bifurcation occurs on the route to chaos. Does the demography drive the disease dynamics?

References: Castillo-Chavez and Yakubu (2000b, 2000c, 2000d), Barrera, Cintron Arias, Davidenko, Denogean, and Franco (2000).

2.10 Project 2.2: A Discrete Time Two-Sex Pair Formation Model

1. Consider the following discrete time two sex pair formation model:

$$\begin{aligned}
x(t+1) &= (\beta_x\mu_x\mu_y + (1-\mu_y)\mu_x + (1-\sigma)\mu_x\mu_y)p(t) \\
&\quad + \mu_x x(t)G(x(t), y(t), p(t)), \\
y(t+1) &= (\beta_y\mu_y\mu_x + (1-\mu_x)\mu_y + (1-\sigma)\mu_x\mu_y)p(t) \\
&\quad + \mu_y y(t)H(x(t), y(t), p(t)), \\
p(t+1) &= \sigma\mu_x\mu_y p(t) + \mu_x x(t)(1 - G(x(t), y(t), p(t))).
\end{aligned}$$

where the functions $G : [0, \infty) \times [0, \infty) \times [0, \infty) \to [0, 1]$ and $H : [0, \infty) \times [0, \infty) \times [0, \infty) \to [0, 1]$ denote the state-dependent probability functions and satisfy the equation

$$\mu_x x(t)(1 - G(x(t), y(t), p(t))) = \mu_y y(t)(1 - H(x(t), y(t), p(t)))$$

and where $\beta_x, \beta_y, \mu_x, \mu_y$, and σ are constants in the interval $[0, 1]$.

(a) Given that

$$G(x(t), y(t), p(t)) = \frac{p(t)}{y(t) + p(t)}$$

where $(x(t), y(t), p(t))$ belong to the set Ω, where

$$\Omega := \{(x(t), y(t), p(t)) \mid 0 \le \frac{x(t)}{y(t)} \le \frac{\mu_y}{\mu_x(1 - G(x(t), y(t), p(t)))}\}$$

find

$$H(x(t), y(t), p(t)).$$

(b) For the given function in (a) find the marriage function $\phi : [0, \infty) \times [0, \infty) \times [0, \infty) \to [0, 1]$ that satisfies the equation

$$\begin{aligned} \phi(x(t), y(t), p(t)) &\equiv \mu_x x(t)(1 - G(x(t), y(t), p(t))) \\ &= \mu_y y(t)(1 - H(x(t), y(t), p(t))). \end{aligned}$$

(c) Show that the marriage function in (b) satisfies the following properties for all $(x(t), y(t), p(t)) \in \Omega$, and the constant $k \in [0, \infty)$:

(i)

$$\phi(x(t), y(t), p(t)) \ge 0$$

(ii)

$$\phi(kx(t), ky(t), kp(t)) = k\phi(x(t), y(t), p(t))$$

(iii)

$$\phi(x(t), 0, p(t)) = \phi(0, y(t), p(t)) = 0.$$

(d) If $\beta_x = \beta_y = \mu_x = \mu_y = \sigma$, use the marriage function in (b) to solve the following equation for the characteristic equation $\lambda = \lambda^*$:

$$-\sigma\mu_x\mu_y + \lambda = \phi\left(\frac{\beta_x\mu_x\mu_y}{\lambda - \mu_x} - 1, \frac{\beta_y\mu_y\mu_x}{\lambda - \mu_y} - 1, 1\right)$$

where

$$\frac{\beta_x\mu_x\mu_y}{\lambda - \mu_x} - 1 > 0 \quad \text{and} \quad \frac{\beta_y\mu_y\mu_x}{\lambda - \mu_y} - 1 > 0.$$

2. Use the marriage function in (b) with $\epsilon = 0$ and $\beta_x = \beta_y = \mu_x = \mu_y = \sigma$ to find a positive fixed point $[\xi_0, \eta_0, 1] \in \Omega$ of the following system (if one exists):

$$\xi(t+1) = \frac{\beta_x \mu_x \mu_y + \mu_x + \mu_x \xi(t)}{\sigma \mu_x \mu_y + \phi(\xi(t), \eta(t), 1)} - 1$$
$$\eta(t+1) = \frac{\beta_y \mu_x \mu_y + \mu_y + \mu_y \eta(t)}{\sigma \mu_x \mu_y + \phi(\xi(t), \eta(t), 1)} - 1$$
$$\varsigma(t+1) = 1$$

3. Use the Jury test to find values of σ (if any exist) for which the fixed point $[\xi_0, \eta_0, 1]$ is stable.

References: Castillo-Chavez and Yakubu (2000e, 2000f).

3

Continuous Single-Species Population Models with Delays

3.1 Introduction

Up to now in our study of continuous population models we have been assuming that $x'(t)$, the growth rate of population size at time t, depends only on $x(t)$, the population size at the same time t. However, there are situations in which the growth rate does not respond instantaneously to changes in population size. One of the first models incorporating a delay was proposed by Volterra (1926) to take into account the delay in response of a population's death rate to changes in population density caused by an accumulation of pollutants in the past. Other causes of response delays which have been mentioned in the biological literature include differences in resource consumption with respect to age structure, migration and diffusion of populations, gestation and maturation periods, delays in behavioral response to environmental changes, and dependence of a population on a food supply that requires time to recover from grazing. In deriving a mathematical model to reflect a particular biological delay mechanism one must consider carefully how this mechanism affects the growth rate. One approach to modeling delays that has been used is formulation of a discrete model (or difference equation) and consideration of the delay in the time between steps. While this is appropriate for populations with a discrete reproduction cycle, such as many fish populations, it does not accurately model populations with continuous growth and time lags. The metered models studied in Section 2.5 allow for a continuous death process but involve a discrete reproduction stage. To describe populations with a continuous reproduction stage we will need to introduce more general classes of models involving *differential-difference equations* or, even more generally, *integro-differential equations*. We will consider models in which the per capita growth rate $x'(t)/x(t)$ at time t is a function of the population size $x(t-T)$, (that is, at T time units previous) and also models with *distributed delay* in which the per capita growth rate is an integral over the past. In addition, we will examine models in which the birth rate at time t is a function of past population size, either at a single time $(t-T)$ or an integral over the past history, and the death rate is a function of the population size at time t. Different biological mechanisms may lead to either of these types of models; the two types are *not* equivalent, although their mathematical analyses follow similar paths.

The assumption of a fixed delay T will lead to a differential-difference equation of the form $x'(t) = f\big(x(t), x(t-T)\big)$. For such an equation one must prescribe initial data for a time interval of length T rather than merely an initial value $x(0)$. In other words, one specifies a function $x_0(t)$ for $-T \leq t \leq 0$ and then requires the initial condition $x(t) = x_0(t)$ for $-T \leq t \leq 0$. The assumption of a distributed delay will lead to an integro-differential equation of the form

$$x'(t) = \int_{-\infty}^{t} f\big(x(t), x(s)\big)p(t-s)ds = \int_{0}^{\infty} f\big(x(t), x(t-u)\big)p(u)du$$

for which one must prescribe initial data on $-\infty \leq t \leq 0$. More precisely if $p(u) \equiv 0$ for $u \geq U$ one must prescribe initial data on $-U \leq t \leq 0$.

Analytic solutions, even for the simplest differential-difference equations, are in general hopeless. For example, suppose we try to solve the equation $x'(t) = -\alpha x(t-1)$, α constant with $x(t) = \epsilon \geq 0$ for $-1 \leq t \leq 0$. Then $x'(t) = -\alpha\epsilon$ for $0 \leq t \leq 1$, which gives $x(t) = \epsilon - \alpha\epsilon t$ for $0 \leq t \leq 1$, and $x(1) = \epsilon(1-\alpha)$. We could use $x(t) = \epsilon(1-\alpha t)$ for $0 \leq t \leq 1$ to obtain $x'(t) = -\alpha \in [1-\alpha(t-1)]$ for $1 \leq t \leq 2$ and then integrate to find $x(t)$ for $1 \leq t \leq 2$. This technique is feasible in principle, but the analytic form of the solution tends to become more and more complicated as we proceed. For this reason, we will concentrate on other ways of attempting to describe solutions. In this example, if $\alpha < 1$ then $x(1) \geq 0$. If $\alpha > 1$, then $x(1) < 0$, and we can see that $x'(t)$ will become positive, suggesting that the solution will oscillate. In fact, if $0 \leq \alpha \leq 1/e$, $x(t)$ remains positive for all t and decreases monotonically to zero. If $1/e \leq \alpha \leq \pi/2$, $x(t)$ oscillates but tends to zero as $t \to \infty$, while if $\alpha > \pi/2$, $x(t)$ oscillates but does not tend to zero, tending instead to a periodic solution. Some additional properties of the solution are established in the exercises below.

Because of the impossibility of obtaining explicit analytic solutions of equations with delays, we will concentrate on qualitative analysis. In Chapters 1 and 2 we obtained some qualitative results for the behavior of solutions of differential equations and difference equations. Even though explicit solution was possible in some cases, the qualitative approach enabled us to obtain information about a larger class of problems. For differential-difference equations the qualitative approach follows a similar line but there are additional technical problems.

Another way of obtaining information about solutions employed in Chapters 1 and 2 was to use a computer algebra system to solve problems numerically and graph the solutions. We will use this approach for differential-difference equations as well, but there are difficulties. A program in Mathematica for solving differential-difference equations is given in the (virtual) Appendix C, but it is somewhat cumbersome. Two programs in Maple given in Appendix C are somewhat less cumbersome but run extremely slowly. For differential-difference equations, the dynamical systems program XPP, or its Windows analogue WinPP (whose acquisition is described in the preface to this book) is considerably more efficient. The solution graphs in this chapter were obtained using WinPP. For example, Figure 3.1 shows the graph of the solution of $x'(t) = -x(t-1)$, with $x(t) = 1$ for $-1 \le t \le 0$.

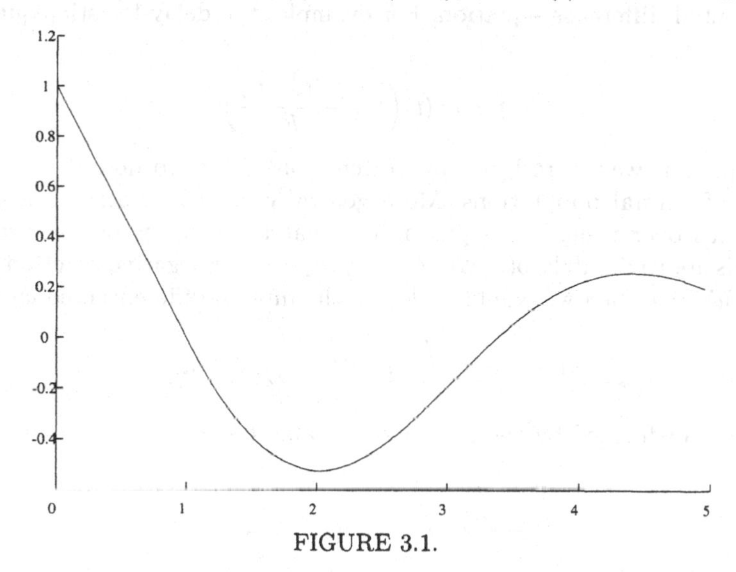

FIGURE 3.1.

Exercises

In Exercises 1 through 5, consider the differential-difference equation $x'(t) = -\alpha x(t-1)$ with initial data $x(t) \equiv 1$ for $-1 \le t \le 0$.

1. Show that $x'(t) \equiv 0$ for $t < 0$ but $x'(t) \equiv -\alpha$ for $0 \le t \le 1$, so that $x'(t)$ is discontinuous at $t = 0$.

2. Find the solution for $1 \le t \le 2$ and show that $x(2) = \frac{1}{2}(\alpha^2 - 4\alpha + 2)$.

3. Show that $x'(t)$ approaches the limit $-\alpha$ as t approaches 1 through values greater than 1, so that $x'(t)$ is continuous at $t = 1$.

4. Use a computer algebra system to sketch the solutions with $\alpha = 0.3$ and $\alpha = 4$.

5*. Show that $x''(t)$ is discontinuous at $t = 1$.

3.2 Models with Delay in Per Capita Growth Rates

If we assume that the per capita growth rate $x'(t)/x(t)$ is a function of $x(t-T)$, as might be appropriate for example in modeling a population whose food supply requires a time T to recover from grazing the causing food supply at time t to depend on the population size at time $(t-T)$, we are led to a model of the form

$$x'(t) = x(t)g\big(x(t-T)\big),$$

a differential-difference equation. For example, the delay logistic equation is

$$x'(t) = rx(t)\Big(1 - \frac{x(t-T)}{K}\Big). \tag{3.1}$$

This equation was introduced by Hutchinson (1948) to describe the dynamics of animal populations. More generally, we could assume a delay distributed over time. If the probability that the delay is between u and $u+\Delta u$ is approximately $p(u)\Delta u$, where $p(u)$ is a nonnegative function with $\int_0^\infty p(u)du = 1$ then we would be led to the integro-differential equation

$$x'(t) = x(t)\int_0^\infty g\big(x(t-u)\big)p(u)du,$$

which is transformed by the change of variable $t - u = s$ to the equivalent form

$$x'(t) = x(t)\int_{-\infty}^t g\big(x(s)\big)p(t-s)ds.$$

The average time delay would then be $\int_0^\infty up(u)du$. One form of continuous delay frequently used in population models is

$$p(u) = \frac{u}{T^2}e^{-u/T},$$

for which it is not difficult to verify that $\int_0^\infty p(u)du = 1$, $\int_0^\infty up(u)du = 2T$, and $p(u)$ has a maximum for $u = T$ (see Exercises 3, 4 below).

Because the analytic solution of differential-difference equations of the form $x'(t) = x(t)g\big(x(t-T)\big)$ is not possible in general, we attempt instead to describe the behavior of solutions in terms of equilibria and asymptotic stability just as we did for differential equations in Chapter 1 and for difference equations in Chapter 2.

Definition. An *equilibrium* of the differential-difference equation

$$x'(t) = x(t)g\bigl(x(t-T)\bigr) \tag{3.2}$$

is a value x_∞ such that $x_\infty g(x_\infty) = 0$, so that $x(t) \equiv x_\infty$ is a constant solution of the differential-difference equation.

Observe that the form of the differential-difference equation (3.2) implies that $x = 0$ is always an equilibrium. The delay logistic equation has two equilibria, $x = 0$ and $x = K$, just as does the (undelayed) logistic differential equation $x'(t) = rx(t)\bigl(1-x(t)/K\bigr)$, which is the case $T = 0$ of the delay-logistic equation (3.1). For the differential equation $x' = xg(x)$, which is the case $T = 0$ of the differential-difference equation $x'(t) = x(t)g\bigl(x(t-T)\bigr)$, an equilibrium x_∞ is asymptotically stable if and only if

$$\bigl(xg(x)\bigr)'\big|_{x=x_\infty} = x_\infty g'(x_\infty) + g(x_\infty) < 0,$$

so that the equilibrium $x = 0$ is asymptotically stable if $g(0) < 0$ and an equilibrium $x_\infty > 0$ is asymptotically stable if $g'(x_\infty) < 0$ (because $g(x_\infty) = 0$). The asymptotic stability of an equilibrium x_∞ of the differential-difference equation $x'(t) = x(t)g\bigl(x(t-T)\bigr)$ requires $\bigl(xg(x)\bigr)'\big|_{x=x_\infty} < 0$ and an additional condition. In order to describe the additional condition required we proceed much as we did in Chapter 1 for differential equations. We begin by linearizing about the equilibrium x_∞ with the aid of Taylor's theorem. That is, we let $u(t) = x(t) - x_\infty$ and obtain the equivalent differential-difference equation

$$\begin{aligned}
u'(t) &= \bigl(x_\infty + u(t)\bigr)g\bigl(x_\infty + u(t-T)\bigr)\\
&= \bigl(x_\infty + u(t)\bigr)\Bigl(g(x_\infty) + g'(x_\infty)u(t-T) + \frac{g''(c)}{2!}u(t-T)^2\Bigr)\\
&= x_\infty g(x_\infty) + g(x_\infty)u(t) + x_\infty g'(x_\infty)u(t-T) + h\bigl(u(t), u(t-T)\bigr)\\
&= g(x_\infty)u(t) + x_\infty g'(x_\infty)u(t-T) + h\bigl(u(t), u(t-T)\bigr),
\end{aligned}$$

where c is between x_∞ and $x_\infty + u(t-T)$ and

$$h\bigl(u(t), u(t-T)\bigr) = g'(x_\infty)u(t)u(t-T) + x_\infty \frac{g''(c)}{2!}u(t-T)^2$$

is "small" when $u(s)$ is small for $t - T \le s \le t$. The *linearization* of the differential-difference equation $x'(t) = x(t)g\bigl(x(t-T)\bigr)$ is defined to be the linear differential-difference equation

$$v'(t) = g(x_\infty)v(t) + x_\infty g'(x_\infty)v(t-T), \tag{3.3}$$

obtained by neglecting higher-order terms collected as $h\bigl(u(t), u(t-T)\bigr)$. Just as with differential equations and difference equations, the importance of the linearization lies in the fact that the behavior of its solutions describes the behavior of solutions of the original equation $x'(t) = x(t)g(x(t-T))$ near the equilibrium. The original equation is equivalent to

$$u'(t) = g(x_\infty)u(t) + x_\infty g'(x_\infty)u(t-T) + h(u(t), u(t-T))$$

and h is small in the sense that for every $\epsilon > 0$ there exists $\delta > 0$ such that $|h(y,z)| < \epsilon(|y| + |z|)$ whenever $|y| < \delta, |z| < \delta$. The following result, which we state without proof, is valid. A proof may be found in Bellman and Cooke (1963).

Theorem 3.1. *If all solutions of the linearization*

$$v'(t) = g(x_\infty)v(t) + x_\infty g'(x_\infty)v(t-T)$$

at an equilibrium x_∞ tend to zero as $t \to \infty$, then every solution $x(t)$ of $x'(t) = x(t)g(x(t-T))$ with $|x(t) - x_\infty|$ sufficiently small for $-T \leq t \leq 0$ tends to the equilibrium x_∞ as $t \to \infty$.

For the equilibrium $x = 0$, the linearization is $v'(t) = g(0)v(t)$. Since $g(0) > 0$ for most models of this type, the equilibrium $x = 0$ is unstable. For an equilibrium $x_\infty > 0$, $g(x_\infty) = 0$ and the linearization is $v'(t) = x_\infty g'(x_\infty)v(t-T)$. The delay logistic equation (3.1), for example, has linearization $v'(t) = -rv(t-T)$ at the equilibrium $x_\infty = K$, since $g(x) = r(1 - x/K)$, $g'(x) = -r/K$, and thus $x_\infty g'(x_\infty) = -r$. In order to determine whether all solutions of a linear differential-difference equation

$$v'(t) = bv(t-T) \tag{3.4}$$

tend to zero as $t \to \infty$, we look for the condition on the parameter λ (possibly complex) that $v(t) = ce^{\lambda t}$ be a solution ($c \neq 0$). Substitution of $v(t) = ce^{\lambda t}$ into (3.4) gives

$$c\lambda e^{\lambda t} = bce^{\lambda(t-T)} = bce^{\lambda t}e^{-\lambda T}.$$

Since $e^{\lambda t} \neq 0$, we may divide both sides by $ce^{\lambda t}$, obtaining

$$\lambda = be^{-\lambda T}. \tag{3.5}$$

This is a transcendental equation for λ having infinitely many roots. A basic result, which we shall state without proof, is that if all roots of the characteristic equation (3.5) have negative real part, then all solutions of the differential-difference equation (3.4) tend to zero as $t \to \infty$. This result is analogous to the corresponding result for differential equations (Section 1.1); a proof may be found in Bellman and Cooke (1963). However, it is much more difficult to analyze the transcendental equation (3.5) in the delay case. The undelayed case is the special case $T = 0$ of (3.4) for which the characteristic equation becomes $\lambda = b$ and the condition that all roots have negative real part is $b < 0$. For the delay case, with $T > 0$, it is possible to show that the condition that all roots of the characteristic equation (3.5) have negative real part is

$$0 < -bT < \frac{\pi}{2}. \tag{3.6}$$

This was first established by Hayes (1950). The condition (3.6) requires $b < 0$, and in addition, that the time lag T not be too large. If the condition (3.6) is violated, the characteristic equation (3.5) has complex roots with real part zero if $-bT = \pi/2$ and positive real part if $-bT > \pi/2$. The appearance of roots $\lambda = iy$ with real part zero when $-bT = \pi/2$ is reflected in the differential-difference equation (3.4) by the appearance of periodic solutions of the form $v(t) = ce^{iyt}$, or $v(t) = c\cos yt$ in real form. Combining this analysis with Theorem 3.1 we see that an equilibrium $x_\infty > 0$ of the differential-difference equation $x'(t) = x(t)g\big(x(t-T)\big)$ is asymptotically stable if $0 < -x_\infty g'(x_\infty)T < \pi/2$. For the delay-logistic equation this stability condition is $0 < rT < \pi/2$. Thus, in addition to the stability condition$\big(xg(x)\big)'\big|_{x=x_\infty} < 0$ for stability of equilibrium for ordinary differential equations, we have an additional requirement that the delay T be sufficiently small.

A graphic display of solution is often helpful for obtaining insights into the behavior of solution, even if the graph does not prove results. For example, here is a display of the solution of the delay logistic equation (3.1) with $r = 1, K = 2, T = 1$.

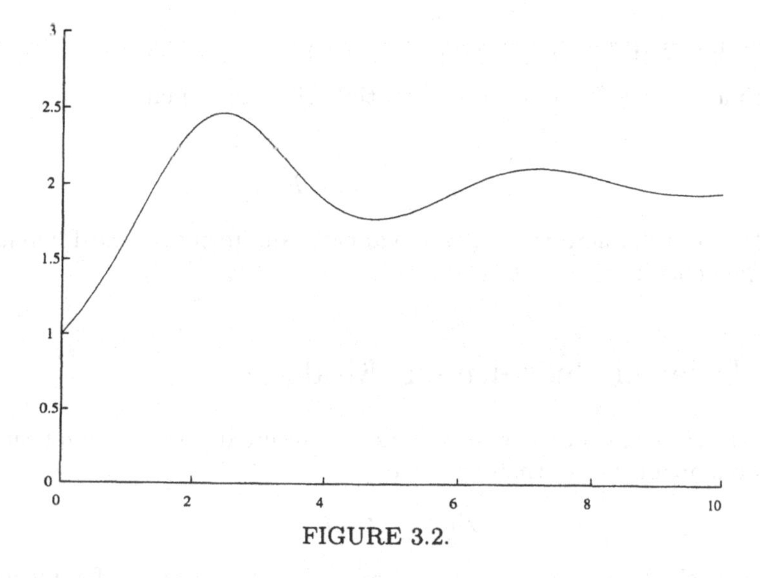

FIGURE 3.2.

Exercises

1. Show that the equilibrium $x = K$ of the delay logistic equation is asymptotically stable if $0 \leq rT < \frac{\pi}{2}$.

2. Show that the equilibrium $x = K$ of the differential-difference equation

$$x'(t) = rx(t) \log\left(\frac{K}{x(t-T)}\right)$$

is asymptotically stable if $0 \leq rT < \frac{\pi}{2}$.

3. Show that the function

$$p(u) = \frac{ue^{-u/T}}{T^2}$$

has a maximum for $u = T$.

4. With the aid of integration by parts show that

$$\int_0^\infty p(u)du = 1 \text{ and } \int_0^\infty up(u)du = 2T$$

for the function $p(u)$ of Exercise 3.

5. Use a computer algebra system to display the solutions of

$$x'(t) = rx(t)\Big(1 - x(t-1)\Big)$$

with $r = 1$, $r = 2$, $r = 3$.

6*. Find all equilibria of the differential-difference equation

$$x'(t) = rx(t)\Big(1 - \frac{x(t-T)}{K}\Big) - H.$$

7*. Find the characteristic equation at each equilibrium of the differential-difference equation of Exercise 5.

3.3 Delayed Recruitment Models

If R is the rate at which new members are recruited into a population and D is the mortality rate, then we have

$$x'(t) = R - D.$$

We will consider models in which $x(t)$ represents the number of adult members, and there is a fixed age T at which members mature and are recruited into the population. If pre-adult mortality depends only on age, there is a constant probability of survival to adulthood. If the birth rate at a given time depends only on the adult population size at that time, then the recruitment rate at time t is a function of the population size at time $(t-T)$. If the death rate at a given time also depends only on the adult population size at that time, then the death rate at time t is a function of the population size at time t. Thus, under the above hypotheses, we have a population model of the form

$$x'(t) = R(x(t-T)) - D(x(t)). \tag{3.7}$$

We will assume that the function $D(x)$ is a monotone increasing function, with $D(0) = 0$ and $\lim_{x\to\infty} D(x) = \infty$. With regard to the function $R(x$, we will consider two different types of recruitment function corresponding to different types of competition for resources among either adults or pre-adults. Under *scramble competition* the available resources are partitioned equally among all individuals. The birth rate is zero when the population size is zero and will also tend to zero for large population sizes, with some maximum birth rate occurring at an intermediate population size. Under *contest competition* a certain number of individuals can be maintained at the expense of the others. The birth rate then increases with population size but is bounded as the population size becomes large. Thus, we make one of the following sets of hypotheses on the recruitment function $R(x)$: Either $R(0) = 0$, R increases to a unique maximum value ρ for $x = \xi$ and then decreases to zero as $x \to \infty$ (scramble competition) *or* $R(0) = 0$, $R'(x) > 0$ for $0 < x < \infty$, and $\lim_{x\to\infty} R(x) = \rho$ (contest competition). We may also admit the possibility of an Allee effect by allowing $R(x) < D(x)$ on some interval $0 < x < \alpha$, with $R(\alpha) = D(\alpha)$, $R'(\alpha) > D'(\alpha)$. We will carry out an equilibrium analysis, following the kind of approach we have used in several settings previously. However, we shall first state some results relating the possible range of population size to the initial data; proofs may be found in Brauer (1986).

Theorem 3.2. *Define M by the condition*

$$D(M) = \rho = \max_{0\leq x\leq\infty} R(x).$$

Then, if the initial data satisfy $0 \leq x(t) \leq M$ for $-T \leq t \leq 0$, the solution $x(t)$ of the differential-difference equation (3.7) *satisfies $0 \leq x(t) \leq M$.*

For studying the possibility of extinction of a population it is useful to have a criterion under which the population size has a positive lower bound.

Theorem 3.3. *In the case of scramble competition define m by the condition*

$$D(m) = R(M).$$

If $M > \xi$, and if the initial data satisfy $0 < m \leq x(t) \leq M$ for $-T \leq t \leq 0$, then the solution $x(t)$ of (3.7) satisfies $m \leq x(t) \leq M$ for $t \geq 0$.

This lower bound result does not cover the possibility of an Allee effect, contest competition, or scramble competition with $M \leq \xi$. For differential-difference equation models, unlike differential equation models, it is possible for the population size to drop to zero without signifying extinction because there may be juvenile members present who will mature into adult members. Extinction is guaranteed, however, if the population size remains zero for a time interval of length T. The start of the qualitative analysis of the delayed-recruitment model (3.7) follows our standard pattern.

Definition. An *equilibrium* of the differential-difference equation

$$x'(t) = R(x(t-T)) - D(x(t))$$

is a value x_∞ such that $R(x_\infty) = D(x_\infty)$ an intersection of the recruitment curve $y = R(x)$ and the mortality curve $y = D(x)$.

The linearization at an equilibrium, obtained in the usual way by letting $u = x - x_\infty$, substituting, expanding by Taylor's theorem and neglecting higher order terms, is

$$u'(t) = -D'(x_\infty)u(t) + R'(x_\infty)u(t-T). \tag{3.8}$$

The following result, analogous to Theorem 3.1 of the preceding section, whose proof may be found in Bellman and Cooke (1963), is valid.

Theorem 3.4. *If all solutions of the linearization (3.8) at an equilibrium x_∞ tend to zero as $t \to \infty$ then every solution $x(t)$ of the differential-difference equation (3.7) with $|x(t) - x_\infty|$ sufficiently small for $-T \leq t \leq 0$ tends to the equilibrium x_∞ as $t \to \infty$.*

In order to describe the behavior of solutions of the linearization, we must study a problem more general than that considered in the preceding section, namely the behavior of solutions of the linear differential-difference equation

$$u'(t) = au(t) + bu(t-T). \tag{3.9}$$

(In the preceding section, we considered the special case $a = 0$.) We look for solutions of the form $u(t) = ce^{\lambda t}$ and obtain the *characteristic equation* $\lambda = a + be^{-\lambda T}$. In order that all solutions of $u'(t) = au(t) + bu(t-T)$ tend to zero as $t \to \infty$, all solutions of the characteristic equation must have negative real part. It is possible to prove the following result [Hayes (1950)].

Theorem 3.5. *Let $p = aT$, $q = bT$. Then all solutions of (3.9) tend to zero as $t \to \infty$ if*

$$p < -q < p \sec z = \frac{z}{\sin z},$$

where z is the solution in $(0, \pi)$ of $z = p \tan z$ if $p \neq 0$, and $z = \pi/2$ so that $z/\sin z = \pi/2$ if $p = 0$.

In fact, if $p < -q < e^{p-1}$ the convergence of solutions to zero is ultimately monotone, while if $p < e^{p-1} < -q < p \sec z$ the convergence of solutions is oscillatory. In particular, if $p < -q < -p$, all solutions tend to zero as $t \to \infty$ for every value of T. The stability region for (3.9) in the (p, q)-parameter plane is shown in Figure 3.3.

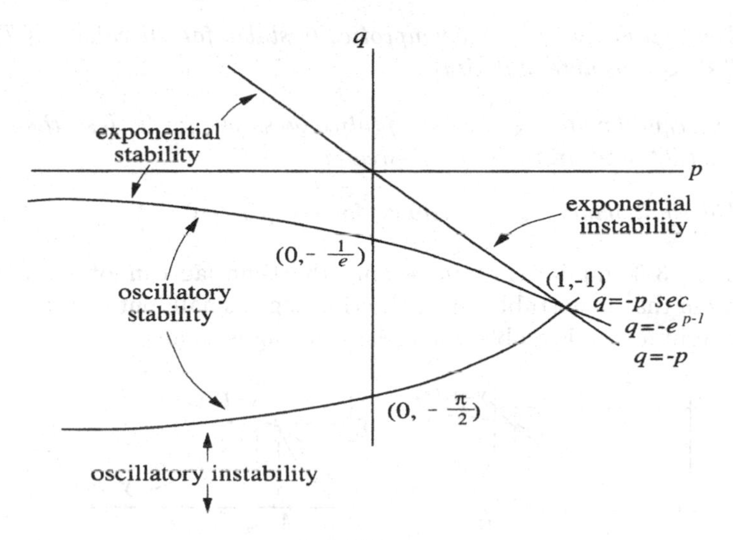

FIGURE 3.3.

For the delayed recruitment model $x'(t) = R\big(x(t-T)\big) - D\big(x(t)\big)$ the linearization is

$$u'(t) = -D'(x_\infty)u(t) + R'(x_\infty)u(t-T),$$

and we have $p = -D'(x_\infty)T < 0$, $q = R'(x_\infty)T$. Combining Theorems 3.4 and 3.5, we obtain the following conclusions (Figure 3.3):

1. At an equilibrium x_∞ with $R'(x_\infty) > 0$, or $q > 0$, we have exponential asymptotic stability if $p < -q$, or $D'(x_\infty) > P'(x_\infty)$ and exponential instability if $p > -q$, or $D'(x_\infty) < R'(x_\infty)$. (This covers contest competition; scramble competition with $M < \xi$, so that $x_\infty < \xi$, the lowest equilibrium when there is an Allee effect; and the case $D(x) > R(x)$ for all $x > 0$, so that $x = 0$ is the only equilibrium. In particular, there can be no oscillation in these situations.)

2. At an equilibrium x_∞ with $R'(x_\infty) < 0$, or $q < 0$, asymptotic stability can be either exponential or oscillatory. There is numerical evidence to indicate that in case of instability there may be periodic solutions, period-doubling, and eventually chaotic behavior, as is the case for difference equations.

3. An equilibrium x_∞ with $|R'(x_\infty)| < D'(x_\infty)$ is asymptotically stable for all T, $0 \le T < \infty$. This property is known as ***absolute stability.***

In fact, it is possible to establish the stronger result Cooke and van den Driessche (1986):

Theorem 3.6. *For an equilibrium x_∞ of the differential-difference equation* (3.7) *one of the following three possibilities must occur:*

i. The equilibrium x_∞ is asymptotically stable for all values of T, $0 \le T < \infty$ (absolute stability).

ii. The equilibrium x_∞ is asymptotically stable for T less than some value T^* and unstable for T greater than T^*.

iii. The equilibrium x_∞ is unstable for every $T \ge 0$.

Theorem 3.6 implies that increasing the time lag cannot stabilize an equilibrium that is unstable when the time lag is small, but can destabilize an equilibrium that is stable when the time lag is small.

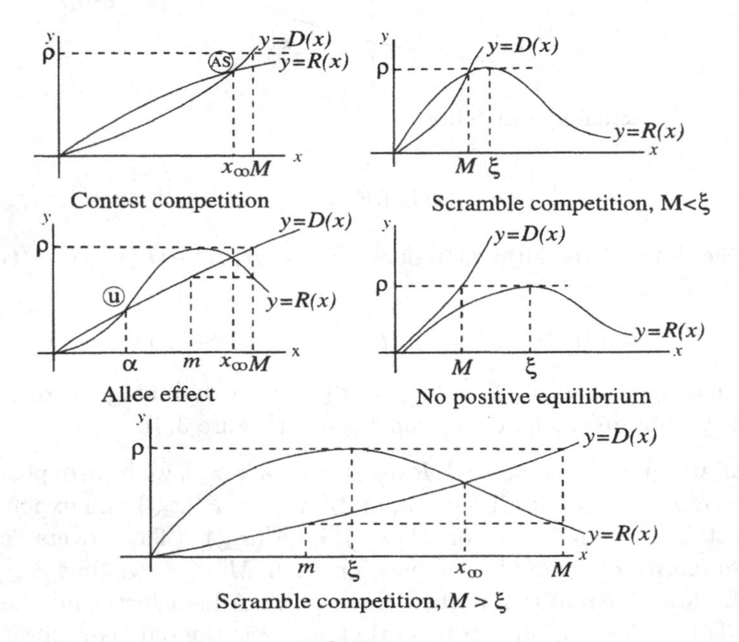

FIGURE 3.4.

In contest competition, since $D(x) \to \infty$ while $R(x)$ is bounded as $x \to \infty$, there must be a largest equilibrium that is asymptotically stable, indeed exponentially asymptotically stable. The smallest asymptotically stable equilibrium (if there should be more than one) serves effectively as a positive lower bound for solutions. In scramble competition with $M < \xi$, there is an exponentially asymptotically stable equilibrium $x_\infty < \xi$, which serves as an effective positive lower bound. When there is an Allee effect with smallest equilibrium α, solutions with initial data less than α tend monotonically to zero, but solutions with initial data greater than α are forced upward, and α serves as a lower bound. If $D(x) > R(x)$ for all $x > 0$ there is no positive lower bound for solutions; indeed all solutions tend to the exponentially asymptotically stable equilibrium $x = 0$.

Example 1. Consider the recruitment function

$$R(x) = rxe^{-x/A}$$

and the mortality function $D(x) = dx$. Then $R(x)$ has a maximum of rA/e attained when $x = A$. Thus, the model

$$x'(t) = rx(t-T)e^{-x(t-T)/A} - dx(t) \tag{3.10}$$

represents scramble competition. There is always an equilibrium at $x = 0$ and there is a positive equilibrium x_∞ given by $re^{-x_\infty/A} = d$, or $x_\infty = A \log r/d$, if and only if $r > d$. Since $R'(x) = re^{-x/A}(1-x/A)$ and $D'(x) = d$, so that $R'(0) = r$, $D'(0) = d$, and $R'(x_\infty) = d(1 - x_\infty/A) = d(1 - \log r/d)$, $D'(x_\infty) = d$. Then Theorem 3.5 shows that the equilibrium $x = 0$ is asymptotically stable if and only if $r < d$, that is, if and only if there is no positive equilibrium. If $r > d$, so that the equilibrium $x = 0$ is unstable, we may analyze the stability of the equilibrium $x_\infty > 0$ by applying Theorem 3.5 with $p = -D'(x_\infty)T = -dT$, $q = R'(x_\infty)T = dT(1 - \log r/d)$. We can deduce that the equilibrium x_∞ is absolutely stable if $-1 < 1 - x_\infty/A < 1$, which reduces to $\log(r/d) < 2$ or $r < de^2$. More generally, the equilibrium x_∞ is asymptotically stable if

$$-dT < -dT\left(1 - \log\frac{r}{d}\right) < -dT \sec z, \tag{3.11}$$

where $z = -dT \tan z$. The condition (3.11) reduces to

$$1 > 1 - \log\frac{r}{d} > \sec z,$$

or

$$0 < \log\frac{r}{d} < 1 - \sec z.$$

This implies that a large ratio r/d tends to make for instability, at least for large values of T. Figure 3.5 shows the solution of the differential–difference equation $x'(t) = 10x(t-1)e^{-x(t-1)} - x(t)$.

Exercises

1. Find all equilibria of

$$x'(t) = \frac{rx(t-T)}{x(t-T)+A} - dx(t)$$

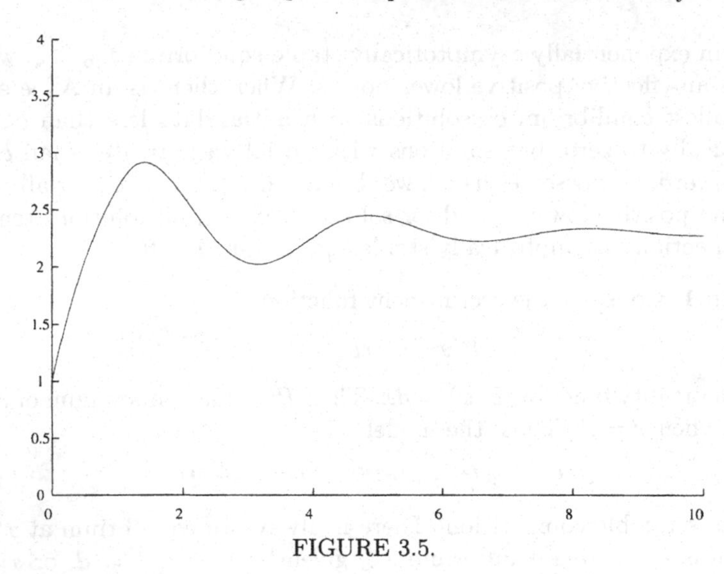

FIGURE 3.5.

and determine for the values of T for which they are asymptotically stable. Use a computer algebra system to graph some solutions for$A = 1$, $r = 2$, $d = 1$.

2. Find all equilibria of

$$x'(t) = \frac{r[x(t-T)]^2}{(x(t-T))^2 + A^2} - dx(t)$$

and determine for the values of T for which they are asymptotically stable. Use a computer algebra system to graph some solutions for $A = 1$, $r = 2$, $d = 1$.

3*. (i) A population is governed by the delayed-recruitment model

$$x'(t) = x(t-T)e^{3-x(t-T)} - x(t).$$

Find all equilibria and determine their stability.

(ii) A fraction p $(0 < p < 1)$ of the population in part (a) is removed per unit time, so that the population is modeled by

$$x'(t) = x(t-T)e^{3-x(t-T)} - x(t) - px(t).$$

Find the relation between the removal fraction p and the largest delay T for which there is an asymptotically stable positive equilibrium.

(iii) Use a computer algebra stystem to graph solutions with $T = 1$ for $p = 0.2$, $p = 0.5$, $p = 0.8$.

4. Use a computer algebra system to graph solutions of the differential-difference equation $x'(t) = rx(t-1)e^{-x(t-1)} - x(t)$ with $r = 2$, $r = 10$.

3.4 Models with Distributed Delay

In Section 3.2 we suggested that the model $x'(t) = x(t)g\big(x(t-T)\big)$ with the per capita growth rate depending on population size with delay T might be generalized to the form

$$x'(t) = x(t)\int_0^\infty g\big(x(t-s)\big)p(s)ds,$$

describing a *distributed delay.* Here $p(s)\Delta s$ represents the probability of a delay between s and $s + \Delta s$, so that $\int_0^\infty p(s)ds = 1$. The *average delay* is then, by definition, $\int_0^\infty sp(s)ds$. An equilibrium of the integro-differential equation

$$x'(t) = x(t)\int_0^\infty g[x(t-s)]p(s)ds$$

is a value x_∞ such that

$$x_\infty \int_0^\infty g(x_\infty)p(s)ds = x_\infty g(x_\infty) = 0,$$

so that $x = 0$ is an equilibrium and equilibria $x_\infty > 0$ are given by $g(x_\infty) = 0$. To linearize about an equilibrium x_∞, we let $u(t) = x(t) - x_\infty$, so that

$$u'(t) = \big(x_\infty + u(t)\big)\int_0^\infty g\big(x_\infty + u(t-s)\big)p(s)ds,$$

and expand using Taylor's theorem, obtaining

$$\begin{aligned}
u'(t) &= \big(x_\infty + u(t)\big)\int_0^\infty \big(g(x_\infty) + g'(x_\infty)u(t-s) + \dots\big)p(s)ds \\
&= \big(x_\infty + u(t)\big)\Big(g(x_\infty) + g'(x_\infty)\int_0^\infty u(t-s)p(s)ds + \dots\Big) \\
&= x_\infty g(x_\infty) + g(x_\infty)u(t) + x_\infty g'(x_\infty)\int_0^\infty u(t-s)p(s)ds + \dots,
\end{aligned}$$

where all quadratic and higher order terms in u have been suppressed. The linearization at the equilibrium is thus

$$v'(t) = g(x_\infty)v(t) + x_\infty g'(x_\infty)\int_0^\infty v(t-s)p(s)ds.$$

As with other types of equations we have studied, such as differential equations, difference equations, and differential-difference equations, the behavior of solutions near an equilibrium is described by the behavior of solutions of the linearization at the equilibrium for integro-differential equations. Therefore, we are led to study the linear integro-differential equation

$$v'(t) = av(t) + b\int_0^\infty v(t-s)p(s)ds$$

with $p(s) \geq 0$ for $0 \leq s < \infty$ and $\int_0^\infty p(s)ds = 1$. In the special case we are considering, $a = g(x_\infty)$ and $b = x_\infty g'(x_\infty)$, so that either $a = 0$ (if $x_\infty > 0$) or $b = 0$ (if $x_\infty = 0$), but we shall consider the general case in which both a and b may be different from zero because it will arise later. It is possible to prove that solutions tend to zero as $t \to \infty$ if $b > 0$ and $a + b < 0$; or if $b < 0$, $a + b < 0$, and the average delay $\int_0^\infty sp(s)ds$ is small enough; but that solutions cannot tend to zero if $a + b \geq 0$. The necessary (but not sufficient) stability condition $a+b < 0$ is in fact the necessary and sufficient condition for asymptotic stability in the undelayed case, corresponding to an ordinary differential equation $x' = xg(x)$ with linearization

$$v' = \left(xg(x)\right)'\big|_{x=x_\infty} v = \left(g(x\infty) + x_\infty g'(x_\infty)\right)v = (a+b)v.$$

To study the behavior of solutions of

$$v'(t) = av(t) + b\int_0^\infty v(t-s)p(s)ds$$

for a specific kernel $p(s)$, we look for solutions $v(t) = ce^{\lambda t}$ and construct a characteristic equation, obtaining

$$\lambda ce^{\lambda t} = ace^{\lambda t} + b\int_0^\infty ce^{\lambda(t-s)}p(s)ds$$

$$\lambda = a + b\int_0^\infty e^{-\lambda s}p(s)ds = a + bL\{p(\lambda)\},$$

where $L\{p(\lambda\}$ denotes the Laplace transform of the function p evaluated at λ. We will consider two specific choices of p, both normalized so that $\int_0^\infty p(s)ds = 1$ and $\int_0^\infty sp(s)ds = T$, making use of the formuls

$$\int_0^\infty e^{-\alpha s}ds = \frac{1}{\alpha}, \quad \int_0^\infty se^{-\alpha s}ds = \frac{1}{\alpha^2}, \quad \int_0^\infty s^2e^{-\alpha s}ds = \frac{2}{\alpha^3}.$$

Our first example is $p_1(s) = 4/T^2 se^{-2s/T}$, with $p_1(0)$ vanishing at zero, rising to a maximum at $s = T/2$ and then falling off exponentially. We have

$$L\{p_1(\lambda)\} = \int_0^\infty e^{-\lambda s}p_1(s)ds = \frac{4}{T^2}\int_0^\infty se^{-(\lambda+\frac{2}{T})s}ds$$

$$= \frac{4}{T^2}\frac{1}{\left(\lambda+\frac{2}{T}\right)^2} = \frac{4}{T^2\lambda^2 + 4T\lambda + 4}$$

The characteristic equation is

$$a+\frac{4b}{\lambda^2T^2+4T\lambda+4}=\lambda,$$

or

$$\lambda^3+\left(\frac{4T-aT^2}{T^2}\right)\lambda^2+\left(\frac{4-4aT}{T^2}\right)\lambda-\frac{4a+4b}{T^2}=0.$$

Our problem now is that we need conditions on the coefficients of a polynomial equation that ensure that all roots have negative real part. This is provided by the *Routh-Hurwitz conditions*, which for a cubic equation $\lambda^3+\alpha\lambda^2+\beta\lambda+\gamma=0$, say that all roots have negative real part if and only if $\alpha>0$, $\gamma>0$, $\alpha\beta>\gamma$. Here, $\alpha=\frac{4-aT}{T}, \beta=\frac{4-4aT}{T^2}, \gamma=-\frac{4(a+b)}{T^2}$, and the stability conditions are $a+b<0$, $aT<4$, $-bt<(2-aT)^2$. Now take $a=g(x_\infty)$, $b=x_\infty g'(x_\infty)$. If $x_\infty=0$ so that $b=0$ these conditions reduce to $g(x_\infty)<0$, which is satisfied in population models only if there is an Allee effect. If $x_\infty>0$ so that $a=0$ the stability conditions reduce to $0<-x_\infty g'(x_\infty)T<4$, quite similar to the single-delay case, where the stability condition was $0<-x_\infty g'(x_\infty)T<\pi/2$.

Our second example is $p_2(s)=1/Te^{-s/T}$, with p_2, decreasing exponentially from $1/T$ to zero, rather than rising to a maximum like $p_1(s)$. We have

$$L\{p_2(\lambda)\}=\int_0^\infty e^{-\lambda s}p_2(s)ds=\frac{1}{T}\int_0^\infty e^{-(\frac{1}{T}+\lambda)s}ds=\frac{1}{\lambda T+1},$$

and the characteristic equation is $a+\frac{b}{\lambda T+1}=\lambda$, or $\lambda^2+\frac{1-aT}{T^2}\lambda-\frac{a+b}{T^2}=0$. The stability condition that both roots of this quadratic equation have negative real part is $1-aT>0$, $-(a+b)>0$. Now take $a=g(x_\infty)$, $b=x_\infty g'(x_\infty)$; if $x_\infty=0$ these conditions reduce to $g(x_\infty)<0$, which is not satisfied. If $x_\infty>0$ the stability condition is just $g'(x_\infty)<0$ since $a=0$, exactly as if there were no delay; there is no requirement that the average delay not be too large. The point is that with distributed delay each delay kernel must be examined in its own right. *It is not true that increasing the average delay always destroys stability.*

Delayed recruitment populations with variable maturation ages can also be described by integro-differential equations, now of the form

$$x'(t)=\int_0^\infty\frac{1}{\pi}B\big(x(t-s)\big)p(s)ds-D\big(x(t)\big).$$

Here $B\big(x(t-s)\big)$ is the number born at time $(t-s)$, reaching age s at time t; $p(s)\Delta s$ is the fraction of those members who reach maturity, doing so at age s to $s+\Delta s$; π is the ratio of the number of members who reach maturity to the number born, so that $P(s)\Delta s/\pi$ is the fraction of members born who reach maturity at age s to $s+\Delta s$; and $D\big(x(t)\big)$ is the number dying at time t. An equilibrium of this integro-differential equation is a solution x_∞ of the equation

$$D(x_\infty) = \frac{1}{\pi}B(x_\infty)\int_0^\infty p(s)ds = \frac{1}{\pi}B(x\infty).$$

The by-now familiar process yields the form of the linearization at an equilibrium as

$$v'(t) = -D'(x_\infty)v(t) + \frac{1}{\pi}B'(x_\infty)\int_0^\infty v(t-s)p(s)ds.$$

This is of the form

$$v'(t) = av(t) + b\int_0^\infty v(t-s)p(s)ds,$$

with characteristic equation $\lambda = a + b\int_0^\infty e^{-\lambda s}p(s)ds$ already described. As before, it can be shown that the equilibrium is asymptotically stable if $(1/\pi)B'(x_\infty) < D'(x_\infty)$ and $B'(x_\infty) > 0$ or $B'(x_\infty) < 0$ and $\int_0^\infty sp(s)ds$ is small enough, but further information depends on the form of $p(s)$.

Exercises

1. Find all equilibria of

$$x'(t) = \frac{1}{\pi}\int_0^\infty \frac{rx(t-s)}{x(t-s)+A}p(s)ds - dx(t)$$

and determine the values of T for which they are asymptotically stable if

(i)

$$p(s) = \frac{4se^{-2s/T}}{T^2}$$

(ii)

$$p(s) = \frac{e^{-s/T}}{T}$$

2. Find all equilibria of

$$x'(t) = \frac{1}{\pi}\int_0^\infty \frac{r\left(x(t-s)\right)^2}{\left(x(t-s)\right)^2 + A^2}p(s)ds - dx(t)$$

and determine the values of T for which they are asymptotically stable for each of the choices of $p(s)$ in Exercise 1.

3. Find all equilibria of

$$x'(t) = \frac{1}{\pi}\int_0^\infty rx(t-s)e^{-x(t-s)}p(s)ds - dx(t),$$

and determine the value of T for which they are asymptotically stable for each of the choices of $p(s)$ in Exercise 1.

3.5 Harvesting in Delayed Recruitment Models

In Section 1.4 we investigated the effect of harvesting on a continuous population model. Here we wish to study whether a continuous model with a time lag, such as a delayed recruitment model of the type introduced in Section 3.3, responds to harvesting in the same way as a model without a time lag. Thus we could consider a population model

$$x'(t) = R\big(x(t-T)\big) - D\big(x(t)\big) - Ex(t) \tag{3.12}$$

to model constant effort harvesting, or a model

$$x'(t) = R\big(x(t-T)\big) - D\big(x(t)\big) - H \tag{3.13}$$

to model constant yield harvesting.

3.5.1 Constant Effort Harvesting

The analysis of the constant effort harvesting model (3.12) is exactly the same as the analysis of the unharvested model with the mortality function $D(x)$ replaced by $D(x) + Ex$. Rather than carrying out the analysis in the general case, we examine only the special case $R(x) = rxe^{-x/A}$, $D(x) = dx$ considered as an example in Section 3.3. By following the analysis of this example we can see that the equilibrium $x = 0$ is asymptotically stable if and only if $r < d + E$, or $E > r - d$. If $0 \leq E < r - d$ there is a positive equilibrium x_∞ that is absolutely stable if $r < (d+E)e^2$, or $E > r/e^2 - d$. In general, increasing the effort tends to stabilize the equilibrium but too high an effort will wipe out the population. If $r > de^2$, so that the stability of the positive equilibrium depends on the delay T, then increasing the harvest effort tends to stabilize the equilibrium but increasing the delay tends to destabilize it. The maximum sustainable yield is found exactly as for the undelayed model in Section 1.5, but the introduction of a time lag into the model raises the possibility of instability of the equilibrium corresponding to the maximum sustainable yield. For the special case $R(x) = rxe^{-x/A}$, $D(x) = dx$, the maximum sustainable yield is achieved when the mortality-harvest curve $y = (d + E)x$ passes through the maximum point $\big(A, rA/e\big)$ of the recruitment curve $y = rxe^{-x/A}$. This requires $\frac{rA}{e} = (d + E)A$, or

$E = r/e - d$. Since $r/e - d > r/e^2 - d$, the equilibrium corresponding to maximum sustainable yield is actually absolutely stable. The behavior of more general compensation models under constant effort harvesting is similar to the above special case. We shall not explore the more complicated situation for depensation models.

3.5.2 Constant Yield Harvesting

The model (3.13) must be modified slightly in order to rule out the possibility that $x'(t) < 0$ when $x(t) = 0$, as negative population sizes have no biological significance. To be realistic (and also to avoid important errors in carrying out numerical approximations), we should specify the model by

$$\begin{aligned} x'(t) &= R(x(t-T)) - D(x(t)) - H & \text{for} \quad x(t) \geq 0 \\ x'(t) &= \max\{R(x(t-T)) - D(x(t)) - H, 0\} & \text{for} \quad x(t) = 0. \end{aligned} \tag{3.14}$$

Because of the time lag, it is possible for $x(t)$ to reach zero without the collapse of the system; even if the adult population is wiped out there may still be immature members who will reach the adult stage. Of course, if $x(t) = 0$ on a time interval of length T there is no possibility of regeneration of the adult population, and we shall consider the population to have become extinct. Under the hypotheses made in Section 3.3 on the recruitment function $R(x)$ and the mortality function $D(x)$ it is possible to show that the bounds on solutions obtained in Section 3.3 are also valid for the constant yield harvesting model (3.14). An equilibrium of the model (3.14) is defined to be a value x_∞ for which

$$R(x_\infty) = D(x_\infty) + H,$$

or the abscissa of an intersection of the recruitment curve $y = R(x)$ and the mortality-harvesting curve $y = D(x) + H$. We may sometimes use the notation $x_\infty(H)$ to emphasize the dependence on H. The effect of increasing the harvest rate H is to move the mortality-harvesting curve upward. For sufficiently large H, the mortality-harvesting curve is above the recruitment curve and there are no equilibria. It is possible to prove that if the model (3.14) has no equilibrium $x_\infty > 0$ then every solution tends to zero as $t \to \infty$ if $H = 0$ and reaches zero in finite time and remains at zero if $H > 0$. There is a critical harvest rate H_c such that there is an equilibrium $x_\infty(H) > 0$ if $H < H_c$ but there is no equilibrium $x_\infty(H)$ if $H > H_c$ (Figure 3.6). When $H = H_c$ the recruitment curve $y = R(x)$ and the mortality-harvesting curve $y = D(x) + H$ are tangent at $x_\infty(H_c)$.

The linearization of (3.14) at an equilibrium x_∞ is

$$u'(t) = -D'(x_\infty)u(t) + r'(x_\infty)u(t-T),$$

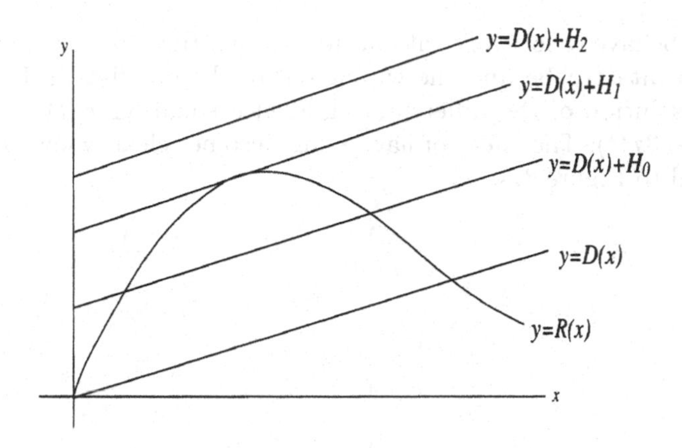

FIGURE 3.6.

which is exactly the same as in the case of no harvesting studied in Section 3.3. The same analysis of the characteristic equation as that in Section 3.3 leads to the following conclusions:

(i) An equilibrium x_∞ with $|R'(x_\infty)| < D'(x_\infty)$ is absolutely stable.

(ii) An equilibrium x_∞ with $-R'(x_\infty) > D'(x_\infty) > 0$ is asymptotically stable for small T and unstable for large T.

(iii) An equilibrium x_∞ with $R'(x_\infty) > D'(x_\infty) > 0$ is unstable for all T.

These conclusions are valid provided $R(x)$ and $D(x)$ satisfy the hypotheses of Section 3.3, namely

$$D(0) = 0, \quad D'(x) \geq 0, \quad \lim_{x \to \infty} D(x) = +\infty$$

and either

$$R(0) = 0, \; R'(x) \geq 0 \text{ for } x \leq \xi, \; R'(x) \leq 0 \text{ for } x \geq \xi, \; \lim_{x \to \infty} R(x) = 0$$

or

$$R(0) = 0, \; R'(x) \geq 0 \text{ for } 0 \leq x < \infty, \; \lim_{x \to \infty} R(x) = r < \infty.$$

Under these hypotheses, $D(x) > R(x)$ for large x, and this implies that if there are any positive equilibria there must be a largest equilibrium x_∞ for which $D'(x_\infty) > R'(x_\infty)$, and this equilibrium must be asymptotically stable at least for sufficiently small T. Because

$$R'(x_\infty(H_c)) = D'(x_\infty(H_c)) > 0$$

the equilibrium $x_\infty(H)$ must be absolutely stable when H is close to H_c. Thus, the effect of harvesting is to stabilize a population system, even one which may behave chaotically without harvesting. However, overharvesting can lead to catastrophe and the wiping out of the population. Figure 3.7 shows the solution of the differential-difference equation $x'(t) = 10x(t-1)e^{-x(t-1)} - 3x(t)$. The effect of harvesting becomes clear when Figure 3.5 is compared to Figure 3.5.

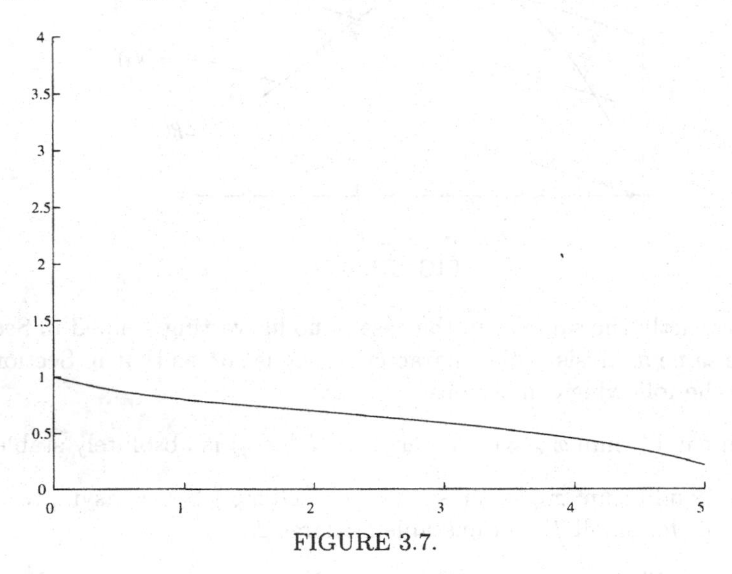

FIGURE 3.7.

Exercises

1. A population under harvesting is governed by the delayed recruitment model

$$x'(t) = x(t-T)e^{3-x(t-T)} - x(t) - H.$$

Find the relation between the harvesting rate H and the largest delay T for which there is an asymptotically stable positive equilibrium. Use a computer algebra system to graph the solutions with $T = 1$ and with $H = 1$, $H = 5$, $H = 10$.

2. Find the critical harvest rate H_c for the model

$$x'(t) = rx(t-T)e^{-x(t-T)/A} - dx(t) - H.$$

3*. For the general model

$$x'(t) = R(x(t-T)) - D(x(t)) - Ex(t)$$

show that the equilibrium $x = 0$ is unstable if $0 \leq E < R'(0) - D'(0)$, and absolutely stable if $E > R'(0) - D'(0)$.

3.6 Nicholson's Blowflies: A Case Study

In 1954 A.J. Nicholson conducted some experiments on the Australian sheep blowfly (*Lucillia cuprina*) in which he observed oscillations of large amplitude in the population size. In these experiments the population was controlled by a single factor, either the rate at which food was supplied to the adult population (in which case the adult population showed oscillations with almost discrete generations and there was a pattern of breeding activity with periods of essentially zero reproduction alternating with relatively long burst of continuous reproductive activity) or the rate at which food was supplied to the larvae (in which case each burst of reproductive activity showed two clearly discrete generations and this reproductive activity occurred in alternate population cycles). It is reasonable to ascribe the pattern of oscillations to some kind of delay mechanism appearing in the per capita growth rate as described in Section 3.2 or in the recruitment rate as described in Section 3.3. Which mechanism is more plausible as an explanation for the oscillations? Suppose we attempt to use the delay logistic equation

$$x'(t) = rx(t)\left(1 - \frac{x(t-T)}{K}\right)$$

to describe this population, as proposed by May (1974). It is convenient to rescale the model in terms of dimensionless variables. We first let $x = Ky$ to transform the model to $y'(t) = ry(t)\big(1 - y(t-T)\big)$. Then we make the change of independent variable $t = Ts$ and let $y(t) = z(s)$, so that $y(t-T) = z(s-1)$ and $y'(t) = (dy/ds)(ds/dt) = (1/T)z'(s)$ to obtain the dimensionless form

$$z'(s) = (rT)z(s)\big(1 - z(s-1)\big)$$

containing a single free parameter rT. For this differential-difference equation there is an equilibrium of $z = 1$ and the linearization at this equilibrium is

$$u'(s) = -rTu(s-1).$$

We know that the equilibrium $z = 1$ is asymptotically stable if $rT < \pi/2$ and that there is a stable periodic solution if $rT > \pi/2$. Numerical simulation for $rT > \pi/2$ indicates relations between the value of the parameter rT, the ratio of maximum to minimum population size, and the period of the periodic solution. From the observed ratio of maximum to minimum population size May estimated $rT \approx 2.1$, which implies a period of 4.54 in dimensionless units, or $4.54T$ in days. Then he deduced from the observed period that the time delay was approximately 9 days, compared to the experimentally observed larva to adult maturation time of 11 days. From these data, he concluded that the agreement between observation and the rather crude model used was sufficiently good to indicate that delay in the per capita growth rate might explain the observed behavior.

In 1980, May's calculations were re-examined by W.S.C. Gurney, S. P. Blythe, and R.M. Nisbet (1980). They observed a maximum to minimum ratio of population sizes in Nicholson's adult-food-limited experiment (the only one for which May made calculations) of 36 ± 17, which would imply $1.880 < rT < 2.035$ (the maximum to minimum population size ratio is difficult to measure accurately, but rT is not very sensitive to changes in this ratio), and that the period of oscillations in the solution is between $4.26T$ and $4.47T$. The fact that the observed period was 38.1 ± 1.5 days implies that the delay T must be between $36./4.47 = 8.2$ days and $39.6/4.26 = 9.3$ days. This is really not very close to the observed delay, as Nicholson observed an egg to reproductive maturity time of between 14.3 and 15.6 days, and as our model describes adult population size this is the appropriate delay. The data also give an independent verification of this delay, since there are clear bursts of reproduction preceding each adult population peak by 14.8 ± 0.4 days. Since egg to adult survival is high (approximately 90 percent) and constant in the adult food-limited experiments, this lag should be the same as the delay T. The discrepancy between a predicted delay of 8.2 to 9.3 days and the observed delay of 14.3 to 15.6 days indicates that the phenomenon observed cannot be explained satisfactorily by use of the delay logistic equation as a model.

A more general model of the form

$$x'(t) = x(t)g\big(x(t-T)\big)$$

would be no more satisfactory than the delay logistic equation because linearization about a positive equilibrium would produce the same linearized equation with the same parameters. Thus we conclude that delay in per capita growth rates does not explain the observed behavior, and we are led to search for a more satisfactory model. In fitting data to experiments more accurate measurements often indicate discrepancies not disclosed by rough measurements. In searching for a model, we will continue to assume that the oscillations are the result of some delay mechanism. While our model will be somewhat more complicated than the delay logistic equation, we would like to have only a small number of controlling parameters. A model with only one controlling parameter in dimensionless variables will always have a linearization about equilibrium either of the form $z'(s) = az(s)$, whose solutions are exponential functions, or of the form $z'(s) = bz(s-1)$, which has been tried and found wanting in our examination of the delay logistic equation. Thus we must expect at least two controlling parameters in a satisfactory model. Because the time required for eggs to mature to adults appears to be related to the oscillatory behavior, it is natural to consider a delayed recruitment model of the form

$$x'(t) = R\big(x(t-T)\big) - D\big(x(t)\big)$$

in which T is the maturation time, $R(x)$ is the number of eggs produced per unit time when the adult population size is x multiplied by the probability of survival to adulthood, and $D(x)$ is the number of adults dying per unit time when the population size is x. Data about egg production, survival to adulthood, and death rates suggest use of the functions

$$R(x) = rxe^{-x/A}, \; D(x) = dx$$

to give the model

$$x'(t) = rx(t-T)e^{-x(t-T)/A} - dx(t), \tag{3.15}$$

both for adult food-limited and for larva food-limited studies, even though the egg production rates and survival probabilities have different forms in the two cases. The changes of variable $x = Ay$, $t = Ts$, with the definition $y(t) = z(s)$ transforms the model to the dimensionless form

$$z'(s) = (rT)z(s-1)e^{-z(s-1)} - (dT)z(s) \tag{3.16}$$

having two controlling parameters rT and dT. We will assume $r > d$ to allow the possibility of a positive equilibrium z_∞ obtained by solving $rTe^{-z} = dT$, so that $z_\infty = \log(r/d) > 0$. The linearization about the equilibrium z_∞ is

$$u'(s) = dT\Big(1 - \log\frac{r}{d}\Big)z(s-1) - dTz(s),$$

and we can describe conditions for the stability of the equilibrium z_∞ in terms of the values of rT and dT. In fact, we can plot in the (dT, rT)-plane the regions of exponential asymptotic stability, oscillatory asymptotic stability, and instability with a stable periodic solution, much as we plotted these regions in the (aT, bT) plane for the equation $u'(t) = au(t) + bu(t-T)$ in Section 3.3. With the aid of numerical simulations we can go further: In the region of instability we can plot the contours along which the ratio of maximum to minumum population size is constant and the contours along which the ratio of period of the solution to delay is constant. The confidence limits for these two observable quantities give four contours that define a confidence region in the (dT, rT)-plane and thus estimate rT and dT. The estimates used earlier (36 ± 17 and 4.6 ± 0.1, respectively) lead us to the estimates $rT = 150 \pm 70$, $dT = 2.9 \pm 0.5$.

We may also estimate r and d independently. During the portion of each cycle when the population size is approaching its minimum the rate of adult recruitment is close to zero. Thus, on these intervals the model is essentially $z'(s) = -dz(s)$, and $z(s)$ behaves like ce^{-ds}. If we plot $\log z$ against s on these intervals, then d is the negative of the slope of the "best straight line" in these plots; this procedure leads to the estimate $d = 0.27 \pm 0.025$, $dT = 4.0 \pm 0.5$, which does not quite agree with the model prediction but at least does not disagree violently. No doubt it would be possible to fudge the data to give agreement. We may estimate r from the fact that the maximum rate of egg production multiplied by the constant survival probability divided by the population size at maximum production is r/e (the maximum value of the function rxe^{-x} attained for $x = 1$). From observation of production rates we estimate $7.4 < r < 14$, $100 < rT < 160$, and this is quite compatible with the model estimate $80 < rT < 220$. We conclude that the delayed recruitment model is a plausible description and may use it to make further predictions about the model. Not only do we obtain population oscillations very similar to those observed experimentally, but we also deduce a pattern of breeding activity with bursts of continuous activity alternating with periods of essentially zero reproduction. If we fit the same model to the observed data for the larva food-limited experiments we obtain $rT = 170 \pm 50$, $dT = 5.5 \pm 1.6$ for one run and $rT = 380 \pm 250$, $dT = 4.7 \pm 0.7$ for a second. Both give separate bursts of reproduction with two clearly discrete generations, different from the prediction of the same model for the adult food-limited experiment but in accordance with the observed differences between the two experiments. This reinforces our acceptance of the model. There is another possible explanation in both models, namely that there is in fact an asymptotically stable equilibrium which is approached (possibly very slowly) in an oscillatory manner. However, for the delay logistic model, if this were the case, the period of oscillation could not be less than $4T$ for any value of the controlling parameter rT, and this is quite incompatible with observation. For the delayed recruitment model the asymptotic stability hypothesis leads to the estimates $rT = 23.5 \pm 4.5$, $dT = 30 \pm 0.7$ for the controlling parameters. The estimate for rT is radically different from the independent estimate, and thus we may reject this explanation for either model. Figure 3.8 shows the solution of the delayed recruitment model with the parameter values $r = 10, d = 0.2,\ T = 15$, which are close to our estimates for the adult food-limited experiment.

When we transformed the delayed recruitment model (3.15) to the dimensionless form (3.16) we were following a procedure which is often useful in constructing and analyzing models. In (3.15) the population size x is measured in number of members or in biomass, and the time t is measured in units of time. Let us suppose that x is measured in grams and t is measured in days. Then $x'(t)$ is measured in gram day^{-1}. Each term of the equation (3.15) must be measured in the same units. Since both x and A are population sizes measured in grams, the quotient x/A is *dimensionless*. This means that a change in units of mass will not alter the value of x/A. Then $e^{-x(t-T)/A}$ is also dimensionless. As $x'(t)$ is measured in gram day^{-1}, each term on the right side of (3.15) must also have these dimensions. Since z is measured in grams, the coefficients r and d must have unit days^{-1}. The change of dependent variable $x = Ay$ gives a new variable y, which is dimensionless; y measures population size in units of size A.

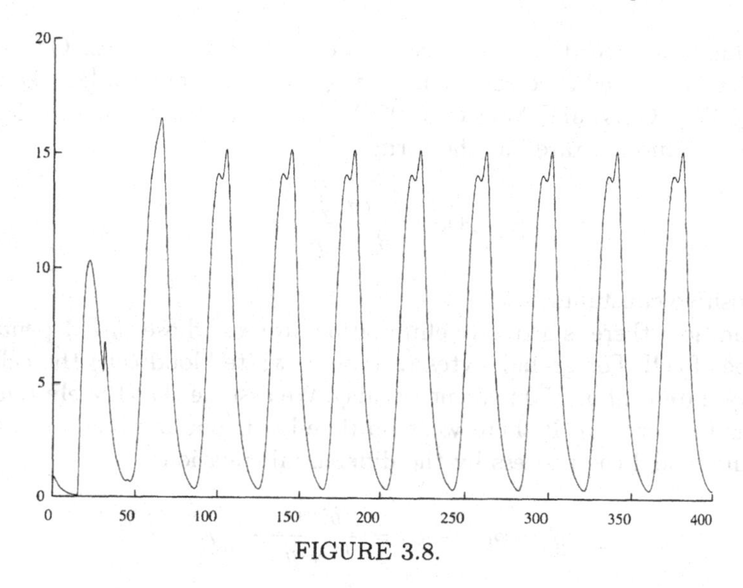

FIGURE 3.8.

Similarly, the change of independent variable $t = Ts$ gives a dimensionless variable s measuring time in units of size T. The model (3.16) is equivalent to (3.15) but is in terms of dimensionless independent and dependent variables. It involves two parameters rT and dT which are dimensionless. The original model (3.15) contained four parameters, r, d, T, A. In general, for every reduction from a variable to a dimensionless variable we may expect to remove one parameter. The dimensionless form (3.16) is more convenient than the form (3.15) for fitting data because its parameters are no longer coupled together.

3.7 Project 3.1: A Model for Blood Cell Populations

Many physiological systems normally display predictable patterns, either remaining almost constant or having regular oscillations. There are diseases, called dynamical diseases, which are marked by changes in these patterns such as changes in the nature of the oscillations. One area in which dynamical diseases have been observed experimentally is the blood system. Some forms of anemia and leukemia have been identified as dynamical diseases. A physiological change may cause a change in parameters of a model, which can alter the qualitative nature of the solutions.

Most types of blood cells are formed from primitive stem cells in the bone marrow, a development process that takes about six days. The cell production process is not well understood, but it appears that the production rate should be small for small density levels and should increase to a maximum and then decrease to zero as cell density increases. One form, which has been used successfully in fitting experimental data [Mackey and Glass (1977); Glass and Mackey (1979), (1988)] is that if the cell density is y, the production rate has the form

$$p(y) = \frac{b\theta^n y}{\theta^n + y^n},$$

with positive constants b, θ, $n > 1$.

In addition, there is a steady elimination process whose rate depends on the type of cell. For granulocytes (a type of white blood cell) the cell destruction rate is about 10 percent per day. We assume that the elimination rate has the form cy. If there were no time lag in the production of cells, we would model the process by the differential equation

$$y' = p(y) - cy = \frac{b\theta^n y}{\theta^n + y^n} - cy. \tag{3.17}$$

Question 1
Show that the change of variable $y = u\theta$ transforms (3.17) to the differential equation

$$u' = \frac{bu}{1 + u^n} - cu. \tag{3.18}$$

This means that θ can be eliminated from the model by choosing θ as the unit of measurement for blood cell density.

Question 2
Find all equilibria of (3.18) and determine which are asymptotically stable and which are unstable. (You will need to distinguish between the cases $b > c$ and $b < c$.)

Question 3
Use a computer algebra system to graph some solutions of (3.18) with the parameter values

$$c = 0.1 \text{ day}^{-1}, \quad b = 0.2 \text{ day}^{-1}, \quad n = 10$$

(which are appropriate for granulocytes in healthy humans).

You should recognize that the model (3.18) cannot produce irregular behavior for any choice of the parameters b and c, and thus cannot explain dynamical disease. In fact, it cannot even produce regular oscillations. By overlooking the time lag in the production process, we have lost an essential feature of the system. This suggests using a differential-difference equation model

$$u'(t) = \frac{bu(t-\tau)}{1+[u(t-\tau)]^n} - cu(t), \tag{3.19}$$

with a delay τ corresponding to the cell development process.

Question 4
Find all equilibria of (3.19), and show that the equilibrium $u = 0$ is asymptotically stable if $b < c$ and unstable if $b > c$.

While it would be possible to obtain a stability criterion for the positive equilibrium of (3.19) if $b > c$, this is quite complicated technically. Instead, we will use a computer algebra system to learn about the behavior of solutions; we may use the resources mentioned in Section 3.1.

Question 5
Use the parameter values

$$c = 0.1 \text{ day}^{-1}, \quad b = 0.2 \text{ day}^{-1}, \quad n = 10, \quad \tau = 6 \text{ days}$$

and sketch some solutions of (3.19).

In chronic myelogeneous leukemia, it is thought that the production time of blood cells increases, and that this may lead to irregular behavior.

Question 6
Use the parameter values

$$c = 0.1 \text{ day}^{-1}, \quad b = 0.2 \text{ day}^{-1}, \quad n = 10, \quad \tau = 20 \text{ days}$$

and sketch some solutions of (3.19).

In order to avoid the complications that arise in the analysis of differential-difference equations, one might try to use a difference equation model to incorporate a time lag. If the production and elimination process occurred at discrete intervals, we might try to use a model

$$\begin{aligned} u_{k+1} &= u_k + \frac{\beta u_k}{1+u_k^n} - \xi u_k \\ &= (1-\xi)u_k + \frac{\beta u_k}{1+u_k^n}. \end{aligned} \tag{3.20}$$

Now, the production delay τ corresponds to one time unit in the difference equation (3.20).

Question 7
Show that to make (3.20) correspond to the differential-difference equation (3.19), we should take $\beta = \tau b$. Show also that we should take $\xi = 1-(1-c)^\tau$. ([Elimination of a fraction c per day gives a remainder of $(1-c)^\tau$ after τ days.)

Question 8
Find all equilibria for (3.20) and obtain conditions in terms of β, ξ and n for their asymptotic stability.

Question 9
Determine whether the positive equilibrium of (3.20) is asymptotically stable for the parameter values corresponding to

$$c = 0.1 \text{ day}^{-1}, \quad b = 0.2 \text{ day}^{-1}, \quad n = 10, \quad \tau = 6 \text{ day}$$

and also for the parameter values corresponding to

$$c = 0.1 \text{ day}^{-1}, \quad b = 0.2 \text{ day}^{-1}, \quad n = 10, \quad \tau = 20 \text{ day}.$$

Use a computer algebra system to sketch some solutions for each of these sets of parameter values, and compare with the sketches obtained in Questions 5 and 6.

Part II

Models for Interacting Species

4

Introduction and Mathematical Preliminaries

4.1 The Lotka-Volterra Equations

In the 1920s Vito Volterra was asked if it were possible to explain the fluctuations which had been observed in the fish population of the Adriatic sea–fluctuations which were of great concern to fishermen in times of low fish populations. Volterra (1926) constructed the model which has become known as the Lotka-Volterra model (because A.J. Lotka (1925) constructed a similar model in a different context about the same time), based on the assumptions that fish and sharks were in a predator-prey relationship.

Here is a description of the model suggested by Volterra. Let $x(t)$ be the number of fish and $y(t)$ the number of sharks at time t. We assume that the plankton, which is the food supply for the fish, is unlimited, and thus that the per capita growth rate of the fish population in the absence of sharks would be constant. Thus, if there were no sharks the fish population would satisfy a differential equation of the form $dx/dt = \lambda x$. The sharks, on the other hand, depend on fish as their food supply, and we assume that if there were no fish the sharks would have a constant per capita death rate; thus, in the absence of fish, the shark population would satisfy a differential equation of the form $dy/dt = -\mu y$. We assume that the presence of fish increases the shark growth rate, changing the per capita shark growth rate from $-\mu$ to $-\mu + cx$. The presence of sharks reduces the fish population, changing the per capita fish growth rate from λ to $\lambda - by$. This gives the

Lotka-Volterra equations

$$\begin{aligned}\frac{dx}{dt} &= x(\lambda - by) \\ \frac{dy}{dt} &= y(-\mu + cx).\end{aligned} \tag{4.1}$$

We cannot solve this system of equations analytically, but we can obtain some information about the behavior of its solutions. Instead of trying to solve for x and y as functions of t, we eliminate t and look for the relation between x and y. In geometric terms, we study the *phase plane*–the (x, y) plane. We look for *orbits*, or *trajectories* of solutions–curves in the phase plane representing the functional relation between x and y with the time t as the parameter. We may eliminate t from the Lotka-Volterra equations by division,

$$\frac{dy/dt}{dx/dt} = \frac{dy}{dx} = \frac{y(-\mu + cx)}{x(\lambda - by)}.$$

We may solve this differential equation by separation of variables:

$$\begin{aligned}\int \frac{-\mu + cx}{x} dx &= \int \frac{\lambda - by}{y} dy \\ -\mu \log x + cx &= \lambda \log y - by + h,\end{aligned}$$

where h is a constant of integration, or

$$-\mu \log x - \lambda \log y + cx + by = h.$$

The minimum value of the function

$$V(x, y) = -\mu \log x - \lambda \log y + cx + by$$

is obtained by setting $\partial V/\partial x = 0$, $\partial V/\partial y = 0$. Then $c - \mu/x = 0, b - \lambda/y = 0$, or $x = \mu/c$, $y = \lambda/b$. This is an equilibrium of the Lotka-Volterra system (i.e., a constant solution $x \equiv \mu/c = x_\infty$, $y \equiv \lambda/b = y_\infty$), which may also be described by the equation

$$\begin{aligned}V(x, y) = h_0 &= -\mu \log x_\infty - \lambda \log y_\infty + cx_\infty + by_\infty \\ &= -\mu \log \frac{\mu}{c} - \lambda \log \frac{\lambda}{b} + \mu + \lambda.\end{aligned}$$

Every orbit of the system is given implicitly by an equation $V(x, y) = h$ for some constant $h \geq h_0$, which is determined by the initial conditions. We make the change of variable $x = x_\infty + u = \mu/c + u$, $y = y_\infty + v = \lambda/b + v$, obtaining

$$V(x, y) = -\mu \log \left(\frac{\mu}{c} + u\right) - \lambda \log \left(\frac{\lambda}{b} + v\right) + c\left(\frac{\mu}{c} + u\right) + b\left(\frac{\lambda}{b} + v\right) = h.$$

We observe that

$$\log\left(\frac{\mu}{c}+u\right)=\log\frac{\mu}{c}+\log\left(1+\frac{cu}{\mu}\right)$$

and if $h-h_0$ is small we may use the approximation $\log(1+x)\approx x-x^2/2$ to approximate this expression by by

$$\log\frac{\mu}{c}+\frac{cu}{\mu}-\frac{c^2u^2}{\mu^2}.$$

Similarly, we may approximate $\log(\lambda/b+v)$ by $\log\lambda/b+bv/\lambda-b^2v^2/\lambda^2$. Then the orbits $V(x,y)=h$ are appproximated by

$$-\mu\log\frac{\mu}{c}-cu+\frac{c^2}{\mu}u^2-\lambda\log\frac{\lambda}{b}-bv+\frac{b^2}{\lambda}v^2+\mu+cu+\lambda+bv=h,$$

or

$$\frac{c^2}{\mu}u^2+\frac{b^2}{\lambda}v^2=h+\mu\log\frac{\mu}{c}+\lambda\log\frac{\lambda}{b}-\mu-\lambda=h-h_0,$$

which represents an ellipse (if $h>h_0$) with the equilibrium (x_∞,y_∞) as its center. This shows that for $h-h_0$ small and positive the orbits are closed curves around the equilibrium; since the solutions run around closed orbits they must be *periodic*. Thus the Lotka-Volterra model predicts the fluctuations that had been observed experimentally. It is possible to show that the period of oscillation is approximately $2\pi/\lambda\mu$, and it is easy to see from the phase portrait that the maximum prey population comes one quarter of a cycle before the maximum predator population (Figure 4.1).

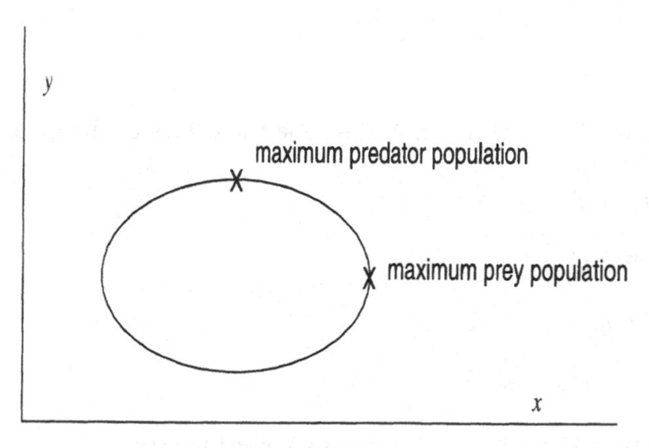

FIGURE 4.1.

The Lotka-Volterra model represented one of the triumphs of early attempts at mathematical modeling in population biology. However, it turns out that there are serious flaws in the model. Any attempt at refinement by introducing self-limiting terms in the per capita growth rates such as in the logistic equation for single populations will lead to qualitatively different behavior of the solutions, orbits that spiral in towards the equilibrium rather than periodic orbits. The price of refinement of the model is loss of agreement with observation. In order to construct a model that predicts periodic solutions we will have to assume nonlinear per capita growth rates for the two species.

Example 1. Describe the orbits of the system

$$\begin{aligned} x' &= y, \\ y' &= -x \end{aligned}$$

Solution. If we consider y as a function of x we have

$$\begin{aligned} \int y dy &= -\int x dx \\ \frac{y^2}{2} &= -\frac{x^2}{2} + c. \end{aligned}$$

Thus every orbit is a circle $x^2 + y^2 = 2c$ with center at the origin. Alternately, we may define the function $V(x, y) = x^2 + y^2$ and calculate

$$\frac{d}{dt}V[x(t), y(t)] = 2x\frac{dx}{dt} + 2y\frac{dy}{dt} = 2xy - 2xy = 0$$

to see that $V(x, y) = x^2 + y^2$ is constant on every orbit.

Exercises

In each of Exercises 1 through 4, describe the orbits of the given system.

1. $x' = xy^2$, $y' = yx^3$

2. $x' = e^{-y}$, $y' = e^x$

3. $x' = \sin y$, $y' = x$

4. $x' = xe^y$, $y' = xye^{-x}$

5. Show that the orbits of the Lotka-Volterra system (4.1) are traversed in a counter-clockwise direction as t increases. [*Hint*: For a point on the orbit with $\lambda/b, x > \mu/c$ we have $x' = 0$ and $y' > 0$. Thus, at this point x is a maxmum and y is increasing.]

4.2 The Chemostat

A chemostat is a piece of laboratory apparatus used to cultivate bacteria. It consists of a reservoir containing a nutrient, a culture vessel in which the bacteria are cultivated, and an output receptacle. Nutrient is pumped from the reservoir to the culture vessel at a constant rate and the bacteria are collected in the receptacle by pumping the contents of the culture vessel out at the same constant rate. The process is called a *continuous culture of bacteria*, in contrast with a *batch culture* in which a fixed quantity of nutrient is supplied and bacteria are harvested after a growth period. We wish to describe the behavior of the chemostat by modeling the number of bacteria and the nutrient concentration in the culture vessel, and we shall sketch the classical theory of the simple chemostat due to Novick and Szilard (1950) and Monod (1950). We will obtain a model for two interacting populations that describes a laboratory realization of a very simple lake. More complicated chemostats, in which two or more cultures are introduced, give multispecies models representing more complicated real world situations. We let y represent the number of bacteria and C the concentration of nutrient in the chemostat, both functions of t. Let V be the volume of the chemostat and Q the rate of flow into the chemostat from the nutrient reservoir and also the rate of flow out from the chemostat. The fixed concentration of nutrient in the reservoir is a constant $C^{(0)}$. We assume that the average per capita bacterial birth rate is a function $r(C)$ of the nutrient concentration and that the rate of nutrient consumption of an individual bacterium is proportional to $r(C)$, say $\alpha r(C)$. Then the rate of change of population size is the birth rate $r(C)y$ of bacteria minus the outflow rate Qy/V. It is convenient to let $q = Q/V$, so that this outflow rate becomes qy. The rate of change of nutrient volume is the replenishment rate $QC^{(0)}$ minus the outflow rate QC minus the consumption rate $\alpha r(C)y$. This gives the pair of differential equations

$$\begin{aligned} \frac{dy}{dt} &= r(C)y - qy \\ \frac{d(CV)}{dt} &= Q(C^{(0)} - C) - \alpha r(C)y. \end{aligned}$$

We divide the second equation by the constant V and let $\beta = \alpha/V$ to give the system

$$\begin{aligned} y' &= r(C)y - qy \\ C' &= q(C^{(0)} - C) - \beta r(C)y. \end{aligned} \qquad (4.2)$$

The system (4.2) describes the chemostat, but we must still make some plausible assumptions on the dependence of the bacterial birth rate (or equivalently the nutrient consumption rate) on the nutrient concentration C. It is reasonable to assume that the function $r(C)$ is zero if $C = 0$ and that it saturates (approaches a limit) when C becomes large. The simplest function with these properties is

$$r(C) = \frac{aC}{C+A},$$

where a and A are constants, and this was the choice originally made by Monod. The explicit chomostat model is now

$$\begin{aligned} y' &= \frac{aCy}{C+A} - qy \\ C' &= q(C^{(0)} - C) - \frac{\beta aCy}{C+A}, \end{aligned} \tag{4.3}$$

where a, A, q, and β are constants. Of these constants, q depends on the flow rate and thus can be adjusted. We may inquire whether it is possible to adjust the flow rate or other parameters in the system so that the chemostat will settle down to an equilibrium. If the flow rate is too large, the whole culture will be washed out; a more interesting equilibrium state would involve both bacteria and nutrient in the culture. Because the system (4.3) cannot be solved analytically, we will have to study it by qualitative methods. This will be done in Section 5.2. The qualitative study of systems of differential equations is much more complicated than the study of a single differential equation. In the remainder of this chapter, we shall describe some useful qualitative results for systems of differential equations, especially for systems of two differential equations for which much more precise results are true than are valid for systems of more than two differential equations. For the most part, we shall state results without proof. Proofs may be found in books on the qualitative theory of differential equations, for example [Brauer and Nohel (1989), Hurewicz (1958), Sánchez (1979), and Waltman (1986)].

4.3 Equilibria and Linearization

One of the main tools in studying continuous models for two interacting populations is linearization at equilibria, just as for models for single species. However, as linearization results can only give information about behavior of solutions near an equilibrium, they will not enable us to examine such questions as the existence of periodic orbits. As might be expected from our discussion of the Lotka-Volterra model, we will want to consider periodic orbits when we study predator-prey systems in Chapter 5, and at that time it will be necessary to make use of some global results not related to linearization. Here we shall describe the main results on linearization for two-dimensional systems. The basic fact is the analogue of the result established for one-dimensional systems (or first order equations), namely that the behavior of solutions near an equilibrium is determined by the behavior of solutions of the linearization at the equilibrium. Since there is a variety of possible behaviors of solutions of the linearization, we will also have to carry out a classification of equilibria according to the behavior of solutions at the equilibrium.

We will consider populations of two interacting species with population sizes $x(t)$ and $y(t)$ respectively. As in our study of continuous single species models, we will assume that $x(t)$ and $y(t)$ are continuously differentiable functions of t whose derivatives are functions of the two population sizes at the same time. Thus, our models will be systems of two first order differential equations,

$$\begin{aligned} x' &= F(x,y) \\ y' &= G(x,y). \end{aligned} \tag{4.4}$$

As in our study of single-species models, the assumptions that lead to this form neglect many factors of importance in many real populations, but the model is a useful first step and may model some real populations quite well. An *equilibrium* is a solution (x_∞, y_∞) of the pair of equations $F(x_\infty, y_\infty) = 0$, $G(x_\infty, y_\infty) = 0$. Thus an equilibrium is a constant solution of the system of differential equations. Geometrically, an equilibrium is a point in the phase plane that is the orbit of a constant solution. If (x_∞, y_∞) is an equilibrium, we make the change of variables $u = x - x_\infty$, $v = y - y_\infty$, obtaining the system

$$\begin{aligned} u' &= F(x_\infty + u, y_\infty + v) \\ v' &= G(x_\infty + u, y_\infty + v). \end{aligned}$$

Using Taylor's theorem for functions of two variables, we may write

$$\begin{aligned} F(x_\infty + u, y_\infty + v) &= F(x_\infty, y_\infty) + F_x(x_\infty, y_\infty)u + F_y(x_\infty, y_\infty) + h_1 \\ G(x_\infty + u, y_\infty + v) &= G(x_\infty, y_\infty) + G_x(x_\infty, y_\infty)u + G_y(x_\infty, y_\infty) + h_2, \end{aligned}$$

where h_1 and h_2 are functions that are small for small u, v in the sense that

$$\lim_{\substack{u\to 0\\ v\to 0}} \frac{h_1(u,v)}{\sqrt{u^2+v^2}} = \lim_{\substack{u\to 0\\ v\to 0}} \frac{h_2(u,v)}{\sqrt{u^2+v^2}} = 0.$$

The *linearization* of the system, obtained by using

$$F(x_\infty, y_\infty) = 0, G(x_\infty, y_\infty) = 0$$

and neglecting the higher order terms $h_1(u,v)$ and $h_2(u,v)$ is defined to be the two-dimensional linear system

$$\begin{aligned} u' &= F_x(x_\infty, y_\infty)u + F_y(x_\infty, y_\infty)v \\ v' &= G_x(x_\infty, y_\infty)u + G_y(x_\infty, y_\infty)v. \end{aligned} \tag{4.5}$$

The coefficient matrix of the system (4.5)

$$\begin{pmatrix} F_x(x_\infty, y_\infty) & F_y(x_\infty, y_\infty) \\ G_x(x_\infty, y_\infty) & G_y(x_\infty, y_\infty) \end{pmatrix}$$

is called the *community matrix* of the system at the equilibrium (x_∞, y_∞). It describes the effect of the size of each species on the growth rate of itself and the other species at equilibrium. Frequently we shall write a system in the form

$$\begin{aligned} x' &= xf(x,y) \\ y' &= yg(x,y), \end{aligned} \tag{4.6}$$

so that $f(x,y)$ and $g(x,y)$ are the per capita growth rates of the two species. The community matrix at equilibrium then has the form

$$\begin{pmatrix} x_\infty f_x(x_\infty, y_\infty) + f(x_\infty, y_\infty) & x_\infty f_y(x_\infty, y_\infty) \\ y_\infty g_x(x_\infty, y_\infty) & y_\infty g_y(x_\infty, y_\infty) + g(x_\infty, y_\infty) \end{pmatrix}.$$

There are four distinct kinds of possible equilibria, as follows.

(i) $(0,0)$, with community matrix

$$\begin{pmatrix} f(0,0) & 0 \\ 0 & g(0,0) \end{pmatrix}.$$

(ii) $(K,0)$ with $K > 0$, $f(K,0) = 0$, having community matrix

$$\begin{pmatrix} Kf_x(K,0) & Kf_y(K,0) \\ 0 & g(K,0) \end{pmatrix}.$$

(iii) $(0,M)$ with $M > 0$, $g(0,M) = 0$, having community matrix

$$\begin{pmatrix} f(0,M) & 0 \\ Mg_x(0,M) & Mg_y(0,M) \end{pmatrix}.$$

(iv) (x_∞, y_∞) with $x_\infty > 0$, $y_\infty > 0$, $f(x_\infty, y_\infty) = 0$, $g(x_\infty, y_\infty) = 0$, having community matrix

$$\begin{pmatrix} x_\infty f_x(x_\infty, y_\infty) & x_\infty f_y(x_\infty, y_\infty) \\ y_\infty g_x(x_\infty, y_\infty) & y_\infty g_y(x_\infty, y_\infty) \end{pmatrix}.$$

We should remark that as from a biological point of view, only non-negative population sizes are of interest, we consider only equilibria having non-negative coordinates and we are concerned with only the first quadrant of the phase plane. In the case (iv) of coexistence of the two species the terms

$f_x(x_\infty, y_\infty)$ and $g_y(x_\infty, y_\infty)$ in the community matrix are self-regulating terms which are normally non-positive.

The terms $f_y(x_\infty, y_\infty)$ and $g_x(x_\infty, y_\infty)$ are *interaction* terms. There are three possible sign combinations, which we shall study separately. If both interaction terms are negative, the two species are said to be in *competition*. If there is one positive and one negative interaction term, the species are said to be in a *predator-prey* relation. Such systems include herbivore-vegetation systems in which we may be interested mainly in the herbivore species but include its food supply in the model for greater realism. The simple chemostat modeled in the previous section is an example of this type. A system in which both interaction terms are positive is called *mutualistic*. We will consider each of these types of interaction in Chapter 5.

An equilibrium (x_∞, y_∞) is said to be *stable* if every solution $(x(t), y(t))$ with $(x(0), y(0))$ sufficiently close to the equilibrium remains close to the equilibrium for all $t \geq 0$. An equilibrium (x_∞, y_∞) is said to be *asymptotically stable* if it is stable and if, in addition, solutions with $(x(0), y(0))$ sufficiently close to the equilibrium tend to the equilibrium as $t \to \infty$. These definitions are the natural analogues of the definitions given previously for first order equations. The analogue of the linearization theorem is also true, although the proof is more complicated; we shall state it without proof.

Theorem 4.1. *If (x_∞, y_∞) is an equilibrium of the system (4.4) and if all solutions of the linearization at the equilibrium (4.5) tend to zero as $t \to \infty$, then the equilibrium (x_∞, y_∞) is asymptotically stable.*

It might appear that stability of an equilibrium is a natural requirement for biological meaning, but we shall require asymptotic stability because of the following perturbation result, analogous to a result given for single-species models, which we state without proof.

Theorem 4.2. *Under the hypotheses of Theorem 4.1:*

(i) If

$$\frac{P(x,y)}{\left((x - x_\infty)^2 + (y - y_\infty)^2\right)^{1/2}}$$

and

$$\frac{Q(x,y)}{\left((x - x_\infty)^2 + (y - y_\infty)^2\right)^{1/2}}$$

tend to zero as $(x, y) \to (x_\infty, y_\infty)$, then solutions of the perturbed system

$$x' = F(x,y) + P(x,y), \quad y' = G(x,y) + Q(x,y)$$

starting close enough to (x_∞, y_∞) *tend to* (x_∞, y_∞) *as* $t \to \infty$.

(ii) If $|P(x,y)| \le A$, $|Q(x,y)| \le A$ *for all* y *and* A *sufficiently small, then solutions of the perturbed system remain within* KA, *for some constant* K, *of solutions of the unperturbed system* $x' = F(x,y)$, $y' = G(x,y)$ *for* $t \ge 0$.

The content of Theorem 4.2 is that an asymptotically stable equilibrium has biological significance, being relatively insensitive to both changes in initial population size and small additional forces.

Example 1. Find the linearization at each equilibrium of the Lotka-Volterra system (4.1)

Solution. The equilibria are the solutions of $x(\lambda - by) = 0, y(-\mu + cx) = 0$. Because the partial derivatives of the functions on the right side of the system are, respectively,

$$\frac{\partial}{\partial x}[x(\lambda - by)] = cy, \quad \frac{\partial}{\partial y}[x(\lambda - by)] = -bx,$$

$$\frac{\partial}{\partial x}[y(-\mu + cx)] = cy, \quad \frac{\partial}{\partial y}[y(-\mu + cx)] = -\mu + cx.$$

the linearization at an equilibrium (x_∞, y_∞) is

$$\begin{aligned} u' &= (\lambda - by_\infty)u - bx_\infty, \\ v' &= cy_\infty u + (-\mu + cx_\infty)v \end{aligned}$$

One equilibrium is $(0,0)$ with linearization

$$u' = \lambda u, \quad v' = -\mu v.$$

A second equilibrium, obtained by solving $\lambda - by = 0$, $-\mu + cx = 0$, is $(\mu/c, \lambda/b)$. The linearization at this equilibrium is

$$\begin{aligned} u' &= -\frac{b\mu}{c}v \\ v' &= \frac{c\lambda}{b}u \end{aligned}$$

Exercises

In Exercises 1 through 8, find the linearization of each given system at each equilibrium

1. $x' = x - y$, $y' = x + y - 2$

2. $x' = y$, $y' = x + y - 1$

3. $x' = y + 1$, $y' = x^2 + y$

4. $x' = y^2 - 8x$, $y' = x - 2$

5. $x' = e^{-y}$, $y' = e^{-x}$

6. $x' = \sin y$, $y' = 2x$

7. $x' = x(\lambda - ax - by)$, $y' = y(\mu - cx - dy)$

8. $x' = x(\lambda - ax + by)$, $y' = y(\mu + cx - dy)$

9. The following two-dimensional nonlinear ordinary differential equation has been proposed as a model for cell differentiation.

$$\begin{aligned} \frac{dx}{dt} &= y - x \\ \frac{dy}{dt} &= \frac{5x^2}{4 + x^2} - y. \end{aligned} \tag{4.7}$$

   (i) Sketch the curves $y = x$ and $y = 5x^2/(4+x^2) = 0$ in the positive quadrant of the $(x, y)-$ plane.

   (ii) Determine the equilibrium points.

   (iii) Linearize the system of differential equations (4.7) at each equilibrium point.

   (iv) Determine the local stability of each positive equilibrium point and classify the equilibrium points.

   (v) Use a computer algebra system to find the numerical solution of system (4.7), with initial points (1.1, 1.2) and (4.5, 3.9), respectively, and plot your solution.

10. [Eldestein-Keshet (1988)]In this problem we shall examine a plant-herbivore model. Let q represent the chemical state of the plant with q low meaning that the plant is toxic and q high meaning that it is good as a food source for the herbivore. Therefore, q is an index of plant quality for the herbivore. Suppose that plant quality is enhanced when herbivory is low or moderate and declines when herbivory is high. Assume that herbivores are small immobile insects with density I. Suppose further that their growth rate depends on the quality of food that they consume. The model equations are

$$\begin{aligned} \frac{dq}{dt} &= k_1 - k_2 q I (I - I_0) \\ \frac{dI}{dt} &= k_3 I \left(1 - \frac{k_4 I}{q}\right). \end{aligned} \tag{4.8}$$

   (i) Explain the equations, and suggest possible meanings for k_1, k_2, k_3, k_4, and I_0.

(ii) Show that the equations can be written in the following dimensionless form:

$$\begin{aligned}\frac{dx}{d\tau} &= 1 - kxy(y-1) \\ \frac{dy}{d\tau} &= \alpha y\left(1 - \frac{y}{x}\right).\end{aligned} \tag{4.9}$$

Determine k and α in terms of the original parameters.

(iii) Show that there is only one equilibrium.

(iv) Determine its stability.

10. Let S, I, and R represent the densities of individuals subject to a disease who are susceptible, infective, and recovered respectively. Suppose that recovered individuals can become susceptible again after some time. The model equations are as follows:

$$\begin{aligned}\frac{dS}{dt} &= -\beta S\frac{I}{N} + \gamma R \\ \frac{dI}{dt} &= \beta S\frac{I}{N} - \nu I \\ \frac{dR}{dt} &= \nu I - \gamma R,\end{aligned}$$

where $N = S + I + R$, and β, γ, and ν are the infection, loss of immunity, and recovery rates, respectively.

(i) Reduce this model to a two-dimensional system of equations.

(ii) Find the equilibrium points. Is there a disease-free equilibrium (an equilibrium with $I = 0$)? Is there an endemic equilibrium (an equilibrium with $I > 0$)?

(iii) Determine the local stability of each of the equilibrium points found in (b).

11. Let $S(t)$ be the number of susceptible individuals at time t, $I(t)$ be the number of infective individuals at time t, $V(t)$ be the number of vaccinated or recovered individuals at time t, and $N(t) = S(t) + I(t) + V(t)$.

Consider the model

$$\begin{aligned}S'(t) &= \mu N - \beta S\frac{I}{N} - \mu S \\ I'(t) &= \beta S\frac{I}{N} - (\mu + \gamma)I \\ V'(t) &= \gamma I - \mu V,\end{aligned} \tag{4.10}$$

where μ is the per capita death rate, γ is the per capita recovery rate, $\beta = pc$ with c the average number of contacts per unit time, and p the probability of transmission per contact by a susceptible with an infective individual.

(i) Find the basic reproductive number R_0.

(ii) Find the equilibria.

(iii) Compute the stability of each equilibria.

Note that the birth rate (also known as recruitment rate) μN into the susceptible class is constant. Why?

12. Consider the model (4.10) but change the recruitment rate from the constant μN to the constant Λ. This means that a certain fixed number of individuals join or arrive into the susceptible class per unit time. The model becomes

$$\begin{aligned} S'(t) &= \Lambda - \beta S\frac{I}{N} - \mu S \\ I'(t) &= \beta S\frac{I}{N} - (\mu+\gamma)I \\ R'(t) &= \gamma I - \mu R, \end{aligned} \tag{4.11}$$

where $N(t) = S(t) + I(t) + R(t)$.

(i) What are the units of Λ, $\beta SI/N$, μ, γ, β, μS?

(ii) Find the equation satisfied by $N(t) = S(t) + I(t) + R(t)$.

(iii) Solve the equation for $N(t)$ and observe that the population size for this model is not constant.

(iv) Show that $N(t) \to \Lambda/\mu$ as $t \to \infty$.

(v) Let $K = \Lambda/\mu$. Consider the limiting system

$$\begin{aligned} S'(t) &= \Lambda - \beta S\frac{I}{K} - \mu S \\ I'(t) &= \beta S\frac{I}{K} - (\mu+\gamma)I \\ R'(t) &= \gamma I - \mu R. \end{aligned} \tag{4.12}$$

To study the dynamics of this limiting system, it is enough to consider the first two equations as $R(t) = K - S(t) - I(t)$.

(i). Find R_0

(ii). Find all equilibria

(iii). Make a stability analysis of the equilibria

Some recent results in dynamical systems guarantee that the limiting system and the original system have the same qualitative dynamics [Castillo-Chavez and Thieme (1995)].

13. Consider the model with vaccination

$$\begin{aligned} S'(t) &= \mu N - \beta S \frac{I}{N} - (\mu + \phi) S \\ I'(t) &= \beta S \frac{I}{N} - (\mu + \gamma) I \\ V'(t) &= \gamma I + \phi S - \mu V. \end{aligned} \tag{4.13}$$

Here V(t) denotes the number of vaccinated or recovered members at time t.

(i) $dN/dt = 0$. What does this imply?

(ii) Discuss why it suffices to study the first two equations.

(iii) Compute $R_0(\phi)$. What is the value of $R_0(0)$? Compare $R_0(\phi)$ with $R_0(0)$.

(iv) Compute the equilibria.

(v) Do the stability analysis for the disease-free equilibrium.

14. If vaccination strategies are incorporated for newborns, we consider that not all new births are susceptible. Suppose that the vaccination rate per capita is p; then a newborn becomes vaccinated with probability p. The modified model is of the new form

$$\begin{aligned} \frac{dS}{dt} &= (1-p)\mu N - \beta S \frac{I}{N} - \mu S = f_1 \\ \frac{dI}{dt} &= \beta S \frac{I}{N} - (\mu + \gamma) I = f_2 \\ \frac{dV}{dt} &= p\mu N + \gamma I - \mu V = f_3 \end{aligned} \tag{4.14}$$

(i) What is the Jacobian matrix of differential equation (4.14) at the disease-free equilibrium points?

(ii) Find the corresponding eigenvalues of the matrix above.

(iii) Find the basic reproductive numbers (R_0). Compare R_0 in model (4.10) with that in model (4.14).

(iv) Study the stability of the disease-free equilibrium points of the model ((4.14)).

4.4 Qualitative Behavior of Solutions of Linear Systems

In the previous section we reduced the analysis of the stability of an equilibrium (x_∞, y_∞) of a system of differential equations

$$\begin{aligned} x' &= F(x,y) \\ y' &= G(x,y) \end{aligned} \tag{4.15}$$

to the determination of the behavior of solutions of the linearization at the equilibrium

$$\begin{aligned} u' &= F_x(x_\infty, y_\infty)u + F_y(x_\infty, y_\infty)v \\ v' &= G_x(x_\infty, y_\infty)u + G_y(x_\infty, y_\infty)v. \end{aligned} \tag{4.16}$$

Next we shall analyze the various possibilities for the behavior of solutions of the two-dimensional linear homogeneous system with constant coefficients

$$\begin{aligned} x' &= ax + by \\ y' &= cx + dy, \end{aligned} \tag{4.17}$$

where a, b, c, and d are constants, in order that we might describe the behavior of solutions of the linearization at an equilibrium. We will then be able to state some refinements of Theorem 4.1 of Section 4.3 that give more specific information about the behavior of solutions near an equilibrium as determined by the community matrix at the equilibrium. We will assume throughout that $ad - bc \neq 0$. This implies that the origin is the only equilibrium of the system (4.17). If this system is the linearization at an equilibrium (x_∞, y_∞) of a nonlinear system (4.15), then this equilibrium is *isolated*, meaning that there is a disc centered at (x_∞, y_∞) containing no other equilibrium of the nonlinear system. It is convenient to use vector-matrix notation: We let $\vec{x}$ denote the column vector $\begin{pmatrix} x \\ y \end{pmatrix}$, $\vec{x}'$ the column vector $\begin{pmatrix} x' \\ y' \end{pmatrix}$, and A the 2×2 matrix

$$\begin{pmatrix} a & b \\ c & d \end{pmatrix}.$$

Then, using the properties of matrix multiplication, we may rewrite the linear system $x' = ax + by$, $y' = cx + dy$ in the form

$$\vec{x}' = A\vec{x}.$$

A linear change of variable $\vec{x} = P\vec{u}$, with P a non-singular 2×2 matrix, (which represents a rotation of axes and a change of scale along the axes), transforms the system to $P\vec{u}' = AP\vec{u}$, or

$$\vec{u}' = P^{-1}AP\vec{u} = B\vec{u}.$$

This is of the same type as the original system $\vec{x}' = A\vec{x}$ and its coefficient matrix $B = P^{-1}AP$ is *similar* to A. If we can solve for $\vec{u}$, we can then reconstruct $\vec{x} = P\vec{u}$, and in fact the qualitative properties of solutions for $\vec{u}$ are preserved in this reconstruction. Thus, we may describe the various possible phase portraits of $\vec{x}' = A\vec{x}$ by listing the various possible canonical forms of A under similarity and constructing the phase portrait for each possibility.

Theorem 4.3. [1] *The matrix*

$$A = \begin{pmatrix} a & b \\ c & d \end{pmatrix},$$

with $\det A = ad - bc \neq 0$, *is similar under a real transformation to one of*

(i) $\begin{pmatrix} \lambda & 0 \\ 0 & \mu \end{pmatrix}$, $\lambda > \mu > 0$ *or* $\lambda < \mu < 0$.

(ii) $\begin{pmatrix} \lambda & 0 \\ 0 & \lambda \end{pmatrix}$, $\lambda > 0$ *or* $\lambda < 0$.

(iii) $\begin{pmatrix} \lambda & 1 \\ 0 & \lambda \end{pmatrix}$, $\lambda > 0$ *or* $\lambda < 0$.

(iv) $\begin{pmatrix} \lambda & 0 \\ 0 & \mu \end{pmatrix}$, $\lambda > 0 > \mu$

(v) $\begin{pmatrix} 0 & \beta \\ -\beta & 0 \end{pmatrix}$, $\beta \neq 0$.

(vi) $\begin{pmatrix} \alpha & \beta \\ -\beta & \alpha \end{pmatrix}$, $\alpha > 0, \beta \neq 0$ *or* $\alpha < 0, \beta \neq 0$.

We now describe the phase portraits for each of these cases in turn.

Case (i). The transformed system is $u' = \lambda u$, $v' = \mu v$, with solution $u = u_0 e^{\lambda t}$, $v = v_0 e^{\mu t}$. If $\lambda < \mu < 0$ then u and v both tend to zero as $t \to \infty$ and $v/u = v_0 e^{(\mu-\lambda)t}/u_0 \to +\infty$. Thus, every orbit tends to the origin with infinite slope (except if $v_0 = 0$, in which case the orbit is on the u-axis), and the phase portrait is as shown in Figure 4.2. If $\lambda > \mu > 0$ the portrait is the same, except that the arrows are reversed.

Case (ii). The system is $u' = \lambda u$, $v' = \lambda v$, with solution $u = u_0 e^{\lambda t}$, $v = v_0 e^{\lambda t}$. If $\lambda < 0$ both u and v tend to zero as $t \to \infty$ and $v/u = v_0/u_0$.

[1] To be proved in Section 4.6, the appendix to this chapter.

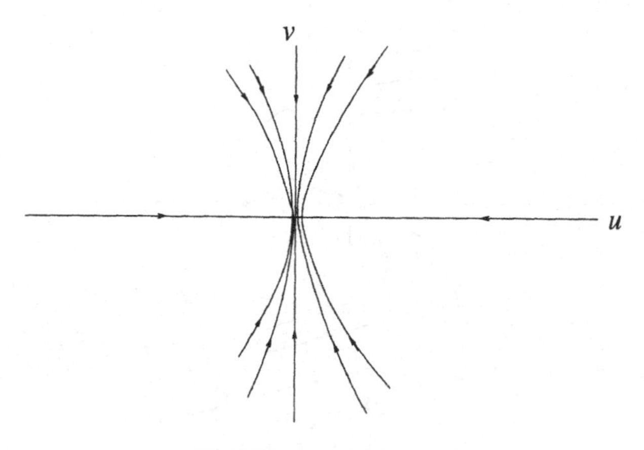

FIGURE 4.2. *(Case i)*

Thus every orbit is a straight line going to the origin, and all slopes as the orbit approaches the origin are possible. The phase portrait is as shown in Figure 4.3. If $\lambda > 0$, the portrait is the same except that the arrows are reversed.

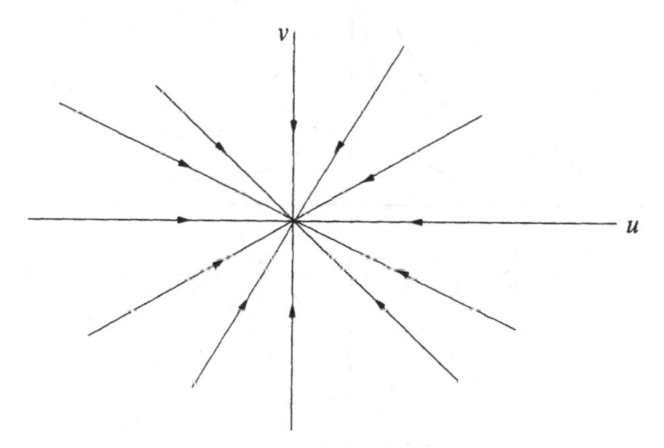

FIGURE 4.3. *(Case ii)*

Case (iii). The system is $u' = \lambda u + v$, $v' = \lambda v$. The solution of the second equation is $v = v_0 e^{\lambda t}$ and substitution into the first equation gives the first order linear equation $u' = \lambda u + v_0 e^{\lambda t}$ whose solution is $u = (u_0 + v_0 t)e^{\lambda t}$. If $\lambda < 0$, both u and v tend to zero as $t \to \infty$ and $v/u = v_0/(u_0 + v_0 t)$ tends to zero unless $v_0 = 0$, in which case the orbit is the u- axis. Because $u = (u_0 + v_0 t)e^{\lambda t}$, $du/dt = \big((u_0\lambda + v_0) + \lambda v_0 t\big)e^{\lambda t}$, and $du/dt = 0$ when $t = -(v_0 + u_0\lambda)/\lambda v_0$. Thus except for the orbits on the u-axis, every orbit has a maximum or minimum u-value and then turns back towards the origin. The phase portrait is as shown in Figure 4.4. If $\lambda > 0$ the portrait is the same except that the arrows are reversed.

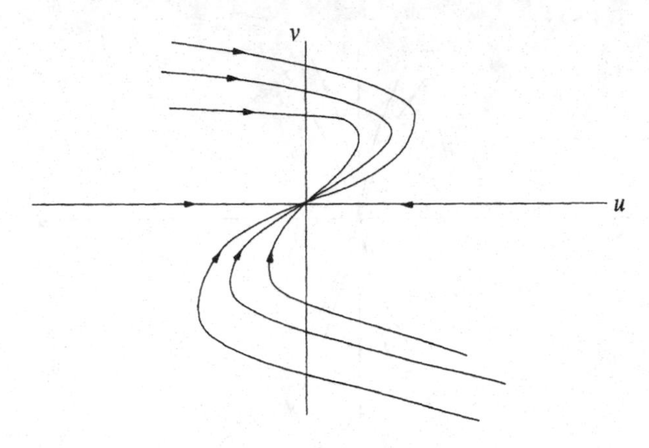

FIGURE 4.4. *(Case iii)*

Case (iv). The solution is $u = u_0 e^{\lambda t}$, $v = v_0 e^{\mu t}$ just as in Case (i), but now u is unbounded and $v \to 0$ as $t \to \infty$, unless $u_0 = 0$, in which case the orbit is on the v-axis. The phase portrait is as shown in Figure 4.5

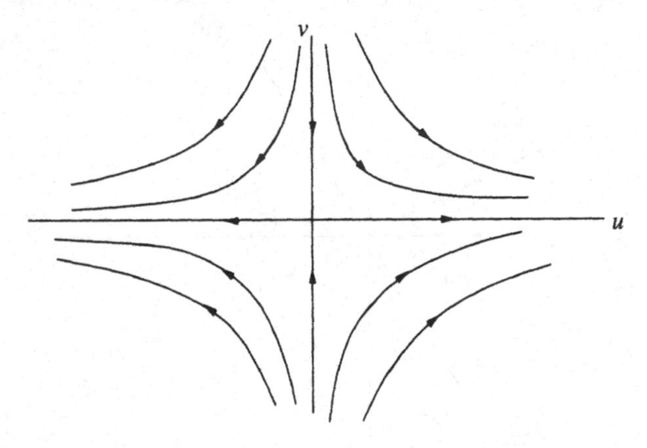

FIGURE 4.5. *(Case iv)*

Case (v). The system is $u' = \beta v$, $v' = -\beta u$. Then $u'' = \beta v' = -\beta^2 u$ and $u = A\cos\beta t + B\sin\beta t$ for some A, B. Thus $v = u'/\beta = -A\sin\beta t + B\cos\beta t$ and $u^2 + v^2 = A^2 + B^2$. Every orbit is a circle, clockwise if $\beta > 0$ and counterclockwise if $\beta < 0$. The phase portrait for $\beta > 0$ is as shown in Figure 4.6.

Case (vi). The system is $u' = \alpha u + \beta v$, $v' = -\beta u + \alpha v$. The change of variables $u = e^{\alpha t} p$, $v = e^{\alpha t} q$, so that $u' = \alpha e^{\alpha t} p + e^{\alpha t} p'$, $v' = \alpha e^{\alpha t} q + e^{\alpha t} q$ reduces the system to $p' = \beta q$, $q' = -\beta p$, which has been solved in Case (v). Thus, $u = e^{\alpha t}(A\cos\beta t + B\sin\beta t)$, $v = e^{\alpha t}(-A\sin\beta t + B\cos\beta t)$, and $u^2 + v^2 = e^{-2\alpha t}(A^2 + B^2)$. If $\alpha < 0$, $u^2 + v^2$ decreases exponentially,

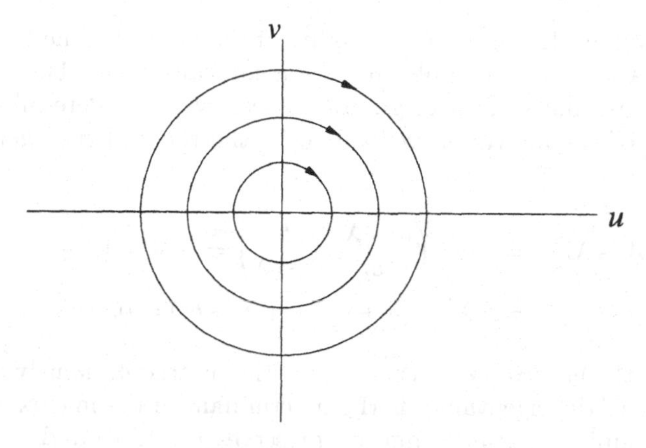

FIGURE 4.6. *(Case v)*

and the orbits are spirals inward to the origin, clockwise if $\beta > 0$ and counterclockwise if $\beta < 0$. The phase portrait is as shown in Figure 4.7. If $\alpha > 0$, the portraits are the same except that the arrows are reversed.

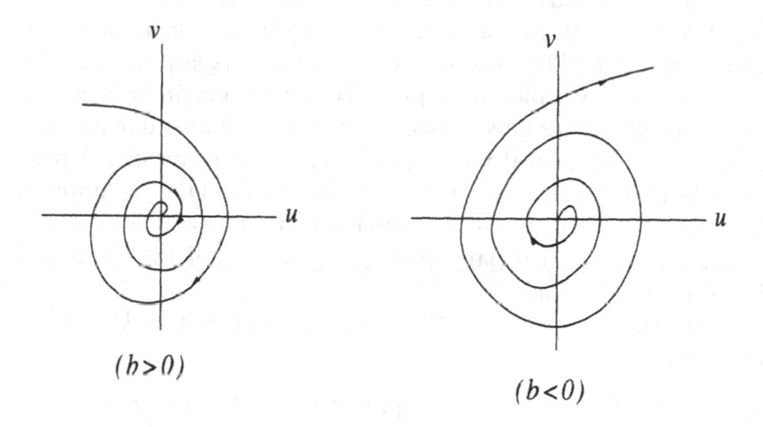

FIGURE 4.7. *(Case vi)*

These six cases may be classified as being of four distinct types. In Cases (i), (ii), and (iii) all orbits approach the origin as $t \to +\infty$ (or as $t \to -\infty$ depending on the signs of λ and μ) with a limiting direction, and the origin is said to be a *node* of the system. In case (iv) only two orbits approach the origin as $t \to +\infty$ or as $t \to -\infty$, and all other orbits move away from the origin. In this case the origin is said to be a *saddle point.* In Case (vi) every orbit winds around the origin in the sense that its angular argument tends to $+\infty$ or to $-\infty$, and the origin is said to be a *vortex, spiral point*, or *focus.* In Case (v), every orbit is periodic; in this case the origin is said to be a *center.* According to Theorem 4.3, asymptotic stability of the origin for the linearization implies asymptotic stability of an equilibrium of a nonlinear system. In addition, instability of the origin for the linearization implies instability of an equilibrium of a nonlinear system. The asymptotic stability or instability of the origin for a linear system is determined by the *eigenvalues* of the matrix A, defined to be the roots of the *characteristic equation*

$$\begin{aligned} \det(A - \lambda I) &= \det\begin{pmatrix} a-\lambda & b \\ c & d-\lambda \end{pmatrix} = (a-\lambda)(d-\lambda) - bc \\ &= \lambda^2 - (a+d)\lambda + (ad-bc) = 0. \end{aligned}$$

The sum of the eigenvalues is the *trace* of the matrix A, namely $a + d$, and the product of the eigenvalues is the determinant of the matrix A, namely $ad - bc$. A similarity transformation preserves the trace and determinant and therefore does not change the eigenvalues. Thus, the eigenvalues of A are λ, μ in Cases (i) and (iv); λ (a double eigenvalue) in Cases (ii) and (iii); the complex conjugates $\pm i\beta$ in Case (v); and $\alpha \pm i\beta$ in Case (vi). Examination of the phase portraits in the various cases shows that the origin is asymptotically stable in Case (i) if $\lambda < \mu < 0$; in Cases (ii) and (iii) if $\lambda < 0$; and in Case (vi) if $\alpha < 0$. Similarly the origin is unstable in Case (i) if $\lambda > \mu > 0$; in Cases (ii) and (iii) if $\lambda > 0$; in Case (iv); and in Case (vi) if $\alpha > 0$. The origin is stable but not asymptotically stable in Case (v) (center). A simpler description is that the origin is asymptotically stable if both eigenvalues have negative real part and unstable if at least one eigenvalue has positive real part. If both eigenvalues have real part zero, the origin is stable but not asymptotically stable. Our assumption that $ad - bc = \det A \neq 0$ rules out the possibility that $\lambda = 0$ is an eigenvalue; thus, eigenvalues with real part zero can occur only if the eigenvalues are pure imaginary, as in Case (v).

Combining this analysis with the linearization result (4.1), we have the following result:

Theorem 4.4. *If (x_∞, y_∞) is an equilibrium of the system 4.15 and if all eigenvalues of the coefficient matrix of the linearization at this equilibrium have negative real part, specifically if*

$$\begin{aligned} \operatorname{tr} A(x_\infty, y_\infty) &= F_x(x_\infty, y_\infty) + G_y(x_\infty, y_\infty) < 0 \\ \det A(x_\infty, y_\infty) &= F_x(x_\infty, y_\infty)G_y(x_\infty, y_\infty) - F_y(x_\infty, y_\infty)G_x(x_\infty, y_\infty) > 0, \end{aligned}$$

then the equilibrium (x_∞, y_∞) is asymptotically stable.

We can be more specific about the nature of the orbits near an equilibrium. In terms of the elements of the matrix A, we may characterize the cases as follows, using the remark that the eigenvalues are complex if and only if

$$\Delta = (a+d)^2 - 4(ad-bc) = (a-d)^2 + 4bc < 0.$$

1. If $\det A = ad - bc < 0$ the origin is a saddle point.

2. If $\det A > 0$ and $\operatorname{tr} A = a + d < 0$, the origin is asymptotically stable, a node if $\Delta \geq 0$ and a spiral point if $\Delta < 0$.

3. If $\det A > 0$ and $\operatorname{tr} A > 0$ the origin is unstable, a node if $\Delta \geq 0$ and a spiral point of $\Delta > 0$.

4. If $\det A > 0$ and $\operatorname{tr} A = 0$ the origin is a center.

It is possible to show that in general the phase portrait of a nonlinear system at an equilibrium is similar to the phase portrait of the linearization at the equilibrium, *except possibly if the linearization has a center.* This is true under the assumption that the functions $F(x, y)$ and $G(x, y)$ in the system (4.15) are smooth enough that Taylor's theorem is applicable, so that the terms neglected in the linearization process are of higher order. If the linearization at an equilibrium has a *node*, then the equilibrium of the nonlinear system is also a node, defined to mean that every orbit tends to the equilibrium (either as $t \to \infty$ or as $t \to -\infty$) with a limiting direction. If the linearization at an equilibrium has a *spiral point*, then the equilibrium of the nonlinear system is also a spiral point, defined to mean that every orbit tends to the equilibrium (either as $t \to \infty$ or as $t \to -\infty$) with its angular variable becoming infinite. If the linearization at an equilibrium has a *saddle point*, then the equilibrium of the nonlinear system is also a saddle point. A saddle point is defined by the characterization that there is a curve through the equilibrium such that orbits starting on this curve tend to the equilibrium but orbits starting off this curve cannot stay near the equilibrium. An equivalent formulation is that there are two orbits tending to the equilibrium as $t \to +\infty$ and two orbits tending away from the equilibrium, or tending to the equilibrium as $t \to -\infty$. These orbits are called *separatrices*; the two orbits tending to the saddle point are the stable separatrices while the two orbits tending away from the saddle point are the unstable separatrices. Other orbits appear like hyperbolas (Figure 4.8).

A *center* is defined to be an equilibrium for which there is an infinite sequence of periodic orbits around the equilibrium and tending to it. If the linearization at an equilibrium has a center then the equilibrium of the nonlinear system *may* be a center, but is not necessarily a center: It could be an asymptotically stable spiral point or an unstable spiral point.

Example 1. If we linearize the Lotka-Volterra system $x' = x(\lambda - by)$, $y' = y(-\mu + cx)$ about the equilibrium (x_∞, y_∞), with $x_\infty = \mu/c$, $y_\infty = \lambda/b$ we obtain the system

$$\begin{aligned} u' &= -\frac{\mu b}{c} v \\ v' &= \frac{\lambda c}{b} u \end{aligned}$$

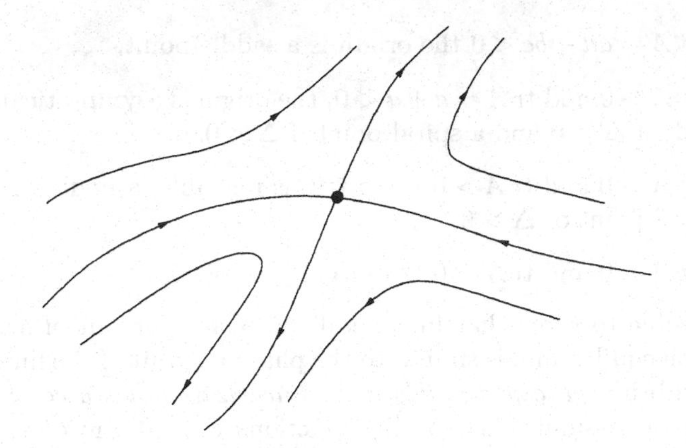

FIGURE 4.8.

with coefficient matrix

$$A = \begin{pmatrix} 0 & -\frac{\mu b}{c} \\ \frac{\lambda c}{b} & 0 \end{pmatrix}$$

with trace zero and determinant $\lambda\mu > 0$. Thus the linearization has a center. To solve the linearization we write

$$\begin{aligned} u'' &= -\frac{\mu b}{c}v' = -\lambda\mu u \\ v'' &= \frac{\lambda c}{b}u' = -\lambda\mu v, \end{aligned}$$

from which we see that both u and v are periodic with frequency $\sqrt{\lambda\mu}$, period $2\pi/\lambda\mu$. We have shown in the preceding section that all solutions of the Lotka-Volterra system near the equilibrium are periodic and thus that the equilibrium is a center, but study of the linearization is not enough to show this. The addition of a small perturbation to the Lotka-Volterra model could change the center to a spiral point, one that may be either asymptotically stable or unstable. Because the Lotka-Volterra model is so sensitive to perturbations it is not really suitable as an population model. When we examine predator-prey systems in more detail we will attempt to refine the model.

Example 2. Determine whether each equilibrium of the system

$$x' = y,\ y' = 2(x^2 - 1)y - x$$

is asymptotically stable or unstable.

Solution. The equilibria are the solution of $y = 0, 2(x^2 - 1)y - x = 0$, and thus the only equilibrium is $(0, 0)$. The community matrix at (x_∞, y_∞) is

$$\begin{pmatrix} 0 & 1 \\ 4x_\infty y_\infty - 1 & 2({x_\infty}^2 - 1) \end{pmatrix}$$

and thus the community matrix at $(0,0)$ is

$$\begin{pmatrix} 0 & 1 \\ -1 & -2) \end{pmatrix},$$

with trace $-2 < 0$ and the determinant $1 > 0$. Thus the equilibrium $(0,0)$ is asymptotically stable.

Example 3. Determine the asymptotic stability or instability of each equilibrium of the Lotka-Volterra system (4.1)
Solution. We have seen in Example 1,Section 4.3 that at the equilibrium $(0,0)$ the community matrix is

$$\begin{pmatrix} \lambda & 0 \\ 0 & -\mu) \end{pmatrix}.$$

Since this has negative determinant, $(0,0)$ is unstable. The community matrix at the equilibrium $(\mu/c, \lambda/b)$ is

$$\begin{pmatrix} 0 & -\frac{b\mu}{c} \\ \frac{c\lambda}{b} & 0 \end{pmatrix}$$

with trace zero. Thus Theorem 4.4 does not settle the question. However, as we have seen in Section 4.1, orbits of the system neither move away from the equilibrium nor approach it. Thus, this equilibrium is neither asymptotically stable nor unstable. [Such an equilibrium is sometimes said to be neutrally stable]

Exercises

In Exercises 1 through 6, for each equilibrium of the given system determine whether the equilibrium is asymptotically stable or unstable

1. $x' = x - y$, $y' = x + y - 2$ (see Exercise 1, Section 4.3)
2. $x' = y$, $y' = x + y - 1$ (see Exercise 2, Section 4.3)
3. $x' = y + 1$, $y' = x^2 + y$(see Exercise 3, Section 4.3)
4. $x' = y^2 - 8x$, $y' = x - 2$ (see Exercise 4 section 4.3)
5. $x' = e^{-y}$, $y' = e^{-x}$ (see Exercise 5, Section 4.3)
6. $x' = \sin y$, $y' = 2x$(see Exercise 6, Section 4.3)
7.* Determine the behavior of orbits of the system

$$\begin{aligned} x' &= x(\lambda - ax - by) \\ y' &= y(\mu - cx - dy). \end{aligned}$$

8.* Determine the behavior of orbits of the system

$$\begin{aligned} x' &= x(2-x) - \frac{xy}{x+1} \\ y' &= \frac{xy}{x+1} - y. \end{aligned}$$

9.* Determine the behavior of orbits of the system

$$\begin{aligned} x' &= x(\lambda - ax - by) \\ y' &= y(-\mu + cx). \end{aligned}$$

10.* Consider the system

$$\begin{aligned} x' &= y \\ y' &= -x - y^3. \end{aligned}$$

(i) Show that $(0,0)$ is the only equilibrium.

(ii) Show that the function $V(x,y) = x^2 + y^2$ decreases along every orbit and tends to zero, showing that $(0,0)$ is asymptotically stable.

11. [Kaplan & Glass(1995)] A chemotherapeutic agent is being used to treat an intracranial tumor. Let x be the number of molecules of the agent in the blood and y the number of molecules that have crossed the blood-brain barrier. When $t = 0$ we have $x = N$ and $y = 0$. The dynamics are described by the differential equations

$$\begin{aligned} \frac{d}{dt}x &= \alpha(y - x) - \gamma x \\ \frac{d}{dt}y &= -\alpha(y - x). \end{aligned}$$

(i) Show that if γ is much larger than α, the roots of the characteristic equation are approximately $-\gamma$ and $-\alpha$.

(ii) Use the result from part (a) to replace the x equation by $dx/dt = -\gamma x$ and solve the equation for y as a function of time for $\alpha = 10^{-3}$ hr^{-1} and $\gamma = 1$hr^{-1}.

(iii) For the values of α and γ in part (b) compute the time when y is a maximum. What is the approximate value of y at this time?

12. [Kaplan & Glass(1995)] Limpets and seaweed live in a tide pool. The dynamics of this system are given by the differential equations

$$\begin{aligned} \frac{ds}{dt} &= s - s^2 - sl \\ \frac{dl}{dt} &= sl - \frac{l}{2} - l^2, \quad l \geq 0, s \geq 0, \end{aligned}$$

where the densities of seaweed and limpets are given by s and l, respectively.

(i) Determine all equilibria of this system.

(ii) For each nonzero equilibrium determined in part (a) evaluate the stability and classify it as a node, focus, or saddle point.

(iii) Sketch the flows in the phase plane.

(iv) What will the dynamics be in the limit as $t \to \infty$ for initial conditions

$$\begin{array}{ll} (i)\ s(0) = 0, & l(0) = 0? \\ (ii)\ s(0) = 0, & l(0) = 15? \\ (iii)\ s(0) = 2, & l(0) = 0? \\ (iv)\ s(0) = 2, & l(0) = 15? \end{array}$$

13. The following system of differential equation is an SIR model for infectious diseases (with vital dynamics):

$$\begin{aligned} \frac{dS}{dt} &= -\beta SI + \mu - \mu S + \theta R \\ \frac{dI}{dt} &= \beta SI - \gamma I - \mu I \\ \frac{dR}{dt} &= 1 - S - I \end{aligned}$$

with $S(0) = S_0$, $I(0) = I_0$, $R(0) = R_0$, where $S = S(t)$ is the fraction of susceptible individuals at time t, $I = I(t)$ is the fraction of infective individuals at time t, and $R = R(t)$ is the fraction of recovered individuals, μ is the per-capita birth and death rate, γ and θ are the rates at which individuals leave the infective and recovered classes, respectively. The *basic reproductive number* (the number of secondary cases produced by one infective individual in a population of susceptibles) in this case is given by the formula $\sigma = \beta/(\mu + \gamma)$.

(i) Choose $\mu = 0.0001$, $\gamma = 0.2$, $\theta = 0.02$, $\beta = 0.15$. Then $\sigma < 1$. Use a computer algebra system to observe that the disease dies out as time goes to infinity.

(ii) Increase β to $\beta = 0.4$, then $\sigma > 1$.

(i) Plot $I(t)$ versus time for initial values $S_0 = 0.2$, $I_0 = 0.8$, $R_0 = 0$.

(ii) Plot $I(t)$ versus $S(t)$ for the following initial conditions (on one graph) and explain the biological meaning suggested by the graph.

$$S_0 = 0.2,\ I_0 = 0.8,\ R_0 = 0;$$
$$S_0 = 0.4,\ I_0 = 0.6,\ R_0 = 0;$$
$$S_0 = 0.6,\ I_0 = 0.4,\ R_0 = 0;$$
$$S_0 = 0.8,\ I_0 = 0.2,\ R_0 = 0.$$

14. Discuss the qualitative behavior of the following epidemiological models (SIR and SIS without vital dynamics):

 (i)

$$\begin{aligned} \frac{dS}{dt} &= -\beta S \frac{I}{N} + \gamma I \\ \frac{dI}{dt} &= \beta S \frac{I}{N} - \gamma I \\ N &= S + I. \end{aligned}$$

 (ii)

$$\begin{aligned} \frac{dS}{dt} &= -\beta S \frac{I}{N} \\ \frac{dI}{dt} &= \beta S \frac{I}{N} - \gamma I \\ \frac{dR}{dt} &= \gamma I \\ N &= S + I + R. \end{aligned}$$

15. [May (1981)] A modification of the logistic equation describing two species interacting mutualistically results in the model

$$\begin{aligned} \frac{dx}{dt} &= rx\left(1 - \frac{x}{K_1 + ay}\right) \\ \frac{dy}{dt} &= ry\left(1 - \frac{y}{K_2 + bx}\right) \end{aligned}$$

 with a, b, K_1, K_2 positive constants. Show that the equilibrium populations are positive if $ab < 1$ and that if this condition is satisfied then the equilibrium is locally asymptotically stable.

16. Sketch the phase-plane behavior of the following systems of ordinary differential equations and classify the stability of the equilibrium point $(0, 0)$.

 (i)

$$\begin{aligned} \frac{dx}{dt} &= 3x - 2y \\ \frac{dy}{dt} &= 4x + y. \end{aligned}$$

(ii)

$$\begin{aligned}\frac{dx}{dt} &= 5x + 2y\\ \frac{dy}{dt} &= -13x - 5y.\end{aligned}$$

(iii)

$$\begin{aligned}\frac{dx}{dt} &= -x - y\\ \frac{dy}{dt} &= \frac{8}{5}x - y.\end{aligned}$$

17. We are going to study an ecological system consisting of two species. The relationship between the two species is one of three kinds. One describes the system as a predator-prey system, a competitive system or a mutualistic system depending on the the nature of the interactions between species. We will use the Lotka-Volterra model to study the system. Let $x(t)$ and $y(t)$ be the population sizes (or densities) at t for the two populations. The key assumption in the Lotka-Volterra model is that the per capital growth rate for each population is a linear function of population sizes. We write the ordinary differential equation model for the system

$$\begin{aligned}\frac{1}{x(t)}\frac{dx}{dt} &= r_x + a_{11}x + a_{12}y \\ \frac{1}{y(t)}\frac{dy}{dt} &= r_y + a_{21}x + a_{22}y\end{aligned} \tag{4.18}$$

where r_x, r_y, and the a_{ij} are constants.

Let $r_x = r_y = 1$, $a_{11} = a_{22} = -1$ and $a_{12} = a_{21} = -1/2$ in the model (4.18). The specific model is

$$\begin{aligned}\frac{1}{x(t)}\frac{dx}{dt} &= 1 - x - \frac{y}{2}\\ \frac{1}{y(t)}\frac{dy}{dt} &= 1 - \frac{x}{2} - y.\end{aligned}$$

Identify the system and make a guess about the fate of population sizes.

18. Consider the system of differential equations

$$\begin{aligned}\frac{dx}{dt} &= xy - y\\ \frac{dy}{dt} &= xy - x.\end{aligned}$$

(i) Compute the equilibria.

(ii) Linearize the equations around each of the equilibria found in part (a). Write the Jacobian matrix and the linear system approximation for each case.

(iii) Draw the phase portrait of each linearized system.

19. Consider the system of differential equations

$$\begin{aligned}\frac{dx}{dt} &= xy - y \\ \frac{dy}{dt} &= xy - x.\end{aligned}$$

(i) Compute the equilibria.

(ii) Linearize the equations around each of the equilibria found in part (a). Write the Jacobian matrix and the linear system approximation for each case.

(iii) Draw the phase portrait of each linearized system.

4.5 Periodic Solutions and Limit Cycles

In the preceding section we analyzed the behavior of solutions starting near an equilibrium. When we studied this question for first order differential equations this information was enough to describe the behavior of all solutions, as every solution was either unbounded or tended to an equilibrium. However, for second order systems there are other possibilities, and our results are valid only locally. We must consider what can happen to a solution that does not begin near an equilibrium; in particular we will want to examine the behavior of solutions of systems that have no asymptotically stable equilibrium. Such systems can arise as models for predator-prey systems. We wish to study the behavior of solutions of a two-dimensional system

$$\begin{aligned}x' &= F(x,y), \\ y' &= G(x,y)\end{aligned} \tag{4.19}$$

by studying the phase portrait in the phase plane. Let $(x(t), y(t))$ be a solution that is bounded as $t \to \infty$. The *positive semi-orbit* C^+ of this solution is defined to be the set of points $(x(t), y(t))$ for $t \geq 0$ in the (x, y) plane. The *limit set* $L(C^+)$ of the semi-orbit is defined to be the set of all points $(\bar{x}, \bar{y})$ such that there is a sequence of times $t_n \to \infty$ with $x(t_n) \to \bar{x}\, y(t_n) \to \bar{y}$ as $n \to \infty$. For example, if the solution $(x(t), y(t))$ tends to an equilibrium (x_∞, y_∞) as $t \to \infty$, then the limit set consists of the equilibrium (x_∞, y_∞). If $(x(t), y(t))$ is a periodic solution so that the semi-orbit C^+ is a closed curve, then the limit set $L(C^+)$ consists of all points of the semi-orbit C^+. It is not difficult to show that the limit set of a bounded semi-orbit is a closed, bounded, connected set. An *invariant* set for the system (4.19) is a set of points in the plane which contains the positive semi-orbit through every point of the set. Thus, for example, an equilibrium is an invariant set, and a closed orbit corresponding to a periodic solution is an invariant set. It is possible to prove, making use of the continuous dependence of solutions of differential equations on initial conditions, that the limit set of a bounded semi-orbit is an invariant set.

The results stated above are valid for autonomous systems of differential equations in all dimensions, but in two dimensions more detailed information on the structure of limit sets is available. The reasons for this involve the topological properties of the plane, especially the Jordan curve theorem which states that a simple closed curve in the plane divides the plane into two disjoint regions–which is not valid in more than two dimensions. The fundamental result on the behavior in the large of solutions of autonomous systems in the plane is the *Poincaré-Bendixson theorem* [Poincaré (1881, 1882, 1885, 1886); Bendixson (1901)], which states that if C^+ is a bounded semi-orbit whose limit set $L(C^+)$ contains no equilibrium points, then either C^+ is a periodic orbit and $L(C^+) = C^+$ or $L(C^+)$ is a periodic orbit, called a *limit cycle*, which C^+ approaches spirally (either from inside or outside). We shall not go into the proof, which may be found in many books on differential equations; a particularly readable elementary source is that of W. Hurewicz (1958).

The properties of limit sets enable us to describe limit sets that do contain equilibrium points. For example, a limit set may consist of a single equilibrium point to which the orbit tends. If a limit set contains a single equilibrium point but contains other points as well, then it must consist of the equilibrium point together with an orbit running from the equilibrium point to itself in order to be connected and invariant. Thus, the equilibrium must be a saddle point and the limit set must be a *separatrix* which runs from the saddle point to itself as t runs from $-\infty$ to $+\infty$ (Figure 4.9).

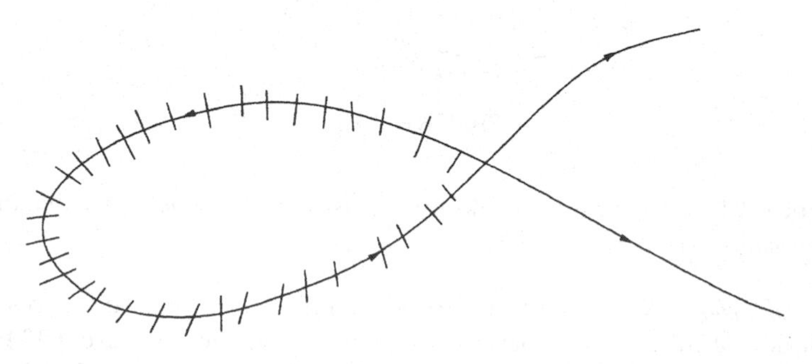

FIGURE 4.9.

If a limit set contains more than one equilibrium then it must also contain orbits joining these equilibria. In essence we can say that a bounded solution tends either to an equilibrium or to a limit cycle, overlooking such "unlikely coincidences" as the possibility of a separatrix running from a saddle point to itself. Thus, if we can show that for a given system all solutions are bounded but that there are no asymptotically stable equilibrium points, we can deduce that there must be at least one periodic orbit. This situation will arise in our study of predator-prey systems. If there is only one periodic orbit then it must be *globally asymptotically stable* in the sense that every orbit tends to it. If there is more than one periodic orbit, each must be asymptotically stable from at least one side: orbits may spiral towards it from the inside, from the outside, or both.

In situations where we know that there is a periodic solution it may be possible to use the concept of the index of a simple closed curve and the index of an equilibrium to help locate the periodic solution. An autonomous system of differential equations (4.19) defines a *vector field* by associating with each point (x, y) a vector with components $(F(x, y), G(x, y))$. A solution of the system has its orbit tangent to this vector field at every point. For any simple, i.e., not intersecting itself, closed curve S we define the *index* of S with respect to the system to to be $1/2\pi$ times the change in the angle of the vector (F, G) as the curve is traversed once in a positive (counter-clockwise) direction. Thus, if S is a periodic orbit it is easy to see that the index of S is $+1$ (Figure 4.10).

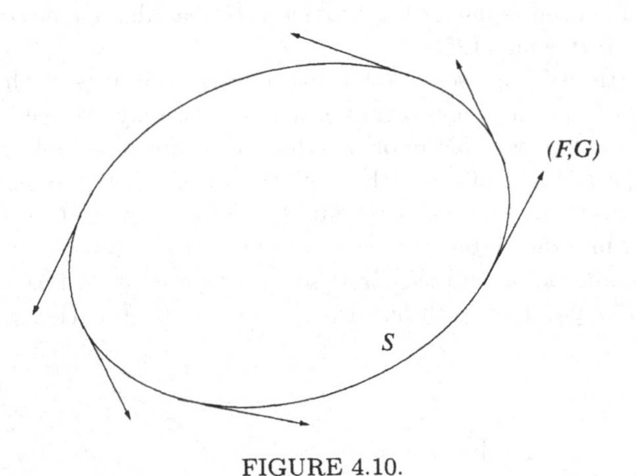

FIGURE 4.10.

Theorem 4.5. *The index of a simple closed curve S whose interior contains no equilibrium points is zero.*

Outline of proof. Write S as a sum of small simple closed curves, with cancellation of all interior portions of their boundaries (Figure 4.11). Over a sufficiently small simple closed curve, the angle of the vector (F, G) cannot

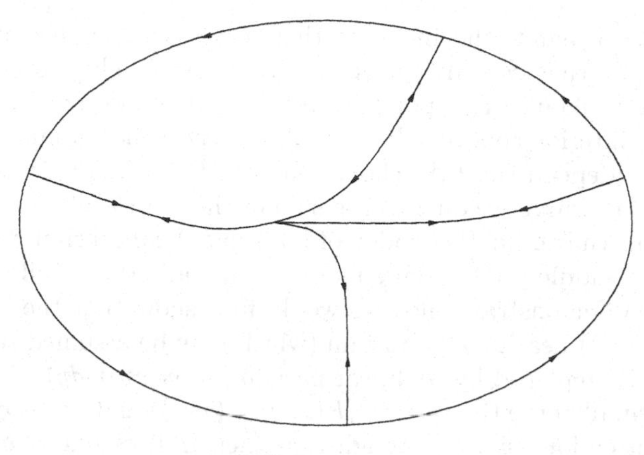

FIGURE 4.11.

change by as much as 2π, by continuity of the vector field. Thus, the index of every sufficiently small simple closed curve is zero. Since the index of S is the sum of the indices of the small curves, the index of S must be zero. The hypothesis that the interior of S contains no equilibrium points is needed to deduce that the vector field is continuous. □

An immediate consequence of Theorem 4.5 is that a periodic orbit must contain an equilibrium in its interior. Another consequence is that it is possible to give an unambiguous definition of the *index of an equilibrium point.* Let P be an equilibrium point and let S be a simple closed curve around P, but whose interior does not contain any other equilibrium points. If $\bar{S}$ is another simple closed curve around P inside S (and thus not containing any other equilibrium points in its interior), the simple closed curve formed by going around S counterclockwise, straight in to $\bar{S}$, clockwise, and then back out to S is a simple closed curve whose interior contains no equilibrium and which therefore has index zero (Figure 4.12).

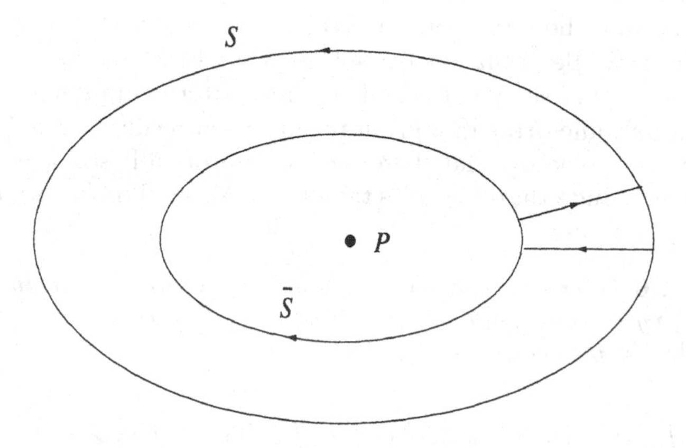

FIGURE 4.12.

On the other hand, the index of this curve is the index of S minus the index of $\bar{S}$; thus S and $\bar{S}$ have the same index. Therefore, we define the index of an equilibrium point P to be the index of any simple closed curve whose interior contains P but contains no other equilibrium. This definition is independent of the choice of curve. For any simple closed curve the index of the curve is equal to the sum of the indices of all equilibria in the interior of the curve. The index of an isolated equilibrium is either $+1$ or -1, with a saddle point having index -1 and all other equilibria having index $+1$. To demonstrate this, we would first show that the vector field $(F(x,y), G(x,y))$ near an equilibrium (which may be assumed to be at the origin) may be replaced by its linear part $(ax+by, cx+dy)$ because for a sufficiently small curve the vectors $(F(x,y), G(x,y))$ and $(ax+by,\ cx+dy)$ must have directions similar enough that their indices must be the same. The remainder of the proof consists of an examination of each type of equilibrium (see, for example, Figure 4.13).

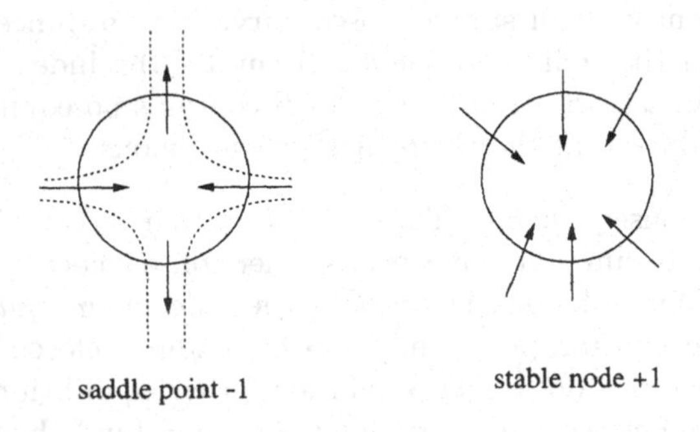

FIGURE 4.13.

We now deduce that the number of equilibria that are not saddle points minus the number of saddle points in the interior of a periodic orbit is one. In particular if a system has several equilibria all but one of which are saddle points, then any periodic orbit must go around the equilibrium which is not a saddle point but cannot go around any saddle point.

We conclude this section by giving some criteria implying that there *cannot* be a periodic orbit in a given region. Such results are of interest in situations where we know that there is an asymptotically stable equilibrium and wish to conclude that all orbits tend to it. We shall make use of Green's theorem in the plane.

Theorem 4.6 (Green). *Let D be a simple connected region in the plane with boundary C, and suppose that $P(x,y)$ and $Q(x,y)$ are continuously differentiable in the closure of D. Then*

$$\int_D\int (Q_x(x,y) - P_y(x,y))dA = \int_C (P(x,y)dx + Q(x,y)dy).$$

Theorem 4.7 (Bendixson (1901)). *Suppose that $F_x(x,y)+G_y(x,y)$ is either strictly positive or strictly negative in a simply connected region D. Then there is no periodic orbit of $x'=F(x,y)$, $y'=G(x,y)$ in D.*

Proof. Suppose that C is a closed orbit in D corresponding to a periodic solution $(x(t),y(t))$ with period T, and let Ω be the interior of C. We apply Green's theorem with $Q(x,y)=F(x,y)$, $P(x,y)=-G(x,y)$, obtaining

$$\begin{aligned}\int_\Omega\int (F_x(x,y)+G_y(x,y))dA &= \int_C (-G(x,y)dx+F(x,y)dy)\\ &= \int_0^T (-y'(t)x'(t)+x'(t)y'(t))dt=0.\end{aligned}$$

But the left side cannot be zero, and we have a contradiction. Thus there can be no perodic orbit in D. □

A useful generalization of Bendixson's theorem is obtained by application of Green's theorem with

$$Q(x,y)=\beta(x,y)F(x,y),\ P(x,y)=-\beta(x,y)G(x,y)$$

for some function $\beta(x,y)$ that is continuously differentiable in D.

Theorem 4.8 (Dulac (1934)). *Let $\beta(x,y)$ be continuously differentiable and suppose that*

$$\frac{\partial}{\partial x}\Big(\beta(x,y)F(x,y)\Big)+\frac{\partial}{\partial y}\Big(\beta(x,y)G(x,y)\Big)$$

is either strictly positive or strictly negative in a simply connected region D. Then there is no periodic orbit of $x'=F(x,y)$, $y'=G(x,y)$ in D.

In the following chapter, we shall be studying models for populations of two interacting species of the form

$$\begin{aligned}x' &= xf(x,y)\\ y' &= yg(x,y).\end{aligned}$$

We may apply Dulac's criterion with $\beta(x,y)=1/xy$, $F(x,y)=xf(x,y)$, $G(x,y)=yg(x,y)$, so that

$$\frac{\partial}{\partial x}(\beta F)+\frac{\partial}{\partial y}(\beta G)=\frac{\partial}{\partial x}\Big(\frac{1}{y}f\Big)+\frac{\partial}{\partial y}\Big(\frac{1}{x}g\Big)=\frac{f_x(x,y)}{y}+\frac{g_y(x,y)}{x}.$$

This gives the following result, which will apply to our models for species in competition and for mutualism, but not to predator-prey models.

Theorem 4.9. *Suppose $f_x(x,y)<0$, $g_y(x,y)<0$ for $x>0$, $y>0$. Then there is no periodic orbit of the system $x'=xf(x,y)$, $y'=yg(x,y)$ in the interior of the first quadrant of the phase plane.*

Example 1. Determine the qualitative behavior of solutions of the system

$$\begin{aligned} x' &= x(1 - \frac{x}{30}) - \frac{xy}{x+10}, \\ y' &= y(\frac{x}{x+10} - \frac{1}{3}) \end{aligned}$$

Solution. Equilibria are solutions of the pair of equations

$$\begin{aligned} x(1 - \frac{x}{30}) - \frac{xy}{x+10} &= 0 \\ y(\frac{x}{x+10} - \frac{1}{3}) &= 0. \end{aligned}$$

There is an equilibrium, $(0, 0)$. If $y = 0$ the first equation gives $x = 0$ or $x = 30$, and thus there is a second equilibrium, $(30, 0)$. If $x \neq 0$, $y \neq 0$, we must solve $1 - x/30 = y/(x + 10), x/(x + 10) = 1/3$, which gives a third equilibrium $(5, 12.5)$. The community matrix at $(0, 0)$ is

$$\begin{pmatrix} 1 & 0 \\ 0 & -1/3 \end{pmatrix}$$

and this equilibrium is unstable because the determinant is negative. The community matrix at $(30, 0)$ is

$$\begin{pmatrix} -1 & -3/4 \\ 0 & -5/12 \end{pmatrix}$$

and this equilibrium is also unstable because the determinant is negative. The community matrix at $(5, 12.5)$ is

$$\begin{pmatrix} 1/9 & -1/3 \\ 5/9 & -0 \end{pmatrix}$$

with positive trace, and $(5, 12.5)$ is also unstable. If we add the two equations of the model we obtain $(x + y)' = x(1 - x/30) - y/3$. Thus $(x + y)$ is decreasing except in the bounded region $y/3 < x(1 - x/30)$. In order that an orbit may be unbounded, we must have $y + z$ unbounded. However, this is impossible since $y + z$ is decreasing whenever $y + z$ is large. Thus every orbit of the system is bounded. Since all equilibria are unstable, the Poincaré-Bendixson theorem implies that there must be a periodic orbit (around $(5, 12.5)$) to which every orbit tends as $t \to \infty$.

Example 2. Determine the behavior as $t \to \infty$ of solutions in the first quadrant of the system

$$x' = x(100 - 4x - 2y)$$

$$y' = y(60 - x - y).$$

Solution. There are equilibria at $(0,0)$, at $(25,0)$, and at $(0,60)$. There is no equilibrium within $x > 0, y > 0$, because such an equilibrium would be found by solving $4x+2y = 100, x+y = 60$ and subtraction of twice the first equation from the second equation gives $2x = -20$. The equilibria $(0,0)$, and $(25,0)$ are unstable because their community matrices are, respectively,

$$\begin{pmatrix} 100 & 0 \\ 0 & 60 \end{pmatrix}$$

and

$$\begin{pmatrix} -100 & -50 \\ 0 & 35 \end{pmatrix}.$$

The equilibrium $(0,60)$ is asymptotically stable because it has community matrix

$$\begin{pmatrix} -20 & 0 \\ -60 & -60 \end{pmatrix}$$

In order to show that every orbit approaches $(0,60)$, we must show that there are no periodic orbits. This follows from Dulac's criterion with $\beta(x,y) = 1/xy$, because

$$\frac{\partial}{\partial x}\left(\frac{100 - 4x - 2y}{y}\right) + \frac{\partial}{\partial y}\left(\frac{60 - x - 2}{x}\right) = -\frac{4}{y} - \frac{1}{x} < 0.$$

Exercises

In each of Exercises 1 through 10, determine the qualitative behavior of solutions of the given system.

1. $x' = 3x(1 - \frac{x}{20}) - \frac{xy}{x+10}$, $y' = y(\frac{x}{x+10} - \frac{1}{2})$
2. $x' = x(1 - \frac{x}{40}) - \frac{2xy}{x+15}$, $y' = 3y(\frac{2x}{x+15} - \frac{6}{5})$
3. $x' = 3x(1 - \frac{x}{20}) - \frac{xy}{x+10}$, $y' = y(\frac{x}{x+10} - \frac{1}{6})$
4. $x' = x(1 - \frac{x}{40}) - \frac{2xy}{x+15}$, $y' = 3y(\frac{2x}{x+15} - \frac{1}{2})$
5. $x' = 3x(1 - \frac{x}{20}) - \frac{xy}{x+20}$, $y' = y(\frac{x}{x+10} - \frac{3}{4})$
6. $x' = x(1 - \frac{x}{40}) - \frac{2xy}{x+15}$, $y' = 3y(\frac{2x}{x+15} - \frac{8}{5})$
7. $x' = x(80 - x - y)$, $y' = y(120 - x - 3y)$
8. $x' = x(60 - x3 - y)$, $y' = y(75 - 4x - y)$
9. $x' = x(40 - x - y)$, $y' = y(90 - x - 2y)$

10. $x' = x(80 - 3x - 2y)$, $y' = y(80 - x - y)$

11.* Show that the equilibrium (x_∞, y_∞) with $x_\infty > 0$, $y_\infty > 0$ of the system

$$\begin{aligned} x' &= rx(1 - \frac{x}{K}) - \frac{axy}{x + A} \\ y' &= sy\left(\frac{ax}{x + A} - \frac{aJ}{J + A})\right) \end{aligned}$$

is unstable if $K > A + 2J$ and asymptotically stable if $J < K < A + 2J$.

12.* Determine the behavior as $t \to \infty$ of solutions of the system

$$\begin{aligned} x' &= rx(1 - \frac{x}{K + ay}) - \frac{axy}{x + A} \\ y' &= sy\left(1 - \frac{y}{M + bx})\right) \end{aligned}$$

with a, b, K, M, r, s positive constants. (*Warning*: The behavior if $ab < 1$ is different from the behavior if $ab > 1$).

13. [Kaplan & Glass(1995)] A theoretical model for mutual inhibition is

$$\begin{aligned} \frac{d}{dt}x &= f(x) = \frac{\left(\frac{1}{2}\right)^n}{\left(\frac{1}{2}\right)^n + y^n} - x \\ \frac{d}{dt}y &= g(x) = \frac{\left(\frac{1}{2}\right)^n}{\left(\frac{1}{2}\right)^n + x^n} - y, \end{aligned}$$

where x and y are positive variables and n is a positive constant greater than one. There is a steady state at $x^* = y^* = 1/2$. Discuss the bifurcation and sketch the flows in the (x, y)-plane as n varies.

14. Consider the system

$$\begin{aligned} x' &= x(ax + by) \\ y' &= y(cx + dy). \end{aligned}$$

(i) Show that every trajectory with $x(0) \geq 0$, $y(0) \geq 0$ satisfies $x(t) \geq 0$, $y(t) \geq 0$ for all $t \geq 0$, (i.e., trajectories starting in the first quadrant remain in the first quadrant forever).

(ii) Use the Dulac criterion with $\beta(x, y) = 1/xy$ to show that there are no periodic orbits if $ac > 0$.

4.6 Appendix: Canonical Forms of 2 × 2 Matrices

In Section 4.4 we stated a theorem giving the possible real canonical forms of a real 2 × 2 matrix. This theorem identified six different cases relating both to the form of the matrix and to the signs of its eigenvalues. If only different canonical form concerns us, the cases reduce to three. Theorem 4.3 of Section 4.4 is a consequence of the following result from linear algebra, whose proof requires an understanding of linear independence, eigenvalues, and eigenvectors but nothing more advanced. We include the proof because most linear algebra texts use a more general approach than ours, one that does require more advanced understanding.

Theorem 4.10. *If* $A = \begin{pmatrix} a & b \\ c & d \end{pmatrix}$ *is a real* 2×2 *matrix with* $\det A = ad - bc \neq 0$ *then there is a real nonsingular* 2×2 *matrix* P *such that* $P^{-1}AP$ *is one of the following:*

(i). $\begin{pmatrix} \lambda & 0 \\ 0 & \mu \end{pmatrix}$, $\lambda \neq 0$, $\mu \neq 0$ *(Cases (i), (ii), (iv) of Theorem 4.3)*

(ii). $\begin{pmatrix} \lambda & 0 \\ 0 & \lambda \end{pmatrix}$, $\lambda \neq 0$, *(Case (iii) of Theorem 4.3)*

(iii). $\begin{pmatrix} \alpha & \beta \\ -\beta & \alpha \end{pmatrix}$, $\beta \neq 0$, *(Cases (v), (vi) of Theorem 4.3)*

Proof. First, consider the case that A has two linearly independent eigenvectors $\vec{u}, \vec{v}$ corresponding to the eigenvalues λ, μ, respectively ($\lambda = \mu$ is permitted). Recall that an eigenvalue of a matrix A is a value of λ such that the linear homogeneous algebraic system $A\vec{u} = \lambda\vec{u}$ has a nontrivial solution and an eigenvector is a corresponding solution vector; λ is an eigenvalue if and only if $\det(A - \lambda I) = 0$. Let P be the matrix whose columns are $\vec{u}$ and $\vec{v}$, $P = (\vec{u}, \vec{v})$. Since $\vec{u}$ and $\vec{v}$ are linearly independent, P is non- singular. We now have

$$P^{-1}AP = P^{-1}(A\vec{u}\,, A\vec{v}) = P^{-1}\big(\lambda\vec{u}\,, \lambda\vec{v}\big).$$

Since $P^{-1}P = I = \begin{pmatrix} 1 & 0 \\ 0 & 1 \end{pmatrix}$, $P^{-1}\vec{u}$ is the first column of I, namely $\begin{pmatrix} 1 \\ 0 \end{pmatrix}$, and $P^{-1}\vec{v}$ is the second column of I, namely $\begin{pmatrix} 0 \\ 1 \end{pmatrix}$. Thus

$$P^{-1}AP = P^{-1}(\lambda\vec{u}\ \lambda\vec{v}) = \begin{pmatrix} \lambda & 0 \\ 0 & \mu \end{pmatrix}.$$

Second, if λ is a double eigenvalue but there is only one linearly independent corresponding eigenvector, take $\vec{v}$ to be any nonzero vector that is not an eigenvector, and let $\vec{u} = A\vec{v} - \lambda\vec{v} \neq \vec{0}$. If $c_1\vec{u}+c_2\vec{v} = \vec{0}$ with c_1 and c_2 not both zero, then c_1 and c_2 must both be different from zero because $\vec{u}$ and $\vec{v}$ are both different from zero. Then $c_1(A\vec{v} - \lambda\vec{v}) + c_2\vec{v} = \vec{0}$, which implies $A\vec{v} = (\lambda - c_2\vec{v}/c_1)$. But then $\lambda - c_2/c_1$ is an eigenvalue, contradicting the assumption that λ is the only eigenvalue. This shows that $c_1 = c_2 = 0$, and thus that $\vec{u}$ and $\vec{v}$ are linearly independent. If $\vec{w}$ is any eigenvector we may write $\vec{w} = a\vec{u} + b\vec{v}$ with $a \neq 0$, because $\vec{u}$ and $\vec{v}$ span R^2 and $\vec{v}$ is not an eigenvector. Then we have

$$\begin{aligned}\vec{0} &= A\vec{w} - \lambda\vec{w} = a(A\vec{u} - \lambda\vec{u}) + b(A\vec{v} - \lambda\vec{v}) \\ &= a(A\vec{u} - \lambda\vec{u}) + b\vec{u} = a\left(A\vec{u} - \left(\lambda - \frac{b}{a}\right)\vec{u}\right).\end{aligned}$$

This would mean that $\lambda - b/a$ is an eigenvalue, but since λ is the only eigenvalue $b = 0$ and therefore $\vec{u}$ is an eigenvector. We now take $P = (\vec{u}\ \vec{v})$, nonsingular because $\vec{u}$ and $\vec{v}$ are linearly independent, and calculate

$$\begin{aligned}P^{-1}AP &= P^{-1}(A\vec{u}\ A\vec{v}) = P^{-1}(\lambda\vec{u}\ \vec{u} + \lambda\vec{v}) \\ &= \lambda P^{-1}(\vec{u}\ \vec{v}) + P^{-1}(0\ \vec{u}) \\ &= \lambda I + \begin{pmatrix} 0 & 1 \\ 0 & 0 \end{pmatrix} = \begin{pmatrix} \lambda & 1 \\ 0 & \lambda \end{pmatrix}.\end{aligned}$$

Third, if there is a pair of complex conjugate eigenvalues $\alpha \pm i\beta$, with $\beta \neq 0$, let $\vec{u} + i\vec{v}$ be an eigenvector corresponding to the eigenvalue $\alpha + i\beta$, with $\vec{u}$ and $\vec{v}$ real. If $\vec{v} = \vec{0}$ then $A\vec{u} = (\alpha+i\beta)\vec{u}$, which is impossible because the left side is real and the right side is not; therefore, $\vec{v} \neq \vec{0}$. A similar argument shows that $\vec{u} \neq \vec{0}$. If $c_1\vec{u}+c_2\vec{v} = \vec{0}$, with $c_1 \neq 0$, $c_2 \neq 0$, we replace $\vec{u}$ by $-c_2\vec{v}/c_1$ in $A(\vec{u}+c\vec{v}) = (\alpha+i\beta)(\vec{u}+i\vec{v})$, obtaining $A\big(-c_2/c_1+i\big)\vec{v} = (\alpha+i\beta)\big(-\frac{c_2}{c_1}+i\big)\vec{v}$, or $A = \vec{v} = (\alpha+i\beta)\vec{v}$. Then $\vec{v}$ would be a real eigenvector, which is not possible, and therefore $\vec{u}$ and $\vec{v}$ are linearly independent. We now define the nonsingular matrix $P = (\vec{u}\ \vec{v})$, and $P^{-1}AP = P^{-1}(A\vec{u}\ A\vec{v})$. From $A(\vec{u} + i\vec{v}) = (\alpha + i\beta)(\vec{u} + i\vec{v}) = (\alpha\vec{u} - \beta\vec{v}) + i(\beta\vec{u} + \alpha\vec{v})$ we have

$$A\vec{u} = \alpha\vec{u} - \beta\vec{v},\ A\vec{v} = \beta\vec{u} + \alpha\vec{v},$$

and thus

$$\begin{aligned}P^{-1}AP^{-1}(A\vec{u}\ A\vec{v}) &= P^{-1}(\alpha\vec{u} - \beta\vec{v}); \beta\vec{u} + \alpha\vec{v}) \\ &= \alpha P^{-1}(\vec{u}\ \vec{v}) + \beta P^{-1}(-\vec{v}\ \vec{u}) \\ &= \alpha I + \beta\begin{pmatrix} 0 & 1 \\ -1 & 0 \end{pmatrix} = \begin{pmatrix} \alpha & \beta \\ -\beta & \alpha \end{pmatrix}.\end{aligned}$$

We have now covered all the cases and have completed the proof of the theorem and of Theorem 4.3 as well. □

4.7 Project 4.1: A Model for Giving up Smoking

1. Let $S(t)$ denote the number of smokers at time t; $P(t)$ the number of potential smokers at time t; $Q(t)$ the number of smokers who quit permanently.

   c: average number of contacts per unit time

   q probability of becoming a smoker given that a P has a contact with a smoker

   $\frac{1}{\mu}$: average time in the system (high school)

   $\frac{1}{\gamma}$: average time as a smoker

   $N(t)$: population size at time t,

   $\beta = qc$. The model is

$$\frac{dP}{dt} = \mu N - \beta P \frac{S}{N} - \mu P$$
$$\frac{dS}{dt} = \beta P \frac{S}{N} - (\mu + \gamma) S$$
$$\frac{dQ}{dt} = \gamma S - \mu Q$$

   (i) Show that $N(t)$ is a constant.

   (ii) Explain each term and coefficient in the model.

   (iii) What are the assumptions of the model?

   (iv) Given the units of μ, β, and γ.

   (v) Give a criterion for invasion in terms of a basic reproductive number R_0.

   (vi) Show that R_0 is a nondimensional quantity.

   (vii) Find all equilibria.

   (viii) Study the stability of these equilibria.

   (ix) Introduce the variables

$$x = \frac{P}{N}, \quad y = \frac{S}{N}, \quad z = \frac{Q}{N}$$

   and write the system in terms of x, y, and z. Do you gain any understanding?

   (x) Alter the model by just one simple modification to make it more realistic.

   (k*) Perform some computer simulations to verify your stability results.

References: Castillo-Garsow, Jordan-Salivia, Rodriguez-Herrera (2000).

4.8 Project 4.2: A Model for Retraining of Workers by their Peers

Suppose we have a system with the following classes of worker at time t:

$$\begin{aligned} R(t) &: \text{reluctant workers} \\ P(t) &: \text{positive workers} \\ M(t) &: \text{master workers} \\ U(t) &: \text{unchangeable (negative) workers} \\ I(t) &: \text{inactive workers} \end{aligned}$$

Assume that $N(t) = R(t) + P(t) + M(t) + U(t) + I(t)$ and that the total number of workers is constant, that is $N(t) = K$ for all t.

The model is

$$\begin{aligned} \frac{dP}{dt} &= qK - \beta P \frac{M}{K} + \delta R - \mu P \\ \frac{dR}{dt} &= (1-q)K - (\delta + \mu)R - \alpha R \\ \frac{dM}{dt} &= \beta P \frac{M}{K} - (\gamma + \mu)M \\ \frac{dU}{dt} &= -\mu U + \alpha R \\ \frac{dI}{dt} &= \gamma M - \mu I \end{aligned}$$

where q, β, δ, μ, and α are constants and $0 \le q \le 1$.

1. Interpret the parameters.
2. Under what conditions will retraining of workers succeed?
3. Will every worker be retrained (e.g., become an M) over the long run?
4. What is the impact of changing $0 \le q \le 1$, $0 < \gamma < \infty$, and $0 < \delta < \infty$? (Compute R_0 and look at the role of these parameters.)
5. Look at the stability of the simplest equilibrium.
6. Do we have a nontrivial equilibrium? Use simulations to guess its stability.
7. What are your conclusions from this model?

4.9 Project 4.3: A Continuous Two-sex Population Model

The two-sex population model of this project begins from a population of males and females that select their partners *more or less* at random. Individuals are either single or in temporarily monogamous heterosexual relationships. The recruitment and death per capita rates are constant but gender specific, while the divorce or separation constant per pair rate is assumed to be gender independent.

Definition of variables and parameters

State variables are denoted by the lower case letters x, y, and p. Here $x(t)$ denotes single males at time t; $y(t)$ denotes single females at time t; and $p(t)$ denotes pairs or married couples at time t. Hence, the total population equals $x + y + 2p$. The constants β_x and β_y denote per capita birth or recruitment rates of females and males, respectively. Furthermore, the constants μ_x and μ_y denote the per capita death rates while σ gives the constant per couple separation rate.

Using these variables and definitions we arrive at the following two-sex model [Keyfitz (1949); Kendall (1949); Fredrickson (1971); McFarland (1972)]:

$$\begin{aligned}
\dot{x} &= (\beta_x + \mu_y + \sigma)p - \mu_x x - \phi(x, y) \\
\dot{y} &= (\beta_y + \mu_x + \sigma)p - \mu_y y - \phi(x, y) \\
\dot{p} &= -(\mu_x + \mu_y + \sigma)p + \phi(x, y).
\end{aligned} \tag{4.20}$$

The population of singles increases because of recruitment, death of a partner, or divorce. The population of singles decreases because of death or marriage. The rate of marriage is *definitely* a nonlinear process. Here it is assumed that it only depends on the population of singles. It is modeled by a nonlinear function $\phi(x, y)$ called the *marriage* or *mating* function. Hence, $\phi(x, y)/x$ is the per capita marriage rate for males and $\phi(x, y)/y$ is the per capita marriage rate for females.

The function $\phi(x, y)$ is nonlinear, defined for $x, y \geq 0$, and it is commonly assumed to have the following properties [Keyfitz (1949); Kendall (1949); Fredrickson (1971); McFarland (1972)]:

1. $\phi(x, 0) = \phi(0, y) = 0$ for all x, $y \geq 0$, and $\phi(x, y) \geq 0$.

2. $\phi(\alpha x, \alpha y) = \alpha\phi(x, y)$ for $\alpha \geq 0$, that is, it is homogeneous of degree one. In other words, if there are x male singles and y female singles then there are $\phi(x, y)$ marriages per unit of time while if there are αx male singles and αy female singles then there are $\alpha\phi(x, y)$ marriages per unit of time.

3. If the number of single males and females increases then the total number of marriages also increases.

Examples of ϕ's. (Generalized means)

$$\phi(a, b) = \rho(\theta x^\alpha + (1-\theta)y^\alpha)^{1/\alpha}$$

for

$$\alpha \geq 0, \quad 0 \leq \theta \leq 1, \quad \rho > 0.$$

Special cases:

$$\begin{aligned}
\phi(x, y) &= \rho\frac{xy}{\theta x + (1-\theta)y} && \text{harmonic mean } (\alpha = -1) \\
\phi(x, y) &= \rho x^\theta y^{1-\theta} && \text{geometric mean } (\alpha \to 0) \\
\phi(x, y) &= \rho \min\{x, y\} && \text{min function } (\alpha \to \infty)
\end{aligned}$$

This model was proposed by D.G. Kendall (1949), N. Keyfitz (1949), A.G. Fredrickson (1971), and D.McFarland (1972) but a complete mathematical analysis was only recently completed by K.P. Hadeler, R. Waldstätter, and A. Wörz-Busekros (1988).

(a) Show that this model does not have (in general) time-independent steady states, that is, constant solutions.

Hint: Suppose such a steady state exists and show that it must then satisfy

$$(\beta_x - \mu_x)p - \mu_x x = 0,$$
$$(\beta_y - \mu_y)p - \mu_y y = 0.$$

Then substitute the values for x and y in the p equation of system (4.20). You should arrive at the condition

$$(\mu_x + \mu_y + \sigma)p = \phi\left(\frac{\beta_x - \mu_x}{\mu_x}, \frac{\beta_y - \mu_y}{\mu_y}\right)p. \tag{4.21}$$

Hence, the parameters of the model have to satisfy equation (4.21). What do you conclude from this?

(b) Since steady states are not generic, we look for exponential solutions, that is, for solutions of the form $ce^{\lambda t}$.

Look for solutions of the form (why is λ the same?)

$$x = \bar{x}e^{\lambda t}$$
$$y = \bar{y}e^{\lambda t}$$
$$p = \bar{p}e^{\lambda t},$$

that is, substitute the above candidates for exponential solutions in the original system (4.20) to arrive at the following nonlinear eigenvalue problem:

$$\begin{aligned}(\beta_x + \mu_y + \sigma)\bar{p} - (\mu_x + \lambda)\bar{x} &= \phi(\bar{x}, \bar{y}) \\ (\beta_y + \mu_x + \sigma)\bar{p} - (\mu_y + \lambda)\bar{y} &= \phi(\bar{x}, \bar{y}) \\ (\mu_x + \mu_y + \sigma + \lambda)\bar{p} &= \phi(\bar{x}, \bar{y})\end{aligned} \tag{4.22}$$

(c) Show that the above nonlinear eigenvalue problem has the trivial solutions

$$\begin{aligned}(1,0,0) &\quad \text{with} \quad \lambda = -\mu_x \\ (0,1,0) &\quad \text{with} \quad \lambda = -\mu_y\end{aligned}$$

and interpret them.

(d) Look for nontrivial solutions. In fact, show that such solutions exist if and only if there exists a λ such that

$$\mu_x + \mu_y + \sigma = \phi\left(\frac{\bar{x}}{\bar{p}}, \frac{\bar{y}}{\bar{p}}\right),$$

or, by substituting $\bar{x}/\bar{p}$, $\bar{y}/\bar{p}$, show that such solutions exist if they satisfy the following nonlinear relationship

$$\mu_x + \mu_y + \sigma + \lambda = \phi\left(\frac{\beta_x}{\mu_x + \lambda} - 1, \frac{\beta_y}{\mu_y + \lambda} - 1\right). \tag{4.23}$$

Hence, we are looking for nonzero λ's that satisfy (4.23). The equation (4.23) is called the *characteristic equation* of this two-sex population.

(e) Assume that

$$\frac{\beta_x - \mu_x - \lambda}{\mu_x + \lambda} > 0, \qquad \frac{\beta_y - \mu_y + \lambda}{\mu_y + \lambda} > 0,$$

Set $\underline{\lambda} = \max(-\mu_x, -\mu_y)$. Then $\mu_x + \lambda > 0$, and $\mu_y + \lambda > 0$ for $\lambda > \underline{\lambda}$. Set $\bar{\lambda} = \min(\beta_x - \mu_x, \beta_y - \mu_y)$. Then $\beta_x - \mu_x - \lambda > 0$ and $\beta_y - \mu_y - \lambda > 0$ for $\lambda < \bar{\lambda}$. Hence $\underline{\lambda} < \lambda < \bar{\lambda}$ (if it exists). Justify the above assumptions and definitions. Show that if $\underline{\lambda} > \bar{\lambda}$ then there is no solution, while if $\underline{\lambda} < \bar{\lambda}$ then there is a solution. In fact, show that if K is defined as

$$K = \phi \cdot \left(\frac{\beta_x}{\mu_x + \lambda} - 1, \frac{\beta_y}{\mu_y + \lambda} - 1\right) \leq \infty,$$

where the possibility that $K = \infty$ cannot be disregarded; this case must also be studied.

Then equation (4.23) has a unique solution if and only if

$$K > \mu_x + \mu_y + \sigma + \lambda.$$

(f) Get a more detailed existence condition by *cleverly* expanding $\phi(x, y)$ in a Taylor series (keeping just the first order terms), that is, by using the fact that it is a homogeneous function of degree one and then letting $y \to \infty$ we have that

$$\phi(x,y) = x\phi(1, \frac{y}{x}) = x\left[\phi(1,0) + \phi_y(1,0)\frac{y}{x}\right] = y\phi_y(1,0).$$

Hence,

$$\begin{aligned}\phi(x,y) &\approx y\phi_y(1,0) \text{ as } x \to \infty \\ \phi(x,y) &\approx x\phi_x(1,0) \text{ as } y \to \infty.\end{aligned}$$

Show that if $\underline{\lambda} = -\mu_x$ then as $x \to \infty$

$$\phi\left(\frac{\beta_x}{\mu_x + \lambda} - 1, \frac{\beta_y}{\mu_y + \lambda} - 1\right) \approx \left[\frac{\beta_y}{\mu_y - \mu_x} - 1\right]\phi_y(1,0).$$

Show that if $\underline{\lambda} = -\mu_y$ then as $y \to \infty$

$$\phi\left(\frac{\beta_x}{\mu_x + \lambda} - 1, \frac{\beta_y}{\mu_y + \lambda} - 1\right) \approx \left[\frac{\beta_x}{\mu_y - \mu_x} - 1\right]\phi_x(1,0).$$

Hence,

$$K = \begin{cases} \frac{\beta_y - \mu_y + \mu_x}{\mu_y - \mu_x}\phi_y(1,0) & \text{for } \underline{\lambda} = -\mu_x \\ \frac{\beta_x - \mu_y + \mu_x}{\mu_y - \mu_x}\phi_x(1,0) & \text{for } \underline{\lambda} = -\mu_y. \end{cases}$$

Finally, verify that the conditions for the existence of nontrivial exponential solutions is

$$\begin{aligned}\mu_x &> \mu_y - \frac{\beta_y\phi_y(1,0)}{\mu_y + \sigma + \phi_y(1,0)} \\ \mu_y &> \mu_x - \frac{\beta_x\phi_x(1,0)}{\mu_x + \sigma + \phi_x(1,0)}.\end{aligned}$$

(g) Look at the proportions of each group (single males, single females, and pairs) and think of how one may study the stability of the exponential solutions that you have just shown to exist.

References: Castillo-Chavez and Huang (1995), Castillo-Chavez, Huang, and Li (1996), Castillo-Chavez and Hsu Schmitz (1993), Castillo-Chavez, Shyu, Rubin, and Umbauch (1992). (1992).

5

Continuous Models for Two Interacting Populations

5.1 Species in Competition

In this chapter we will consider populations of two interacting species with population sizes $x(t)$ and $y(t)$, respectively, modeled by a system of two first order differential equations:

$$\begin{aligned} x' &= F(x,y) \\ y' &= G(x,y). \end{aligned} \tag{5.1}$$

There are different kinds of biological interaction represented mathematically by different signs for the rates of change of the growth rates $F(x,y)$ and $G(x,y)$ with respect to the population sizes x and y. We begin with the case of species in *competition*, perhaps competing for the same nutrient resource. Competition means that an increase in the size of either population tends to decrease the growth rate of the other population, so that

$$F_y(x,y) < 0, \qquad G_x(x,y) < 0. \tag{5.2}$$

We also assume that an increase in the size of either population tends to inhibit the growth rate of that po[pulation, so that

$$F_x(x,y) < 0, \qquad G_y(x,y) < 0. \tag{5.3}$$

The classical experimental result on species in competition is the *principle of competitive exclusion* of G.F. Gause (1934), that two species competing for the same resource cannot coexist. The experimental evidence is somewhat equivocal and there is considerable doubt about the universality of the principle.

We will discuss primarily models in which the per capita growth rates are linear of the form

$$\begin{aligned} x' &= x(\lambda - ax - by) \\ y' &= y(\mu - cx - dy), \end{aligned} \tag{5.4}$$

where λ, μ, a, b, c, d are positive constants, but we will also indicate extensions to models with nonlinear growth rates.

For competition models with linear per capita growth rates, there are four equilibria, as follows

I. $(0,0)$, with community matrix

$$\begin{pmatrix} \lambda & 0 \\ 0 & \mu \end{pmatrix},$$

necessarily an unstable node.

II. $(K,0)$, with $K = \lambda/a$. The community matrix is

$$\begin{pmatrix} -\lambda & \frac{b\lambda}{a} \\ 0 & \mu - \frac{c\lambda}{a} \end{pmatrix},$$

with trace $\frac{1}{a}(a\mu - a\lambda - c\lambda)$ and determinant $\frac{\lambda}{a}(c\lambda - a\mu)$.

III. $(0,M)$, with $M = \frac{\mu}{d}$. The community matrix is

$$\begin{pmatrix} \lambda - \frac{b\mu}{d} & 0 \\ -\frac{\mu c}{d} & -\mu \end{pmatrix},$$

with trace $\frac{1}{d}(d\lambda - d\mu - b\mu)$ and determinant $\frac{\mu}{d}(b\mu - d\lambda)$.

IV. (x_∞, y_∞), with $x_\infty > 0$, $y_\infty > 0$, $ax_\infty + by_\infty = \lambda$, $cx_\infty + dy_\infty = \mu$, or

$$x_\infty = \frac{d\lambda - b\mu}{ac - bc}, \qquad y_\infty = \frac{a\mu - c\lambda}{ad - bc}.$$

The community matrix is

$$\begin{pmatrix} -ax_\infty & -bx_\infty \\ -cy_\infty & -dy_\infty \end{pmatrix},$$

with trace $-ax_\infty - dy_\infty < 0$ and determinant $x_\infty y_\infty (ad - bc)$.

We will distinguish four cases, corresponding to the four possible sign combinations for $b\mu - d\lambda$ and $a\mu - c\lambda$, but ignoring the possibilities $b\mu - d\lambda = 0$ (which would give $x_\infty = 0$) and $a\mu - c\lambda = 0$ (which would give $y_\infty = 0$). An equilibrium $(K, 0)$ or $(0, M)$ describes a situation in which one species survives but the other species loses the struggle for existence and becomes extinct. An equilibrium (x_∞, y_∞) with $x_\infty > 0, y_\infty > 0$ describes coexistence of the two species.

Because

$$\begin{aligned} \Delta &= (-ax_\infty - dy_\infty)^2 - 4(ad - bc)x_\infty y_\infty \\ &= a^2x_\infty^2 + 2adx_\infty y_\infty + d^2y_\infty^2 - 4adx_\infty y_\infty + 4bcx_\infty y_\infty \\ &= a^2x_\infty^2 - 2adx_\infty y_\infty + d^2y_\infty^2 + 4bcx_\infty y_\infty \\ &= (ax_\infty - dy_\infty)^2 + 4bcx_\infty y_\infty > 0, \end{aligned}$$

and because it is easy to verify that $\Delta \geq 0$ for equilibria with either or both species absent, it is not possible for the system to have a spiral point or a center.

Case 1. $d\lambda - b\mu > 0$, $a\mu - c\lambda > 0$. In this case $d/b > \mu/\lambda > c/a$, and $ad - bc > 0$. Thus, there is an equilibrium (x_∞, y_∞) in the first quadrant of the phase plane. Because the determinant of the community matrix is positive while the trace is negative, this equilibrium is asymptotically stable and must be an asymptotically stable node. As the community matrices for the equilibrium $(K, 0)$ and $(0, M)$ both have negative determinants, these equilibria are saddle points. The phase portrait is as shown in Figure 5.1: In this figure and in the other figures in the section, the symbol $\overset{\leftarrow}{\downarrow}$ means that orbits are directed downward and to the left, etc. Equilibria are labelled •, with Ⓢ indicating that the equilibrium is asymptotically stable. In this case, every orbit tends to (x_∞, y_∞) as $t \to \infty$, indicating coexistence of the two species for all initial population sizes.

Case 2. $c\lambda - a\mu > 0, b\mu - d\lambda > 0$. In this case $c/a > \mu/\lambda > /b$, and $ad - bc < 0$. Again, there is an equilibrium (x_∞, y_∞) in the first quadrant, but because the determinant of the community matrix is negative, this equilibrium is a saddle point. The equilibria $(K, 0)$ and $(0, M)$ are both asymptotically stable nodes because their community matrices have positive determinant and negative trace. The phase portrait in Figure 5.2 shows the asymptotically stable separatrices at (x_∞, y_∞) as dotted curves.

These separatrices divide the first quadrant into two regions, one containing initial points for which orbits tend to $(K, 0)$, and the other containing initial points for which orbits tend to $(0, M)$. In this case, coexistence of the two species is impossible, and one species will always win the competition for survival. The winner of this competition is determined by the initial population sizes. Although coexistence would be the result of initial population sizes on the separatrices, this possibility is too sensitive to disturbances to be biologically meaningful.

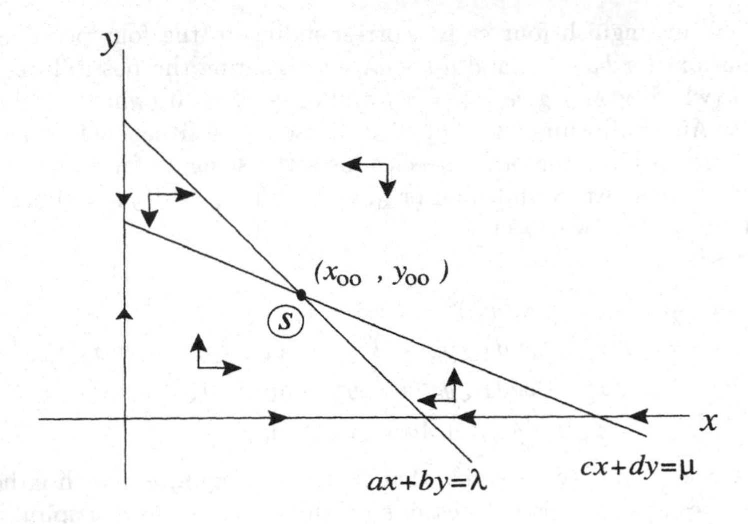

FIGURE 5.1.

Case 3. $c\lambda - a\mu > 0, d\lambda - b\mu > 0$. In this case there is no equilibrium in the interior of the first quadrant. It is easy to verify by examination of the community matrices that the equilibrium at $(K, 0)$ is an asymptotically stable node, while the equilibrium at $(0, M)$ is a saddle point (Figure 5.3).

All orbits tend to $(K, 0)$ as $t \to \infty$, corresponding to extinction of the y species and survival of the x species for all initial population sizes.

Case 4. $a\mu - c\lambda > 0, b\mu - d\lambda > 0$. Again, there is no equilibrium in the interior of the first quadrant. It is again easy to determine the stability of the equilibria on the axes. Now $(K, 0)$ is a saddle point while $(0, M)$ is an asymptotically stable node (Figure 5.4). All orbits tend to $(0, M)$ as $t \to \infty$, corresponding to extinction of the x species and survival of the y species for all initial population sizes.

The numbers K and M are the respective carrying capacities of the x and y species, the equilibrium population size that each species would reach in the absence of the other species. Sometimes competition between species is described as "qualified" or "unqualified", according to the inhibiting effect of each species on the other. We will say that the competition is *unqualified* if $x_\infty/K + y_\infty/M \leq 1$, that is, if competition tends to reduce the total population size (where the carrying capacities are the units of population size). This is intended to describe a situation in which both species depend on a common source of food and the competition makes survival more difficult for both species. Because

$$x_\infty = \frac{d\lambda - b\mu}{ad - bc}, y_\infty = \frac{a\mu - c\lambda}{ad - bc},$$

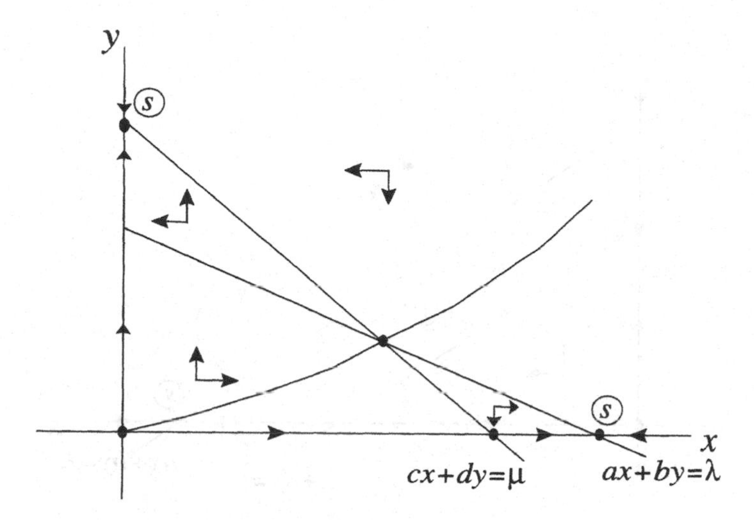

FIGURE 5.2.

the condition $x_\infty/K + y_\infty/M \leq 1$ is equivalent to

$$\frac{a}{\lambda}\frac{d\lambda - b\mu}{ad-bc} + \frac{d}{\mu}\frac{a\mu - c\lambda}{ad-bc} \leq 1,$$

which is equivalent to

$$a\mu(d\lambda - b\mu) + d\lambda(a\mu - c\lambda) \leq (ad-bc)\lambda\mu.$$

If $ad - bc > 0$, this reduces to $ab\mu^2 - (ad+bc)\lambda\mu + cd\lambda^2 \geq 0$, or $(cd - a\mu)(d\lambda - b\mu) \geq 0$. Thus, the system is in Case 3 or 4. If $ad - bc < 0$ the inequalities are reversed, so that $(cd - a\mu)(d\lambda - b\mu) < 0$, which shows that the system is in Case 2. This shows that the assumption of unqualified competition rules out the possibility of coexistence of the two species (Case 1). The *principle of competitive exclusion*, originally formulated by Gause in 1934 on the basis of experimental evidence, states that coexistence under unqualified competition is impossible. We have shown that the principle of competitive exclusion is a consequence of the assumption of linear per capita growth rates. The principle itself is quite controversial; there is experimental evidence suggesting that coexistence under unqualified competition is possible. Our argument shows only that the principle of competitive exclusion is a consequence of the specific model hypothesized, not that the principle has biological validity. It is important to remember that if a mathematical model predicts behavior which contradicts valid observations, one must conclude that there is a flaw in the model.

The condition $x_\infty/K + y_\infty/M \leq 1$ of unqualified competition says geometrically that the equilibrium (x_∞, y_∞) lies below the line $x/K + y/M = 1$ joining the equilibria $(K, 0)$ and $(0, M)$. The condition that the equilibrium

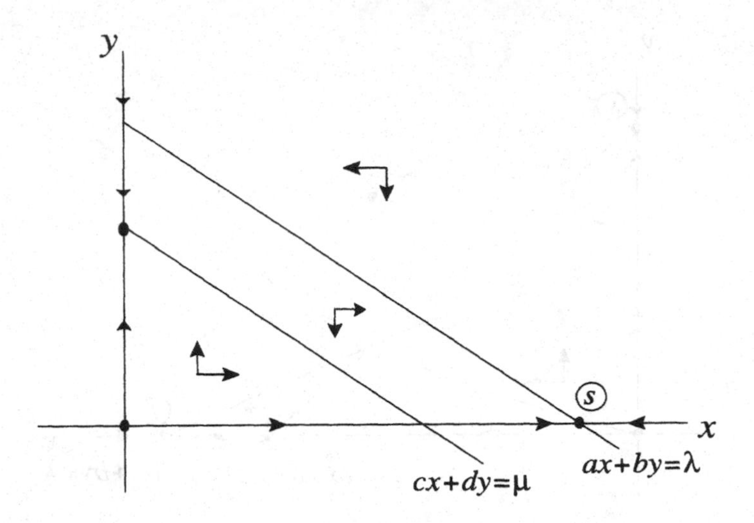

FIGURE 5.3.

(x_∞, y_∞) be asymptotically stable is that the crossing of the curves $x' = 0$ and $y' = 0$ at (x_∞, y_∞) must be as in Figure 5.5,

It may be seen by looking at the phase portraits in Cases 1 and 2 that if the *isoclines* $x' = 0$ and $y' = 0$ are straight lines, these two conditions are incompatible. However, it is possible to construct models of the form $x' = xf(x,y)$, $y' = yg(x,y)$ for species in competition for which coexistence and unqualified competition are compatible. An example of such a model, which has been proposed, [Ayala, Gilpin, and Ehrenfeld (1973)] is

$$\begin{aligned} x' &= \lambda x\left(1 - \left(\frac{x}{K_1}\right)^{\theta_1} - \frac{b}{\lambda}y\right) \\ y' &= \mu y\left(1 - \left(\frac{y}{K_2}\right)^{\theta_2} - \frac{c}{\mu}x\right). \end{aligned}$$

In describing two species in competition by a model of the form

$$\begin{aligned} x' &= xf(x,y) \\ y' &= yg(x,y), \end{aligned}$$

it is reasonable to assume that $f_x(x,y) < 0$ and $g_y(x,y) < 0$ for $x > 0$, $y > 0$, and also that $f_y(x,y) < 0$, $g_x(x,y) < 0$ for $x > 0$, $y > 0$. This assumption says that an increase in the population size of either species reduces the growth rate of the population size of both species. In order to locate equilibria, we examine the *isoclines* of the two species, namely the curves $f(x,y) = 0$ and $g(x,y) = 0$, respectively. Just as in the case of linear per capita growth rates, there are four types of equilibria: (I) $(0,0)$; (II) $(K,0)$, where K now is defined by $f(K,0) = 0$; (III) $(0,M)$, where M now is defined by $g(0,M) = 0$; (IV) (x_∞, y_∞), with $x_\infty > 0$, $y_\infty > 0$, where $f(x_\infty, y_\infty) = 0$, $g(x_\infty, y_\infty) = 0$.

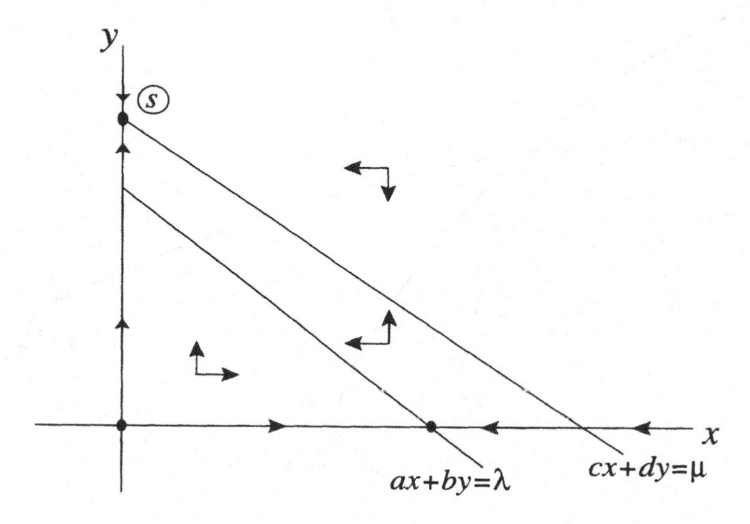

FIGURE 5.4.

At an intersection (x_∞, y_∞) of the isoclines $f(x,y) = 0$ and $g(x,y) = 0$, the slope of the x isocline is $-f_x(x_\infty, y_\infty)/f_y(x_\infty, y_\infty)$, and the slope of the y-isocline is $-g_x(x_\infty, y_\infty)/g_y(x_\infty, y_\infty)$. Because of the hypothesis $f_x < 0$, $f_y < 0$, $g_x < 0$, $g_y < 0$ each of these slopes is negative. The condition that the slope of the x isocline is less (more negative) than the slope of the y isocline is

$$-\frac{f_x(x_\infty, y_\infty)}{f_y(x_\infty, y_\infty)} < -\frac{g_x(x_\infty, y_\infty)}{g_y(x_\infty, y_\infty)},$$

or

$$f_x(x_\infty, y_\infty)g_y(x_\infty, y_\infty) - f_y(x_\infty, y_\infty)g_x(x_\infty, y_\infty) > 0.$$

The community matrix at the equilibrium is

$$\begin{pmatrix} x_\infty f_x(x_\infty, y_\infty) & x_\infty f_y(x_\infty, y_\infty) \\ y_\infty g_x(x_\infty, y_\infty) & y_\infty g_y(x_\infty, y_\infty) \end{pmatrix},$$

with trace

$$x_\infty f_x(x_\infty, y_\infty) + y_\infty g_y(x_\infty, y_\infty) < 0,$$

and determinant

$$x_\infty y_\infty \big(f_x(x_\infty, y_\infty)g_y(x\infty, y\infty) - f_y(x\infty, y\infty)g_x(x_\infty, y_\infty)\big).$$

Thus the determinant is positive, so that the equilibrium is asymptotically stable, if and only if the slope of the x-isocline is less than the slope of the y-isocline; otherwise the equilibrium is a saddle point. There are four cases,

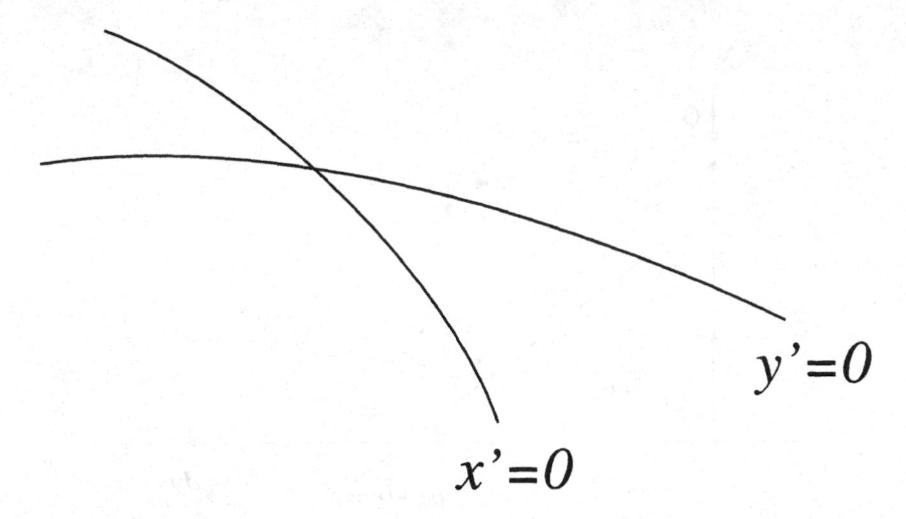

FIGURE 5.5.

exactly as for linear per capita growth rates, except that the isoclines are no longer necessarily straight lines. Otherwise, all the analysis carries over to this more general situation. Isoclines which are concave upwards allow the possibility of an equilibrium below the line $x_\infty/K + y_\infty/M = 1$, so that the competition is unqualified. Such an equilibrium may be either asymptotically stable or unstable (Figure 5.6). Thus, the principle of competitive exclusion is not a consequence of these more general hypotheses.

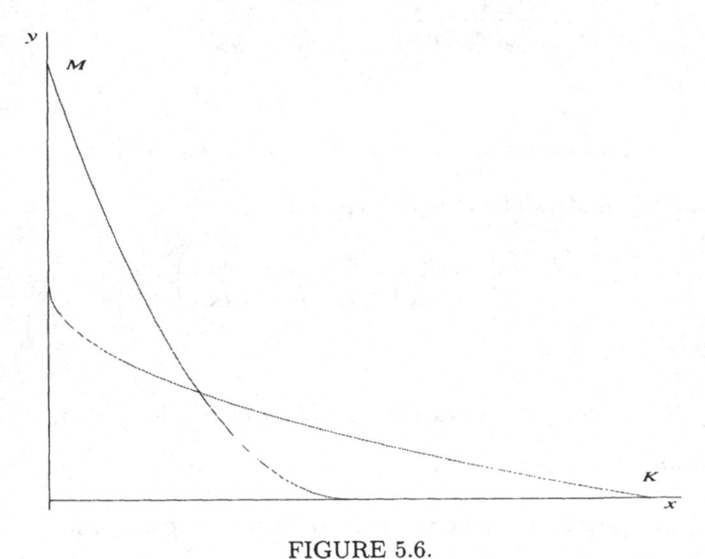

FIGURE 5.6.

Every orbit of a system $x' = xf(x, y)$, $y' = yg(x, y)$ for which $f(x, y) < 0$ when $x > K$, and $g(x, y) < 0$ when $y > M$ must be bounded because

x and y must both be decreasing functions of t for x and y, respectively, large enough. According to the Poincaré-Bendixson theorem, every orbit then must tend either to an equilibrium or to a limit cycle. For competition models with $f_x(x,y) < 0, g_y(x,y) < 0$ for $x > 0, y > 0$, we have seen in Section 4.5 that the Dulac criterion rules out the possibility of periodic solutions. Therefore every orbit must tend to an equilibrium, and by analyzing the nature of each equilibrium we have actually described the global behavior of solutions, not merely the behavior of solutions starting near an equilibrium.

Example 1. Determine the outcome of a competition modelled by the system

$$\begin{aligned} x' &= x(100 - 4x - y) \\ y' &= y(60 - x - 2y) \end{aligned}$$

Solution: A coexistence equilibrium is found by solving the system

$$\begin{aligned} 4x + y &= 100 \\ x + 2y &= 60 \end{aligned}$$

By eliminating one of the variables we can obtain the equilibrium (20, 20). The community matrix at an equilibrium (x_∞, y_∞) is

$$\begin{bmatrix} -4x_\infty & -y_\infty \\ -x_\infty & -2y_\infty \end{bmatrix}$$

which in this case is

$$\begin{bmatrix} -80 & -20 \\ -20 & -40 \end{bmatrix}$$

with negative trace and positive determinant. Thus, the coexistence equilibrium is asymptotically stable and every orbit approaches this equilibrium. The two species coexist.

Exercises

In each of Exercises 1 through 4, determine the outcome of the competition modelled by the given system.

1. $x' = x(80 - x - y),\ y' = y(120 - x - 3y)$ [see Exercise 7, Section 4.5]

2. $x' = x(60 - x3 - y),\ y' = y(75 - 4x - y)$ [see Exercise 8, Section 4.5]

3. $x' = x(40 - x - y),\ y' = y(90 - x - 2y)$ [see Exercise 9, Section 4.5]

4. $x' = x(80 - 3x - 2y),\ y' = y(80 - x - y)$ [see Exercise 10, Section 4.5]

5. What is the outcome of a competition modelled by the system

$$\begin{aligned} x' &= x(2 - x - x^2 - y) \\ y' &= y(16 - 2x - x^2 - y)? \end{aligned}$$

6.* What is the outcome of a competition modelled by the system

$$\begin{aligned} x' &= x(80 - x - y) \\ y' &= y(120 - x - 3y - 2y^2)? \end{aligned}$$

5.2 Predator-prey Systems

We have already discussed the Lotka-Volterra system

$$x' = x(\lambda - by), \qquad y' = y(-\mu + cx), \tag{5.5}$$

which unrealistically predicts population oscillations that have been observed in real populations (see Section 4.1). The reason for describing this prediction as "unrealistic" is that the model is extremely sensitive to perturbations. A change in initial population size would produce a change to a different periodic orbit, while the addition of a perturbing term to the system of differential equations could produce the same type of change or could produce a *qualitative* change in the behavior of orbits, which might either spiral in to an equilibrium or spiral out from an equilibrium.

Let us try to construct a more realistic model by assuming that in the absence of predators the prey species would obey a logistic model. This would suggest a system

$$x' = x(\lambda - ax - by), \qquad y' = y(-\mu + cx). \tag{5.6}$$

There are two possible direction fields, depending on the sign of $c\lambda - a\mu$, as shown in Figure 5.7.

The community matrix at the interior equilibrium (x_∞, y_∞) (if there is one) is

$$\begin{pmatrix} -ax_\infty & -bx_\infty \\ cy_\infty & 0 \end{pmatrix},$$

with determinant $bcx_\infty y_\infty > 0$ and trace $-ax_\infty < 0$; thus (x_∞, y_∞) is asymptotically stable. The equilibrium (x_∞, y_∞) may be a spiral point, so that oscillations are possible, but only damped oscillations. The Dulac criterion shows that there cannot be a periodic orbit. It is easy to verify that if $c\lambda - a\mu > 0$ the equilibrium $(\lambda/a, 0)$ is a saddle point, while if $c\lambda - a\mu < 0$, the equilibrium $(\lambda/a, 0)$ is an asymptotically stable node. Thus in each case there is exactly one asymptotically stable equilibrium to which every orbit tends.

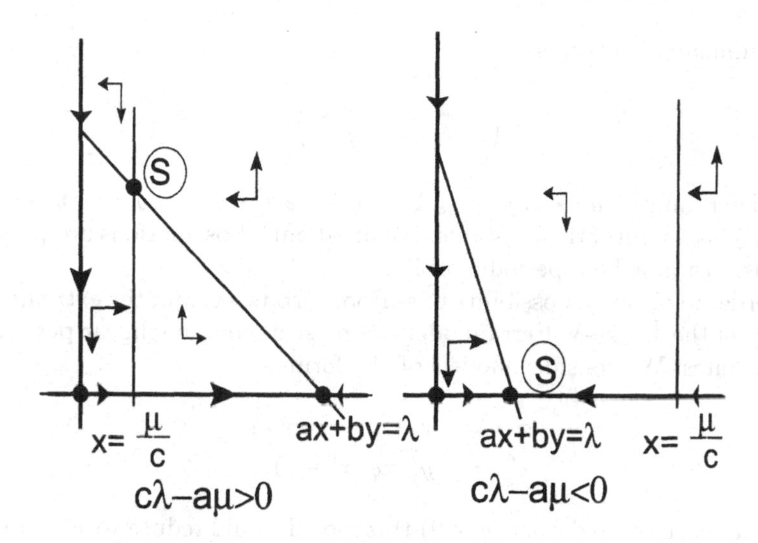

FIGURE 5.7.

Another plausible modification of the Lotka-Volterra system is

$$x' = x(\lambda - ax - by), \qquad y' = y(\mu + cx - dy), \tag{5.7}$$

with μ allowed to be either positive or negative. The direction field is as in Figure 5.8.

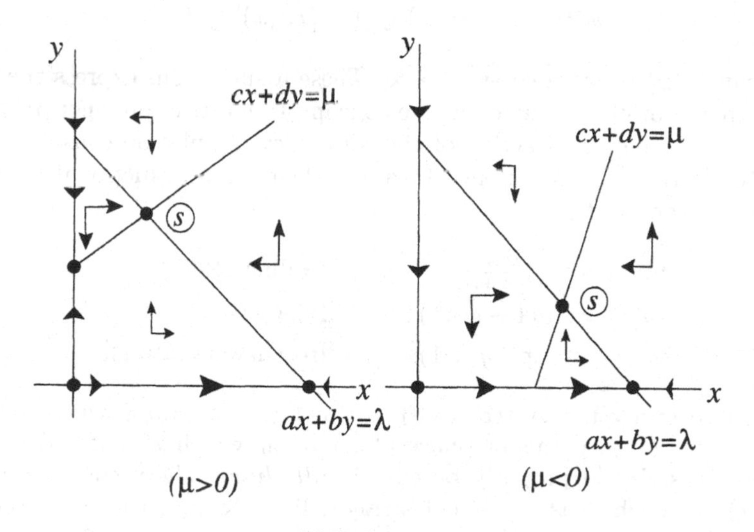

FIGURE 5.8.

(We shall omit the cases in which there is no interior equilibrium, but the reader may easily sketch these also.) At the interior equilibrium (x_∞, y_∞)

the community matrix is

$$\begin{pmatrix} -ax_\infty & -by_\infty \\ cx_\infty & -dy_\infty \end{pmatrix},$$

with determinant $(ad + bc)x_\infty y_\infty > 0$ and trace $-ax - dy < 0$; therefore, (x_∞, y_∞) is asymptotically stable. Again, damped oscillations are possible but there cannot be a periodic orbit.

In order to obtain a possibility of periodic orbits without the extreme sensitivity of the Lotka-Volterra model, we must assume nonlinear per capita growth rates. We consider models of the form

$$\begin{aligned} x' &= xf(x) - xy\phi(x) \\ y' &= y(cx\phi(x) - e). \end{aligned} \tag{5.8}$$

In the absence of predators ($y \equiv 0$) this model would reduce to $x' = xf(x)$. Thus, $xf(x)$ represents the growth rate of the prey species by itself. The term $xy\phi(x)$ is called the *predator functional response*; $x\phi(x)$ is the number of prey consumed per predator in unit time. The constant c is the conversion efficiency of prey into predators, and the term $cxy\phi(x)$ is called the *predator numerical response*. The constant e is the predator mortality rate. Models of the form (5.8) are known as Rosenzweig-MacArthur models, having been proposed by M.L. Rosenzweig and R.H. MacArthur (1963).

In (5.8) it is reasonable to assume

$$\phi(x) \geq 0, \quad \phi'(x) \leq 0, \quad [x\phi(x)]' \geq 0, \tag{5.9}$$

and that $x\phi(x)$ is bounded as $x \to \infty$. These assumptions express the idea that as prey population increases the consumption rate of prey per predator increases but that the fraction of the total prey population consumed per predator decreases. Some explicit forms for the predator functional response that have been used are

$$\begin{aligned} x\phi(x) &= \tfrac{\alpha x}{x+A} && \text{[Holling (1965)]} \\ x\phi(x) &= a(1 - e^{-cx}) && \text{[Ivlev (1961)]} \\ x\phi(x) &= ax^q \ (q < 1) && \text{[Rosenzweig (1971)].} \end{aligned}$$

In order to study the system (5.8), we plot the *isoclines*, which are the curves in the (x, y)-plane, or phase plane, along which $x' = 0$ and $y' = 0$. These curves should properly be called *null clines*. The predator isocline is $c\phi(x) = e$, which is a vertical straight line. To plot the prey isocline $y = f(x)/\phi(x)$, we proceed as follows: sketch the graph of $xf(cx)$ and of $y_i x\phi(x)$ for a sequence of values y_i. If $y_i x\phi(x)$ intersects $xf(x)$ for $x = x_i$, then (x_i, y_i) is a point on the prey isocline. The prey isocline may then be sketched by projecting to another graph in the (x, y)-plane (Figure 5.9).

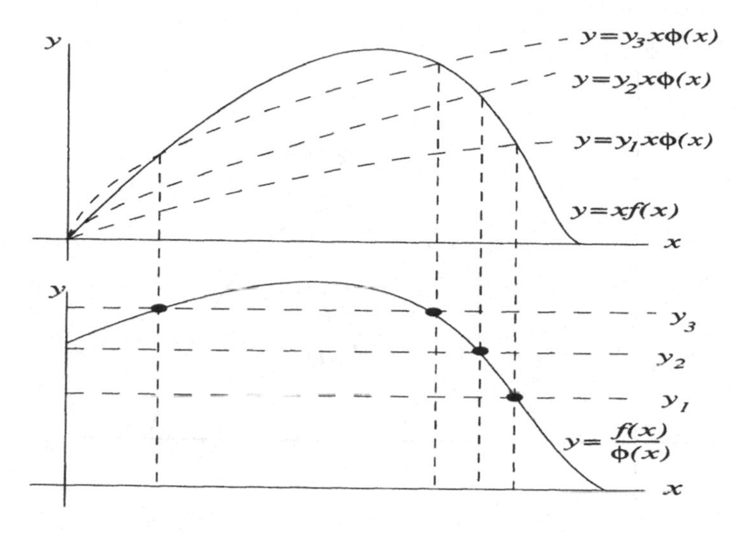

FIGURE 5.9.

Observe that the prey isocline may have a maximum if the graph of $x\phi(x)$ is concave down, so that $y_i x\phi(x)$ and $xf(x)$ may intersect in more than one point for some values of y_i. There are, in fact, biological reasons for expecting the prey isocline to have a maximum [Rosenzweig (1969)]. It is reasonable for the function $x\phi(x)$ to increase with x but approach a maximum when the number of prey is large. As the prey population increases it should be able to survive in the face of an increase in the predator population because of this saturation, but when the prey population goto still larger self-limiting effects tend to take over to bring the prey isocline down. It is possible to incorporate an *Allee effect*, in which the prey population is required to maintain a minimum size for survival,by having the prey isocline intersect the x axis at two points (Figure 5.10)

The Lotka-Volterra model (5.5) and the refinements (5.6), (5.7) with linear per capita growth rates, do not allow any bending of the isoclines. Also, the refinements (5.6), (5.7) satisfy $f_x(x,y) < 0$, $g_y(x,y) \leq 0$ in the first quadrant of the phase plane. Thus, by the Dulac criterion they cannot have periodic orbits.

For a Rosenzweig-MacArthur model (5.8) with an equilibrium (x_∞, y_∞) given by $\phi(x_\infty) = e/c$, $y_\infty = f(x_\infty)/\phi(x_\infty) = cf(x_\infty)/e$, the community matrix at equilibrium is

$$\begin{pmatrix} x_\infty f'(x_\infty) - x_\infty y_\infty \phi'(x) & -x_\infty \phi(x_\infty) \\ c y_\infty [x\phi(x)]'_{x_\infty} & 0 \end{pmatrix},$$

whose determinant is $c x_\infty y_\infty \phi(x_\infty)[x\phi(x)]'_{x_\infty} > 0$ and whose trace is $x_\infty \big(f'(x_\infty) - y_\infty \phi'(x)\big)$.

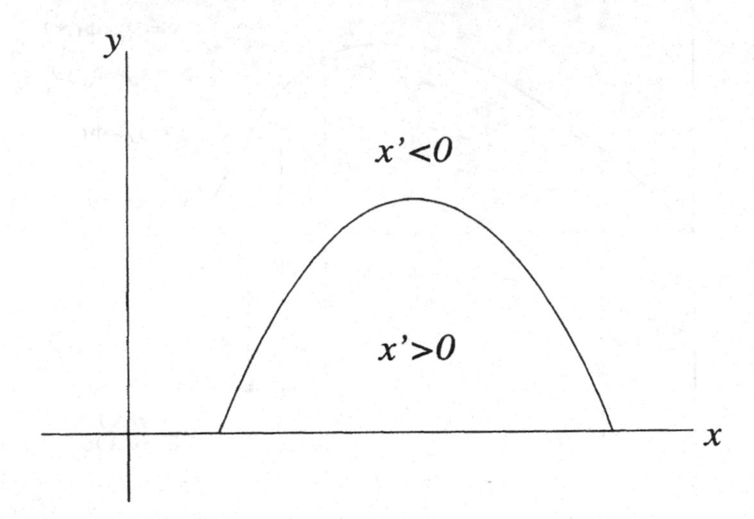

FIGURE 5.10.

The slope of the prey isocline at the equilibrium is

$$\frac{\phi(x_\infty)f'(x_\infty) - f(x_\infty)\phi'(x_\infty)}{(\phi(x_\infty))^2} = \frac{f'(x_\infty) - y_\infty\phi'(x_\infty)}{\phi(x)},$$

which has the same sign as the trace of the community matrix. Thus, the equilibrium is asymptotically stable if the prey isocline has negative slope at equilibrium and unstable if the prey isocline has positive slope at equilibrium (Figure 5.11).

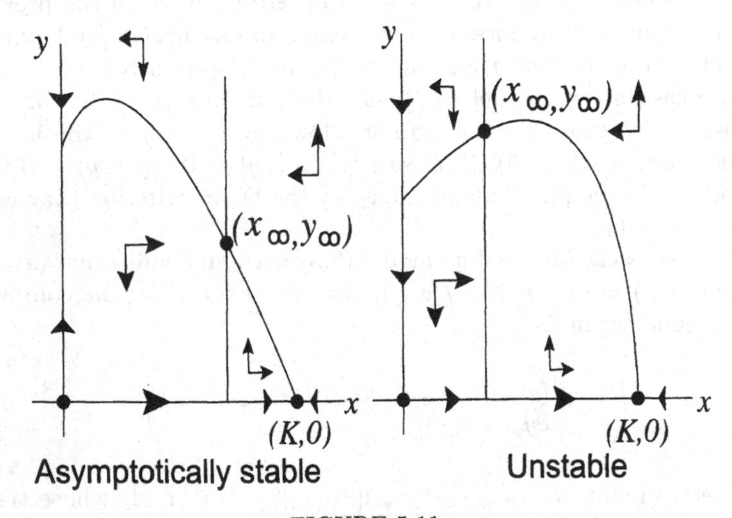

FIGURE 5.11.

No orbit can cross either the positive x axis or the positive y axis. An orbit starting on the y axis tends to the origin while an orbit starting on the x axis tends to the equilibrium on the positive x axis. Every orbit starting in the interior of the first quadrant, if bounded, remains in the interior of the first quadrant and must, by the Poincaré-Bendixson theorem, tend to the equilibrium (x_∞, y_∞), or to the equilibrium $(K, 0)$ on the positive x-axis, or to a limit cycle around the equilibrium (x_∞, y_∞). We will prove shortly in a more general setting that every orbit must remain bounded. If there is no equilibrium (x_∞, y_∞), that is, if $c\phi(K) < e$, where $f(K) = 0$, then the equilibrium $(K, 0)$ is asymptotically stable and every orbit tends to it (Figure 5.12).

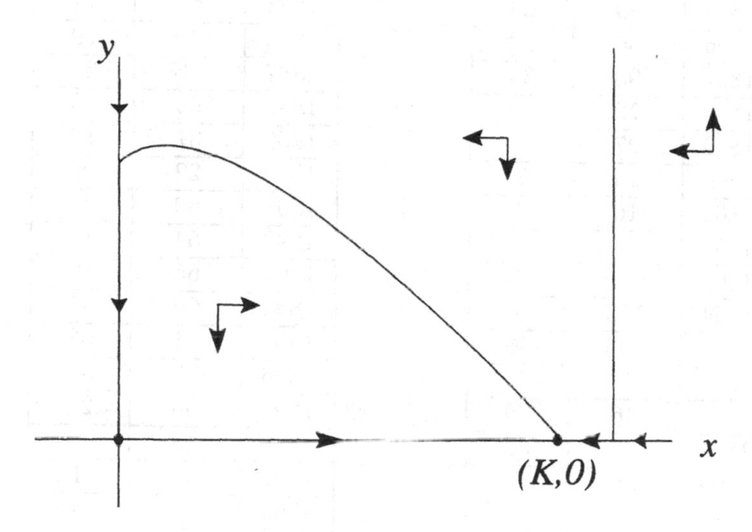

FIGURE 5.12.

If the prey isocline has negative slope at (x_∞, y_∞), then (x_∞, y_∞) is asymptotically stable and every orbit tends to it; $(K, 0)$ is a saddle point (Figure 5.13). If the prey isocline has positive slope at (x_∞, y_∞), then (x_∞, y_∞) is unstable and there must be a limit cycle around (x_∞, y_∞) to which every orbit tends (Figure 5.14). We ignore the pathological (and very sensitive to perturbations) possibility of an infinite sequence of periodic orbits around (x_∞, y_∞) and assume that there can only be a single limit cycle. Then every orbit spirals either outward or inward to this limit cycle. Thus we have a more plausible model than the original Lotka-Volterra model to explain the observed periodic behavior.

Year	Hares	Lynx
1848	21	44
1849	12	20
1850	24	9
1851	50	5
1852	80	5
1853	80	6
1854	90	11
1855	69	23
1856	80	32
1857	93	34
1858	72	23
1859	27	15
1860	14	7
1861	16	4
1862	38	5
1863	5	5
1864	153	16
1865	145	36
1866	106	77
1867	46	68
1868	23	37
1869	2	16
1870	4	8
1871	8	5
1872	7	7
1873	60	11
1874	46	19
1875	50	31
1876	103	43
1877	87	27
1878	68	18
1879	17	15
1880	10	9
1881	17	8
1882	16	8
1883	15	27
1884	46	52
1885	55	74
1886	137	79
1887	137	34
1888	95	19
1889	37	12
1890	22	8
1891	50	9
1892	54	13
1893	65	20
1894	60	37
1895	81	56
1896	95	39
1897	56	27
1898	18	15
1899	5	4
1900	2	6
1901	15	9
1902	2	19
1903	6	36
1904	45	59
1905	50	61
1906	58	39
1907	20	10

TABLE 5.1. Hare & Lynx population sizes

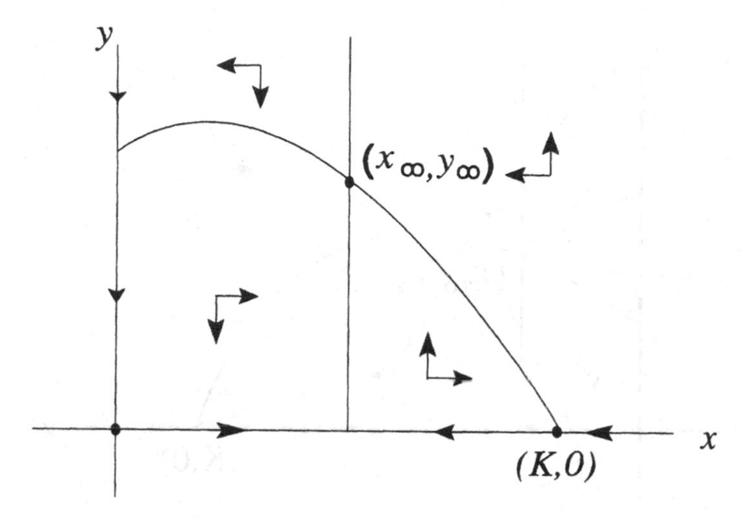

FIGURE 5.13.

Another classical example of interacting populations in which oscillations have been observed is the data collected by the Hudson's Bay Company in Canada during the period 1821-1940 on furs of the snowshoe hare (lepus americanus) and Canadian lynx (lynx canadiensis) brought to the company by trappers. Table 5.1 shows some data for the period 1848-1907, with the hare data estimated from the data shown in graph form in [MacLulich (1937)] and the lynx data taken from [Elton and Nicholson (1942)], and with population numbers given in thousands. In each case, the data has been obtained by analyzing a variety of reports. This data is also depicted in Figure 5.15 in the phase (hare-lynx) plane.

There are many problems with this data as a measure of population sizes. One problem is that trapping data may not accurately describe the total population sizes, and another problem is the varying delays between catching animals and bringing their skins to a trading post. For example, it is pointed out in [Elton & Nicholson (1942)] that the data in [MacLulich (1937)] should be shifted forward by a year. In the earlier studies, it was found that the population sizes of hares, lynx, and also several other species, fluctuated in time with a period of approximately 10 years. The lynx oscilations were ascribed to the fact that hare is the main food for lynx. However, the relation was not viewed as a predator-prey interaction until the systems view was introduced into theoretical ecology, notably in the text by Odum (1953), which was the first to display the lynx and hare data in the same graph. This appears to be another example of a predator-prey system with the oscillations being explained as a limit cycle. However, a graph of the hare and lynx populations in the hare (prey)-lynx (predator) plane shows orbits going clockwise in some cycles, whereas the limit cycle predicted by a Rosenzweig-MacArthur model is always counterclockwise.

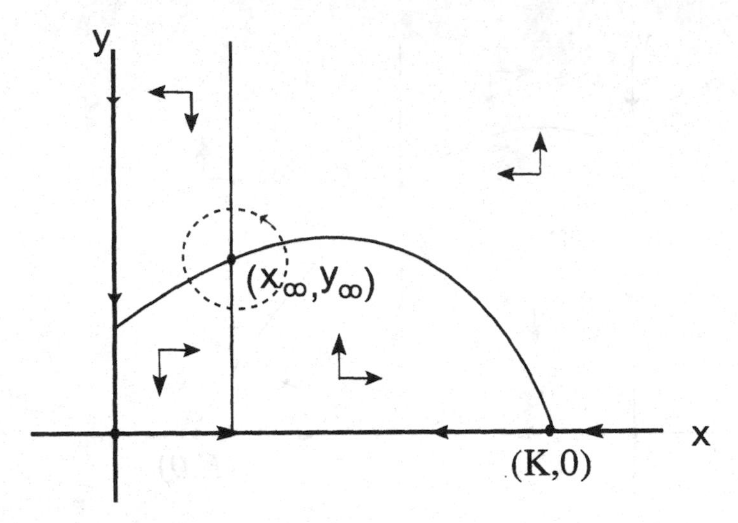

FIGURE 5.14.

This does not show that hares eat lynx [Gilpin (1973)]! It indicates that something is wrong with the model. Various suggestions have been made to explain the anomaly. One possibility is that the numbers of animals of the two species caught are not necessarily proportional to the actual population sizes. The trappers should be viewed as predators who choose which type of animal to pursue rather than as external experimenters who measure population size. Lynx are useful for their fur, which is more valuable than rabbit fur, but rabbits are also useful as food for trappers. Thus, when rabbits are abundant and food is plentiful, hunters can conveniently pursue both hares and lynx, as lynx tend to go where there are rabbits. When rabbits are scarce, trappers must concentrate on rabbits for food and therefore will catch fewer lynx. An important lesson to be learned is that population models outside the laboratory are seldom simple and comprehensive, while experimental data are also subject to substantial errors. Predictions of qualitative behavior may be possible, but quantitative data are unreliable, especially over long periods of time.

There is an even more basic flaw in the predator-prey model as an explanation for the lynx-hare cycle: Hare population size oscillates even in the absence of lynx, probably due to climactic variations and to epidemics which recur, killing most of the hares. A two-species predator-prey model, whether of Lotka-Volterra or Rosenzweig-MacArthur type can not exhibit oscillations when only one species is present.

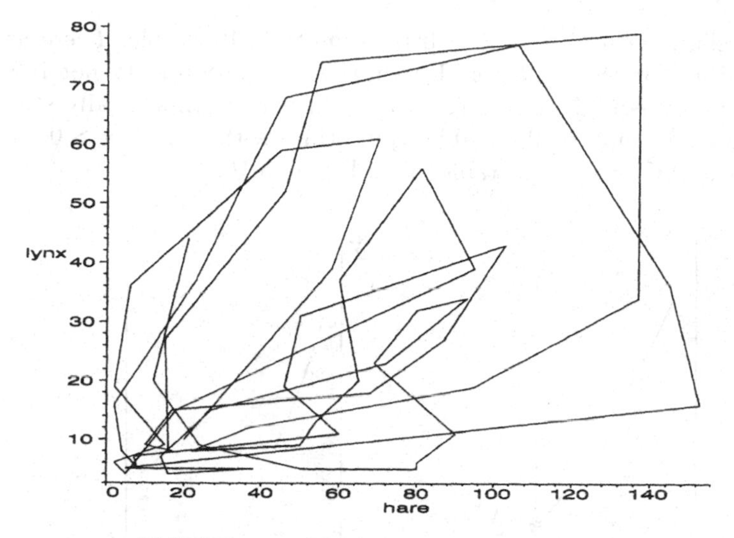

FIGURE 5.15. The hare-lynx phase plane

Example 1. In Section 4.2 we formulated the model for the simple chemostat

$$\begin{aligned} C' &= q(C^{(0)} - C) - \frac{\beta a C y}{C + A} \\ y' &= \frac{aCA}{C + A} - qy \end{aligned} \tag{5.10}$$

with a, A, q, and β constants. This is a Rosenzweig-MacArthur model with the nutrient concentration C playing the role of the prey population and the bacteria population size y playing the role of the predator population. In the model (5.8) we have

$$\begin{aligned} xf(x) &= q(C^{(0)} - C) \\ x\phi(x) &= \frac{\beta a C}{C + A} \\ c &= \frac{1}{\beta} \\ e &= q. \end{aligned}$$

For (5.10), the prey isocline is the curve

$$y = \frac{q(C + A)(C^{(0)} - c)}{a\beta C}$$

which is unbounded as $C \to 0$, and it is not difficult to verify that $dy/dc < 0$ at every point on the prey isocline, and that the prey isocline meets the C-axis for $C = C^{(0)}$ while the predator isocline is the line $C = Aq/(a - q)$. Because the prey isocline is monotone decreasing, an intersection of the two isoclines with $C > 0$, $y > 0$ is asymptotically stable. There are two possibilities for the behavior. Either the two isoclines do not intersect, and the only equilibrium is $(C^{(0)}, 0)$, which is asymptotically stable, or there is an asymptotically stable equilibrium with $C > 0$, $y > 0$, and the equilibrium $(C^{(0)}, 0)$ is unstable, as in Figure 5.16.

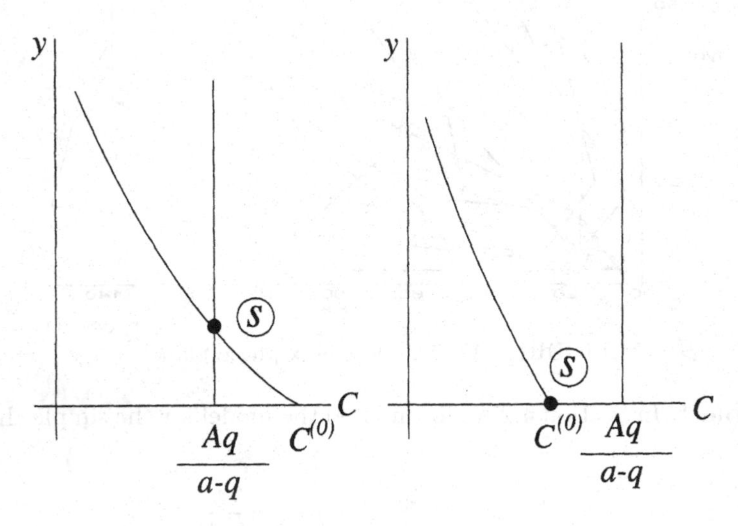

FIGURE 5.16.

The condition that the chemostat have an equilibrium with a positive bacteria population is

$$C^{(0)} > \frac{Aq}{a - q}.$$

In terms of the parameter q, which depends on the flow rate and is thus adjustable, this coexistence condition is

$$q < \frac{aC^{(0)}}{A + C^{(0)}}.$$

Example 2. Determine the qualitative behavior of a predator-prey system modelled by

$$\begin{aligned} x' &= x(1 - \frac{x}{30}) - \frac{xy}{x + 10} \\ y' &= y(\frac{x}{x + 10} - \frac{1}{3}) \end{aligned}$$

Solution. We have studied this system in Example 1, Section 4.5 and shown that every orbit approaches a periodic orbit around the (unstable) equilibrium (5, 12.5). Thus the two species coexist with oscillations.

Example 3. Determine the equilibrium behavior of a predator-prey system modeled by

$$\begin{aligned} x' &= x(1-\frac{x}{30}) - \frac{xy}{x+10} \\ y' &= y(\frac{x}{x+10} - \frac{3}{5}). \end{aligned}$$

Solution. Equilibria are solutions of the pair of equations

$$\begin{aligned} x\left(1-\frac{x}{30}-\frac{y}{x+10}\right) &= 0 \\ y\left(\frac{x}{x+10}-\frac{3}{5}\right) &= 0. \end{aligned}$$

There is an equilibrium $(0,0)$, and if $y=0$ we must have $x=30$. There is a coexistence equilibrium obtained by solving

$$\begin{aligned} \frac{x}{x+10} - \frac{3}{5} &= 0 \\ 1-\frac{x}{30}-\frac{y}{x+10} &= 0 \end{aligned}$$

The first of these equations gives $x=15$, and then the second equation gives $y=12.5$. Thus, there are three equilibria: $(0,0), (30,0)$, and $(14,12.5)$. The community matrix at an equilibrium (x_∞, y_∞) is

$$\begin{pmatrix} 1-\frac{x_\infty}{15}-\frac{10y_\infty}{(x_\infty+10)^2} & -\frac{x_\infty}{x_\infty+10} \\ \frac{10y_\infty}{(x_\infty+10)^2} & \frac{x_\infty}{x_\infty+10}-\frac{3}{5} \end{pmatrix}$$

At $(0,0)$, the community matrix is

$$\begin{pmatrix} 1 & 0 \\ 0 & -\frac{3}{5} \end{pmatrix}$$

and thus $(0,0)$ is unstable. At $(30,0)$ the community matrix is

$$\begin{pmatrix} -1 & -\frac{3}{4} \\ 0 & \frac{1}{5} \end{pmatrix}.$$

Since the determinant of this matrix is negative, $(30,0)$ is also unstable. The community matrix at $(15,12.5)$ is

$$\begin{pmatrix} -\frac{1}{5} & -\frac{3}{5} \\ \frac{1}{5} & 0 \end{pmatrix}$$

with negative trace and positive determinant. Thus $(15,12.5)$ is asymptotically stable and every orbit approaches this equilibrium.

Exercises

In each of Exercises 1 through 6, determine the outcome of the predator-prey relationship modelled by the given system

1. $x' = 3x(1-\frac{x}{20}) - \frac{xy}{x+10}$, $y' = y(\frac{x}{x+10} - \frac{1}{2})$ [see Exercise 1, Section 4.6]

2. $x' = x(1-\frac{x}{40}) - \frac{2xy}{x+15}$, $y' = 3y(\frac{2x}{x+15} - \frac{6}{5})$ [see Exercise 2, Section 4.6]

3. $x' = 3x(1-\frac{x}{20}) - \frac{xy}{x+10}$, $y' = y(\frac{x}{x+10} - \frac{1}{6})$ [see Exercise 3, Section 4.6]

4. $x' = x(1-\frac{x}{40}) - \frac{2xy}{x+15}$, $y' = 3y(\frac{2x}{x+15} - \frac{1}{2})$ [see Exercise 4, Section 4.6]

5. $x' = 3x(1-\frac{x}{20}) - \frac{xy}{x+20}$, $y' = y(\frac{x}{x+10} - \frac{3}{4})$ [see Exercise 5, Section 4.6]

6. $x' = x(1-\frac{x}{40}) - \frac{2xy}{x+15}$, $y' = 3y(\frac{2x}{x+15} - \frac{8}{5})$ [see Exercise 6, Section 4.6]

7. Show that the equilibrium (x_∞, y_∞) with $x_\infty > 0, y_\infty > 0$ of the predator-prey system modeled by

$$\begin{aligned} x' &= rx(1-\frac{x}{K}) - \frac{axy}{x+A} \\ y' &= sy(\frac{ax}{x+A} - \frac{aJ}{J+A}) \end{aligned}$$

is unstable if $K > A + 2J$, and asymptotically stable if $J < K < A + 2J$ [see Exercise 11, Section 4.5].

8. Use a computer algebra system to plot the hare and lynx populations given in Table 5.1 as functions of time separately but in the same graph

9. Use a computer algebra system to plot the hare and lynx data given in Table 5.1 over some time intervals of 10-12 years in the phase plane.

5.3 Laboratory Populations: Two Case Studies

Such ideas as competitive exclusion and oscillations in predator-prey systems have been suggested by real-world observations, but the data is never completely unequivocal. There are too many factors which are ignored in simple models for a close fit between model and data. For example, it is rare for a population system to involve only two species without any interaction with other species. However, laboratory experiments may allow more control and give better opportunities to test the validity of models. In this section, we describe two experiments conducted by G. F. Gause (1934a, 1934b), one on two species in competition and one on a predator-prey system.

The possibility of competitive exclusion is illustrated by some experiments conducted by Gause (1934a) on paramecium aurelia and paramecium caudatum (two kinds of protozoa). First, Gause measured the population sizes of each separately and fit the results to logistic models. Then he measured the sizes of the two populations together. The results are shown in Table 5.2, with time measured in days and population sizes measured in numbers of individuals per 0.5 cc.

			Competition	
Time	**P. aurelia**	**P. caudatum**	**P. aurelia**	**P. caudatum**
0	2	2	2	2
1	3	5	4	8
2	29	22	29	20
3	92	16	66	25
4	173	39	141	24
5	210	52	162	-
6	210	54	219	-
7	240	47	153	-
8	-	50	162	21
9	-	26	150	15
10	240	69	175	12
11	219	51	260	9
12	255	57	276	12
13	252	70	285	6
14	270	53	225	9
15	240	59	222	3
16	249	57	220	0

TABLE 5.2. Paramecium population sizes

For comparison of the population sizes of the two species, the volumes are used rather than the numbers of individuals. Because P. caudatum has larger volume than P. aurelia, the volume of P. caudatum is taken as the unit and the number of P. aurelia is multiplied by 0.39. The general shape of the individual growth curves looks rather like logistic growth, although there appear to be oscillations around the carrying capacity, which are probably caused by small variations in temperature and the composition of the growth medium. Gause therefore fit the single species data to logistic models. We use slightly different values for the parameters than those obtained by Gause in order to fit his data better. We use intrinsic growth rate 1.12 and carrying capacity 95 for P. aurelia and intrinsic growth rate 0.84 and carrying capacity 60 for P. caudatum. Figure 5.17 shows the data and model curve for P. aurelia and Figure 5.18 shows the data and model curve for P. caudatum.

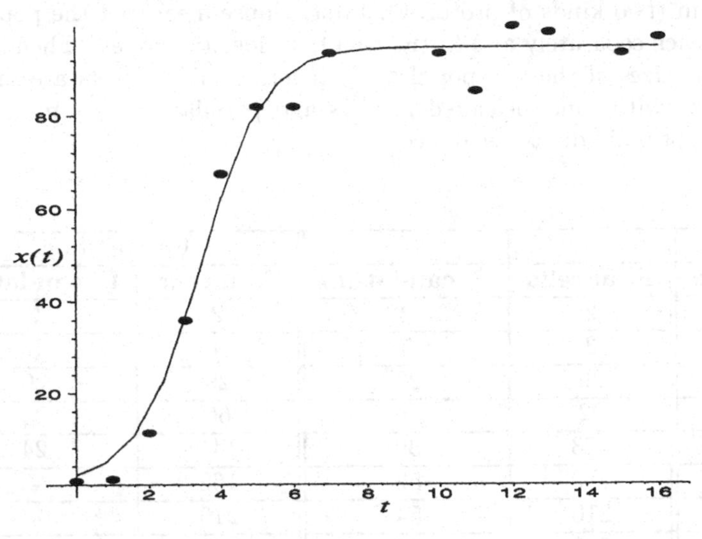

FIGURE 5.17. P. aurelia

The data for the mixed populations, shown as a phase portrait in Figure 5.19, indicates clearly a P. caudatum population dying out and a P. aurelia population approaching its single-species carrying capacity. We may fit the data to a model of the form (5.4) with x the density of P. aurelia and y the density of P. caudatum and parameter values $\lambda = 1.12$, $\mu = 0.84$, $a = 1.12/95$, $d = 0.84/60$, and b and c to be determined. By comparing the conversion of food to biomass for each species, Gause estimated values for $b = 1.8103/105, c = 0.4846/64$ in the initial stages, but he considered these parameters to be functions of time. In the model (5.4), these coefficients are constants, and a reasonable fit with the data is obtained with the parameter values $b = (1.55)(1.12)/95, c = (0.65)(0.84/60$. Using these parameter values, we may calculate that $c\lambda - a\mu > 0, d\lambda - b\mu > 0$. This suggests that the system is in what we called Case 3 in Section 5.1, and thus that P. aurelia should survive and P. caudatum should become extinct. The behavior of the model (5.4) agrees quite well with the experimental data but the approach to extinction of P. caudatum is considerably slower than the experimental data indicates. The experiment is consistent with the principle of competitive exclusion. More elaborate experiments have been carried out in [Park (1948)], and these also support the principle of competitive exclusion. However, other experiments, such as those reported in [Ayala, Gilpin, and Ehrenfeld (1973)] cast doubt on the universality of the principle.

In the real world, there are many situations in which species avoid competition by using somewhat different food supplies or by searching for food

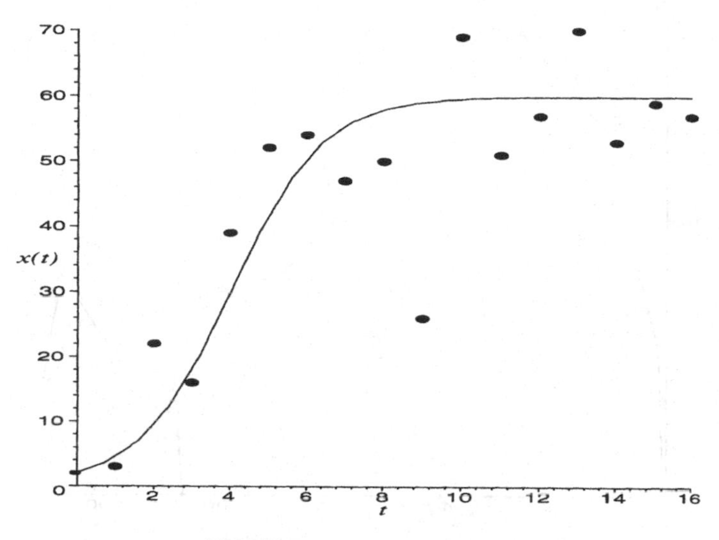

FIGURE 5.18. P. caudatum

in different locations. This suggests that the principle of competitive exclusion may be broadly applicable but that species may arrange to coexist in spite of it.

Another set of experiments performed by Gause (1934a, 1934b) involved Paramecium aurelia (protozoa) and Saccharomyces exiguus (a yeast on which the protozoa feed). Gause obtained the data given in Table 5.3, with time measured in days. Because the original data was displayed only in a graph, the population sizes are estimates read from the graph.

A phase portrait display of this data is shown in Figure 5.20. The phase portrait looks more like a limit cycle than a Lotka-Volterra model, as there is some decrease in amplitude from the initial state, and Gause recognized that the Lotka-Volterra model was not a perfect fit. However, if we try to fit the data to a Lotka-Volterra model (5.5), we may use the estimates $\lambda = 0.65, \mu = 0.32$ obtained by Gause. In order to estimate the other two parameters b and c, we may use the fact that in the Lotka-Volterra model the center of the closed orbits is the equilibrium point $(\mu/c, \lambda/b)$ and then we may use the values $b = 0.0108$ and $c = 0.0058$ obtained by estimating that the equilibrium point is $(55, 60)$. Perhaps a model of Rosenzweig-MacArthur type would give a better fit, but to fit data to such a model is a much more difficult matter.

Some of Gause's experiments led to neither coexistence in equilibrium nor coexistence with oscillations of the predator and prey species. Instead, the predators consumed all the prey and then died from lack of food. This problem also arose in more elaborate experiments conducted by Huffaker (1958); in some cases the system went through a few cycles and then collapsed. In experiments on predator-prey systems, it may be necessary to

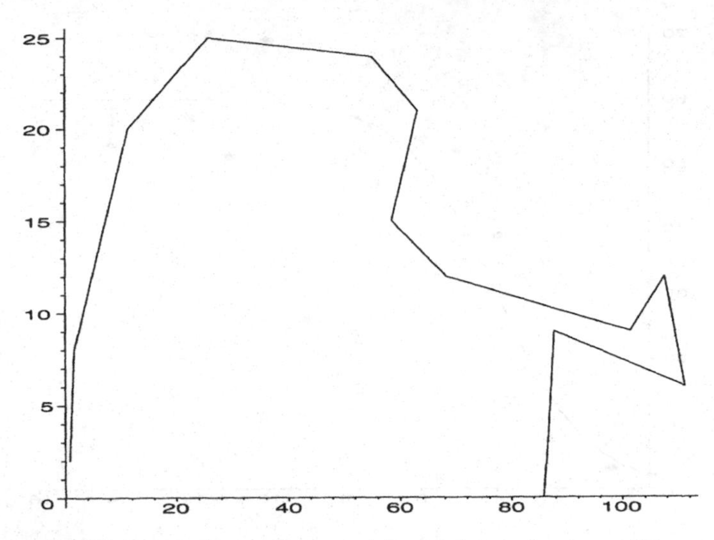

FIGURE 5.19. P. aurelium and P. caudatum in competition

perturb the system by providing a refuge for prey or allowing some immigration to make coexistence possible.

Exercises

1. Use a computer algebra system to plot solutions of the system (5.4) with the parameters derived here and initial data $x(0) = 2, y(0) = 2$.
2. Use a computer algebra system to plot solutions of the system (5.5) with the parameters derived here and an initial point on the periodic orbit indicated by Figure 5.20.

5.4 Kolmogorov Models

Two-species models whose per capita growth rates are functions of the population size are described generally by models of the form

$$\begin{aligned} x' &= xf(x,y) \\ y' &= yg(x,y), \end{aligned} \tag{5.11}$$

sometimes called *Kolmogorov* models. We model predator-prey behavior by the assumptions:

(i) $f_y(x,y) < 0$, $g_x(x,y) > 0$, $g_y(x,y) \leq 0$.

(ii) For some (prey carrying capacity) $K > 0$, $f(K,0) = 0$ and $f(x,y) < 0$ if $x > K$; for some (minimum prey population to support predators) $J > 0$, $g(J,0) = 0$.

Time	S. exiguus	P. aurelia
0	155	90
1	40	175
2	20	120
3	10	60
4	25	10
5	55	20
6	120	15
7	110	55
8	50	130
9	20	70
10	15	30
11	20	15
13	70	20
15	135	30
16	135	80
17	50	170
18	15	90
19	20	30

TABLE 5.3. Predator - prey population sizes

Because $g_x(x,y) > 0$, $g_y(x,y) \leq 0$ the predator isocline $g(x,y) = 0$ slopes upward to the right from $(J,0)$ (or is vertical if $g_y(x,y) = 0$), as in the Rosenzweig-MacArthur model). The interpretation of a condition $g_y(x,y) \equiv 0$ is that the predators do not interfere with one another in searching for prey. The Rosenzweig-MacArthur model is a special case of the Kolmogorov model with $f(x,y) = f(x) - y\phi(x), g(x,y) = cx\phi(x) - e$. Then $f_y(x,y) = -\phi(x) < 0$, $g_x(x,y) = [x\phi(x)]' > 0$, K is determined by $f(K) = 0$, and J by $cJ\phi(J) = e$. An example of a Kolmogorov model which is not a Rosenzweig-MacArthur model is the Leslie model (1948) with a predator "carrying capacity" proportional to prey population size, that is,

$$g(x,y) = b\left(1 - \frac{y}{ax}\right).$$

Theorem 5.1. *Under the hypotheses* (i), (ii) *above, every solution of a Kolmogorov model with* $x(0) > 0$, $y(0) > 0$ *remains bounded for* $0 \leq t < \infty$.

Proof. The proof of this theorem is quite difficult and may be omitted without seriously affecting comprehension. Since $x'(t) < 0$ whenever $x > K$, it is impossible to have $x(t) \to \infty$. A solution would become unbounded only if $y(t) \to \infty$ in the region where $x' < 0$ or $f(x,y) < 0$, and $y' > 0$ or $g(x,y) > 0$. Pick a constant $\alpha > 0$ and choose any $x_0 > K$, y_0 above

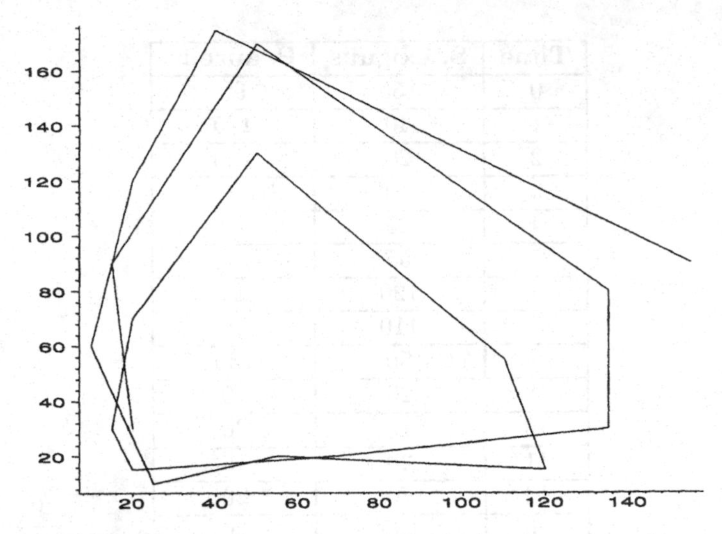

FIGURE 5.20. S. exiguus and P. aurelia phase portrait

the maximum height of the curve $f(x, y) = -\alpha$. Let $(x(t), y(t))$ be the solution with $x(0) = x_0$, $y(0) = y_0$. We will show that this solution crosses $g(x, y) = 0$ and enters the region where $x' < 0$, $y' < 0$, and thus is not unbounded. Any solution with $x(0) < x_0, y(0) < y_0$ must also remain bounded because it cannot cross the orbit of $(x(t), y(t))$. So long as this orbit remains in the region in which $x' < 0$, $y' > 0$, $(x(t), y(t))$ remains above the curve $f(x, y) = -\alpha$, and therefore we have $f(x(t), y(t)) \leq -\alpha$. Also, $(d/dt)g(x(t), y(t)) = g_x(x(t), y(t))x'(t) + g_y(x(t), y(t))y'(t) \leq 0$, since $g_x(x, y) > 0$, $x' < 0$, $g_y(x, y) \leq 0$, $y' > 0$. Thus, $g(x(t), y(t)) \leq g(x_0, y_0)$ for $t \geq 0$. We now let $\beta = g(x_0, y_0)/\alpha > 0$ and define $V(x, y) = x^\beta y$. Then

$$
\begin{aligned}
\frac{d}{dt}V(x(t), y(t)) &= \beta x^{\beta-1}yx'(t) + x^\beta y'(t) \\
&= \beta x^{\beta-1}yxf(x, y) + x^\beta yg(x, y) \\
&= x^\beta y\big(f(x, y) + g(x, y)\big) \\
&\leq x^\beta y\big(-\beta\alpha + g(x_0, y_0)\big) \leq 0.
\end{aligned}
$$

Thus $V(x(t), y(t))$ is a decreasing function, and $V(x(t), y(t)) \leq V(x_0, y_0)$ for $t \geq 0$. We now have

$$
y(t) \leq \frac{x_0^\beta y_0}{[x(t)]^\beta} \leq \left(\frac{x_0}{J}\right)^\beta y_0
$$

in the region $x \geq J$, which contains the region where $x' < 0$, $y' > 0$. Thus the solution $(x(t), y(t))$ cannot have $y(t) \to \infty$ but must proceed to the region where $x' < 0$, $y' < 0$ and remain bounded. □

An immediate consequence of Theorem 5.1, together with the Poincaré-Bendixson theorem, is the theorem of Kolmogorov (1936): Every orbit of a predator-prey system of Kolmogorov type tends either to a stable equilibrium or to a stable limit cycle as $t \to \infty$. We may now return to the problem of explaining the periodic oscillations that originally led to the Lotka-Volterra model (Section 4.1) and that suggest that a predator-prey model with nonlinear per capita growth rates is a more suitable description.

Models of Kolmogorov type are the most general models for situations in which the per capita growth rate of each species depends only on the population sizes of both species. The effect of the factors x and y in the equations for x' and y', respectively, is to ensure that neither species generate spontaneously; an orbit for which $x = 0$ at some time must remain on the y-axis of the phase plane for all later times. We are also assured that every orbit starting in the first quadrant $(x > 0, y > 0)$ of the phase plane remains in the first quadrant.

5.5 Mutualism

There are situations in which the interaction of two species is mutually beneficial, for example, plant-pollinator systems. The interaction may be *facultative*, meaning that the two species could survive separately, or *obligatory*, meaning that each species will become extinct without the assistance of the other.

If we model a mutualistic system by a pair of differential equations with linear per capita growth rates

$$\begin{aligned} x' &= x(\lambda - ax + by) \\ y' &= y(\mu + cx - dy), \end{aligned}$$

the mutualism of the interaction is modeled by the positive nature of the interaction terms cx and by. In a facultative interaction, the constants λ and μ are positive, while in an obligatory relation the constants λ and μ are negative. In each type of interaction there are two possibilities, depending on the relation between the slope a/b of the x isocline and the slope c/d of the y isocline (Figure 5.21). If $ad > bc$ the mutualistic effects are smaller than the self-limiting terms in the per capita growth rates and the slope of the x-isocline is greater than the slope of the y-isocline.

In both facultative and obligatory interactions, if $ad < bc$ there is a region of the phase plane in which solutions become unbounded, and this suggests that either we must restrict models of this form by requiring $ad > bc$, or we must consider models with nonlinear per capita growth rates. For models with linear per capita growth rates and $ad > bc$ it is easy to verify that in the facultative case the only asymptotically stable equilibrium is the intersection (x_∞, y_∞) of the lines $ax - by = \lambda$, $-cx + dy = \mu$ with $x_\infty > 0, y_\infty > 0$ and every orbit tends to this equilibrium. In the obligatory case, the only asymptotically stable equilibrium is the origin and every orbit

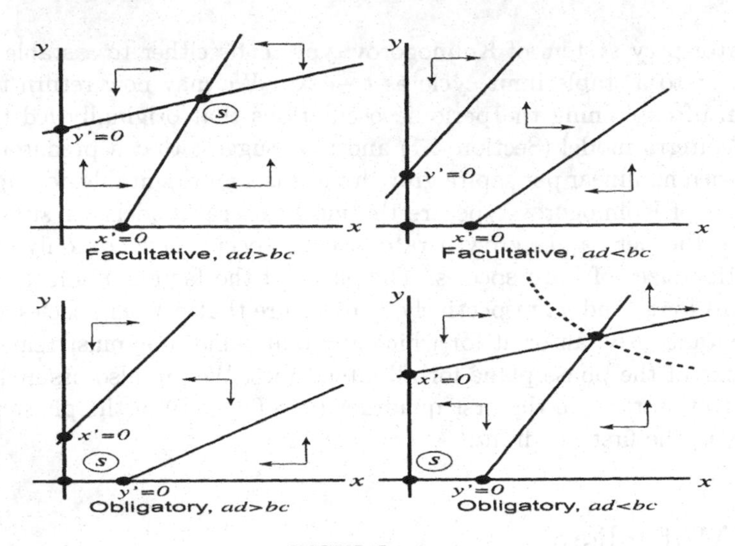

FIGURE 5.21.

tends to the origin. Thus, in the obligatory case neither species survives. While our model may be acceptable in the facultative case, it is clear that the possibility of obligatory mutualism is not described by this model. If we consider the obligatory case with $ad < bc$ there is an equilibrium (x_∞, y_∞) with $x_\infty > 0$, $y_\infty > 0$, which may be shown to be a saddle point whose stable separatrices separate the phase plane into a region of mutual extinction and a region of unbounded growth. Such a separation is plausible biologically, but we must alter the model so as to rule out the possibility of unbounded growth in order to give a more realistic model.

We shall describe a model of Kolmogorov type to describe mutualistic systems, either facultative or obligatory and show how general hypotheses that are biologically reasonable lead to qualitative predictions about the behavior of solutions. We assume a model of the form

$$\begin{aligned} x' &= xf(x,y) \\ y' &= yg(x,y). \end{aligned}$$

To describe the mutualistic effect of the y species on the x species, we assume

$$f_x(x,y) < 0, \qquad f_y(x,y) \geq 0$$

for $x \geq 0, y \geq 0$. The case $f_y(x,y) \equiv 0$ is known as *commensalism.* Under commensalism the population size of the x species is independent of the population size of the y species. Our treatment will include the possibility of commensalism, but obviously we will not assume that the model is commensal for both species.

Because $f_x(x,y) < 0$, the x-isocline $f(x,y) = 0$ may be written in the form $x = \phi(y) \geq 0$ with $\phi'(y) \geq 0$ on some interval $\alpha \leq y < \infty$. If $\alpha = 0$

and $\phi(0) = K > 0$ so that $f(K, 0) = 0$, the x species is facultative and K is the carrying capacity. If $\alpha > 0$ or $\alpha = 0$ but $\phi(0) = 0$, the x species is obligatory. In this case the x species would die out in the absence of the y species.

We assume also that

$$\phi(\infty) = K^* < \infty$$

so that the mutualistic effect of the y population cannot be so strong as to allow the x species population size to become unbounded. Since ϕ is an increasing function, $K^* \geq K$; the number K^* may be viewed as an increased carrying capacity for the x species produced by the mutualistic effect of the y species. Specifically, we assume

$$g_x(x, y) \geq 0, \qquad g_y(x, y) < 0$$

for $x \geq 0, y \geq 0$, so that the y isocline $g(x, y) = 0$ may be written in the form $y = \psi(x) \geq 0$ with $\psi'(x) \geq 0$ for $\beta \leq x < \infty$ $(\beta \geq 0)$. The y species is facultative if $\beta = 0$ and $\psi(0) = M > 0$, so that $g(0, M) = 0$. If $\beta > 0$ or $\beta = 0$ but $\psi(0) = 0$, the y species is obligatory. We also assume

$$\psi(\infty) = M^* < \infty.$$

At an interior equilibrium (x_∞, y_∞), the community matrix is

$$\begin{pmatrix} x_\infty f_x(x_\infty, y_\infty) & x_\infty f_y(x_\infty, y_\infty) \\ y_\infty g_x(x_\infty, y_\infty) & y_\infty g_y(x_\infty, y_\infty) \end{pmatrix},$$

with trace

$$x_\infty f_x(x_\infty, y_\infty) + y_\infty g_y(x_\infty, y_\infty) > 0$$

and determinant

$$x_\infty y_\infty \left(f_x(x_\infty, y_\infty) g_y(x_\infty, y_\infty) - f_y(x_\infty, y_\infty) g_x(x_\infty, y_\infty)\right),$$

which is positive if and only if the crossing of the isoclines at (x_∞, y_∞) is so that the x-isocline is above the y-isocline to the right of x_∞. Also the quantity

$$\begin{aligned} \Delta &= \left(x_\infty f_x(x_\infty, y_\infty) - y_\infty g_y(x_\infty, y_\infty)\right)^2 + \\ &\quad 4 x_\infty y_\infty f_y(x_\infty, y_\infty) g_x(x_\infty, y_\infty) \geq 0. \end{aligned}$$

Thus, the equilibrium (x_∞, y_∞) is either a saddle point or an asymptotically stable node, depending on the crossing of the isoclines. If we assume that the isoclines are not tangent at any equilibrium, then interior equilibria must alternate between saddle points and asymptotically stable nodes. The stable separatrices at a saddle point separate the domains of attractions of the adjacent nodes.

Because the x isocline approaches the vertical line $x = K^*$ asymptotically and the y isocline approaches the horizontal line $y = M^*$ asymptotically, there must be a "last" equilibrium–the one with the largest values of x and y,–and this equilibrium must be an asymptotically stable node. Also, if the y isocline is above the x isocline for small x there must be at least one interior equilibrium.

We can now describe the qualitative behavior of mutualistic systems of Kolmogorov type for both facultative and obligatory interactions. In the accompanying figures we will use **(SP)** to describe a saddle point, **(U)** to designate an unstable node, and **(AS)** to designate an asymptotically stable node.

Case I. Both species facultative. In this case $f(K,0) = 0, g(0,M) = 0$, and $f(0,0) > 0$. Thus, the origin is an unstable node and $(K,0), (0,M)$ are saddle points. Since the y isocline is above the x isocline for small x, the first interior equilibrium is an asymptotically stable node, and since the last equilibrium is an asymptotically stable node, there must be an odd number (at least one) of interior equilibria. Every orbit starting in the interior of the first quadrant tends to an interior node (Figure 5.22).

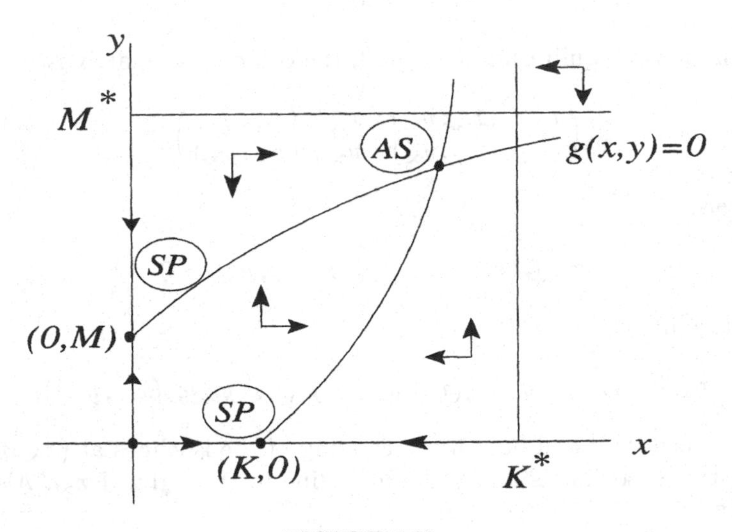

FIGURE 5.22.

Case II. Both species obligatory. In this case the x isocline meets the y axis at $(0,\alpha)$ and the y isocline meets the x axis at $(\beta,0)$. Since $f(0,0) < 0, g(0,0) < 0$, the origin is an asymptotically stable node. Since the x-isocline is above the y-isocline for small x, the first interior equilibrium is a saddle point. Thus there is an even number, possibly zero, of interior equilibria. If there are two interior equilibria, the first one is a saddle point whose stable separatrices separate the region of initial states for which orbits tend to the origin (extinction of both species) from the region of initial states for which orbits tend to the asymptotically stable interior node (coexistence) (Figure 5.23). If there is no interior equilibrium the origin is an asymptotically stable node and all orbits tend to the origin (extinction of both species) (Figure 5.24).

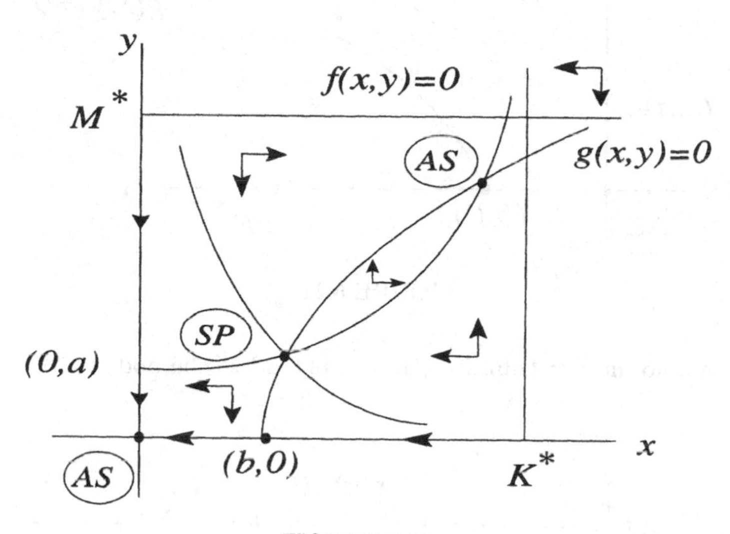

FIGURE 5.23.

Which of the two situations occurs depends on the parameters of the model. Changes in external conditions may alter the parameters of the system and cause a transition from one state to another. Thus coexistence may be destroyed, either by such a transition or by a perturbation which moves an orbit across the separatrix into the extinction region.

Case III. Facultative-obligatory mutualism. We take the x species to be obligatory, so that the x isocline meets the y axis at $(0, \alpha)$, and the y species to be facultative, so that the y isocline meets the y axis at $(0, M)$. There are two distinct subcases, depending on the relative sizes of α and M, but in each of these subcases, $f(0,0) < 0$, $g(0,0) > 0$ and the origin is a saddle point.

If $\alpha > M$, then $f(0, M) < 0$ and $(0, M)$ is an asymptotically stable node. Since the x isocline is above the y isocline for small x, the first interior equilibrium is a saddle point. As in the obligatory-obligatory case (Case II), there must be an even number (possibly zero) of interior equilibria. If there are two interior equilibria, the first is a saddle point whose stable separatrices separate the region of initial states for which orbits tend to the node $(0, M)$ (extinction of the x species) from the region of initial states for which orbits tend to the interior stable node (coexistence) (Figure 5.25).

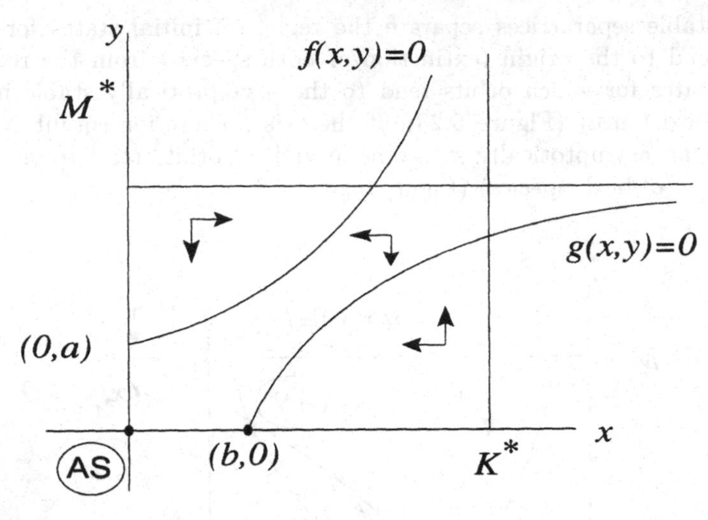

FIGURE 5.24.

If there are no interior equilibria all orbits tend to the node (0,M) (Figure 5.26).

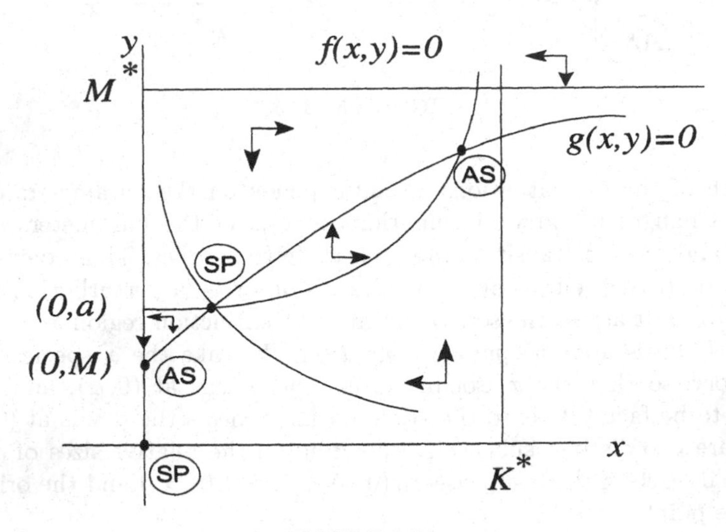

FIGURE 5.25.

If $\alpha < M$, then $f(0, M) > 0$, and $(0, M)$ is a saddle point. Since the y isocline is above the x isocline for small x, the first interior equilibrium is an asymptotically stable node. As in the facultative-facultative case (Case I) there must be an odd number of interior equilibria and every orbit starting in the interior of the first quadrant of the phase plane must tend to an asymptotically stable interior node (Figure 5.27).

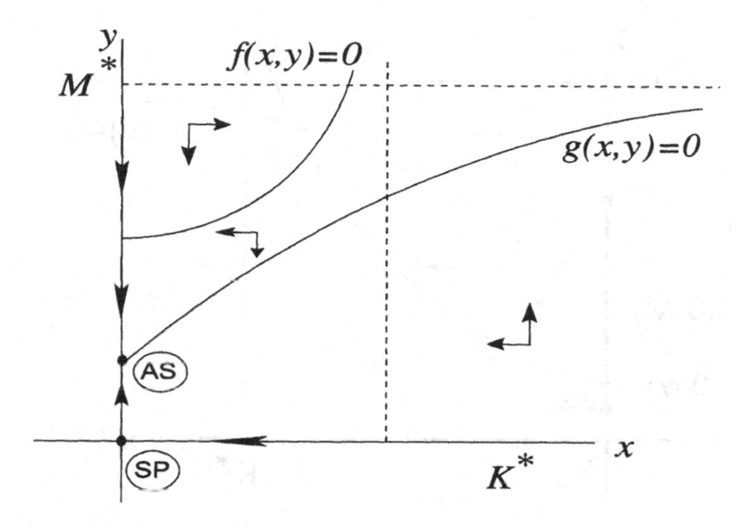

FIGURE 5.26.

We now have a general description of what a two-species mutualistic model should be and how its solutions should behave. There have been relatively few attempts to suggest explicit models for mutualism, possibly because experimental data to fit a model are difficult to obtain. While in the model we have proposed both isoclines be curved, all the results that we have deduced on behavior of solutions would remain true if one of the isoclines were a straight line.

Some forms that have been used include:

$$\begin{aligned} f(x,y) &= \frac{a_1 y}{1+b_1 x} - c, \\ g(x,y) &= d_2(K-y) + \frac{a_2 x}{1+b_2 x} \end{aligned} \qquad \text{[Soberon and DelRio (1981)]},$$

$$\begin{aligned} f(x,y) &= r\left(1 - \frac{x}{K\left(1-e^{x(1-y/M)}\right)}\right), \\ g(x,y) &= s\left(1 - \frac{y}{L\left(1-e^{y(1-x/N)}\right)}\right) \end{aligned} \qquad \text{[Dean (1983)]}.$$

If we think of a logistic model with parameters r and K, we could use

$$f(x,y) = r\left(1 - \frac{x(y+A)}{KA+K^*y}\right)$$

to model enhancement of K, or

$$f(x,y) = \left(\frac{rA+r^*y}{y+A}\right)\left(1-\frac{x}{K}\right)$$

to model enhancement of r, or

$$f(x,y) = \left(\frac{rA+r^*y}{y+A}\right)\left(1 - \frac{x(y+B)}{KB+K^*y}\right)$$

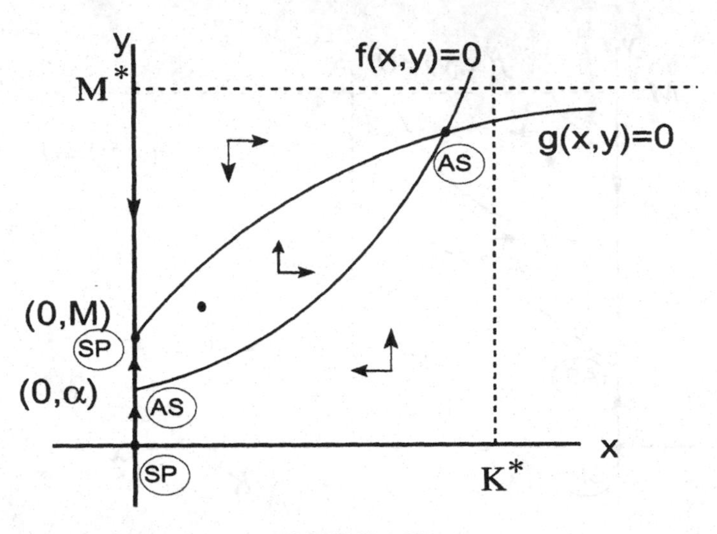

FIGURE 5.27.

to model simultaneous enhancement of r and K through mutualism, with similar forms for $g(x, y)$.

Exercises

In Exercises 1 through 4, find all equilibria of the given mutualistic system and determine their stability. Are there any unbounded orbits?

1. $x' = x(-20 - x + 2y)$, $y' = y(-50 + x - y)$

2. $x' = x(-20 - x + 2y)$, $y' = y(-50 + x - y)$

3. $x' = x(y/(1 + x) - 1)$, $y' = y(10 - y + x/(1 + y))$

4. $x' = x(1 - x/10) \cdot ((2y + 1)/(1 + y))$, $y' = y(1 - y/20)$

5.* Determine the behavior as $t \to \infty$ of solutions of the system

$$x' = rx(1 - \frac{x}{K + ay}), y' = sy(1 - \frac{y}{M + bx})$$

with a, b, K, M, r, and s positive constants. [see Exercise 12, Section 4.5. [*Warning*: The behavior if $ab < 1$ is different from the behavior if $ab > 1$].

5.6 The Spruce Budworm: A Case Study

The spruce budworm is an insect that inhabits spruce and fir trees in eastern North America. In most years it is relatively scarce, but it exhibits outbreaks in which the population grows by a factor of as much as 1000. In an outbreak year, budworms may devour the new needles produced by an evergreen forest, ultimately killing 80 percent of the mature trees in the forest. Such outbreaks may destroy forests. When the food supply is destroyed (this may take 7 to 14 years from the start of the outbreak) the budworm population declines rapidly to a low level and the forest may then begin to recover. We have described a simple model for the budworm population in Exercise 13, Section 1.4 and Project 1.1. Here we will recall this model and refine it to incorporate the forest as well. The dynamics of the spruce budworm proceed on a much faster time scale than the dynamics of the forest, and our model will incorporate two time scales. The process is to begin with the fast (budworm) variable, assuming that the slow (forest) variables remain constant. The next step is to formulate a slow variable model and analyze it in order to decide how to refine the fast variable model to incorporate dependence of its parameters on the slow variables. This will allow us to simulate outbreaks caused by a change in the behavior of the fast variable system caused by parameter changes. Our description is based on the paper [Ludwig, Jones, and Holling (1978)].

We let B devote the spruce budworm population size and assume the fast variable model

$$B' = r_B\left(1 - \frac{B}{K_B}\right) - \beta\frac{B^2}{\alpha^2 + \beta^2}. \tag{5.12}$$

This is the model of Exercise 13, Section 1.4 with different notation. Here r_B is the intrinsic growth rate and K_B is the carrying capacity in a logistic growth model. In addition, there is predation by birds and parasites that saturates for high budworm populations at a level β, and α is the budworm population at which predation is half the maximum. The model involves the four parameters r_B, K_B, β, α. Later, when we incorporate the slow variables, we will allow K_B and α to depend on these slow variables.

It is convenient to scale the equation (5.12) to reduce the number of parameters. We use α as the unit of budworm population size and introduce $u = B/\alpha$ as a new dependent variable. This transforms (5.12) to the equation

$$\begin{aligned}
\alpha u' &= r_B u\alpha\left(1 - \frac{\alpha u}{K_B}\right) - \beta\frac{u^2}{1+u^2} \\
\frac{\alpha}{\beta}u' &= r_B\frac{\alpha}{\beta}u\left(a - \frac{\alpha u}{K_B}\right) - \frac{u^2}{1+u^2}.
\end{aligned}$$

Now we define the new parameters

$$R = \frac{\alpha}{\beta}, \qquad Q = \frac{K_B}{\alpha}$$

and rewrite the differential equation as

$$\frac{\alpha}{\beta}u' = Ru\left(1 - \frac{u}{Q}\right) - \frac{u^2}{1+u^2}. \tag{5.13}$$

Equilibria of (5.13) are $u = 0$ and the intersections of the line $y = R(1 - u/Q)$ and the curve $y = u/(1+u^2)$. It will turn out that during an outbreak Q remains nearly constant but R may vary considerably. Thus, in order to understand the dynamics we will need to investigate how the equilibria depend on R. The curve $y = u/(1+u^2)$ starts at the origin, increases to a maximum at $(1, 1/2)$, and then decreases with an inflection part at $(\sqrt{3}, \sqrt{3}/4)$ and approaches zero as $u \to \infty$. The line $y = R(1 - u/Q)$ passes through the points $(0, R)$ and $(Q, 0)$ in the (u, y)-plane. A "typical" situation is as shown in Figure 5.28.

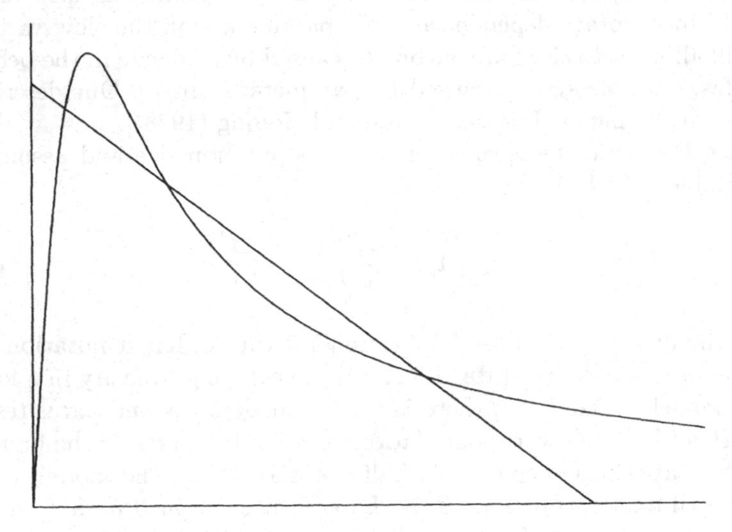

FIGURE 5.28.

There are four equilibria: an unstable equilibrium at $u = 0$, two asymptotically stable equilibria u_- and u_+, separated by an unstable equilibrium $u = u_c$. The domain of attraction of the small equilibrium u_- is the interval $(0, u_c)$ and the domain of attraction of the small equilibrium u_+ is the interval (u_c, ∞). If R increases until u_- and u_c coalesce, as in Figure 5.29, then only the equilibrium u_+ remains. Thus, if a system is in equilibrium at u_- and R is increased there could be a jump to u_+. Conversely, if R is decreased until u_+ and u_c coalesce, as in Figure 5.30, then only the equilibrium u_- remains. Thus, if a system is in equilibrium at u_+ and R is decreased there could be a crash to u_-.

We think of the parameter R as representing the resources of the forest supply as food for budworms, and the equilbria u_- and u_+ as corresponding to budworm limitation by predators and food supply, respectively. As forest

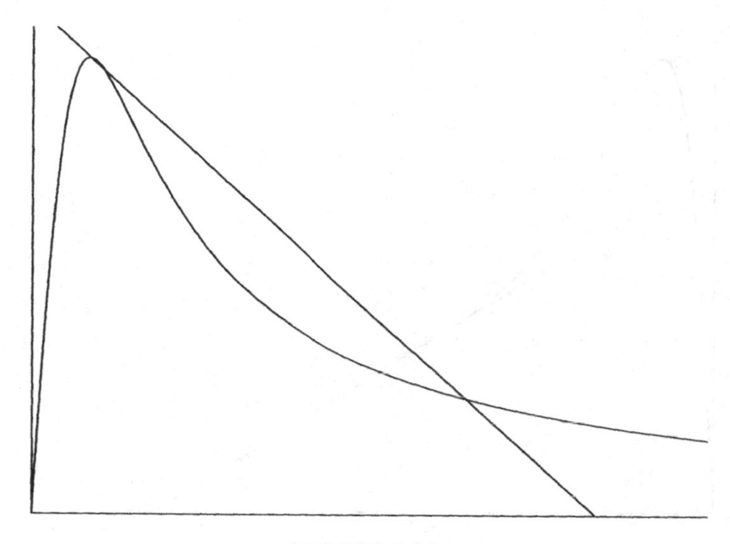

FIGURE 5.29.

conditions improve budworm growth exceeds control by predators and there is an outbreak. If this outbreak destroys the forest, the predators may regain control and cause a budworm crash. The outbreak and crash will generally occur at different population levels.

In order to build in the possibility of forest destruction, we need to form a model to describe the state of the forest assuming a fixed budworm population. We will use as forest variables the *total surface area of branches* S and a variable E, which may be viewed as an *energy reserve*, to describe the foliage and health of the trees. We choose a logistic form for the equation governing S,

$$S' = r_S S\left(1 - \frac{SK_E}{K_S E}\right). \tag{5.14}$$

The factor K_E/E is included because S may decrease under stress through the death of branches or whole trees. Normally, however, E will be close to its carrying capacity K_E and S will approach its carrying capacity K_S. We assume that the energy reserve E also satisfies an equation of logistic type

$$E' = r_E E\left(1 - \frac{E}{K_E}\right) - p\frac{B}{S}. \tag{5.15}$$

The term pB/S describes the stress on the trees excited by budworm consumption of foliage. Since B has units of number of budworms per acre and S has units of branch surface area per acre, B/S is the number of budworms per unit of branch surface area.

The next step is to determine the qualitative behavior of the slow variable system (5.14), (5.15), treating B as constant, by analyzing its equilibria.

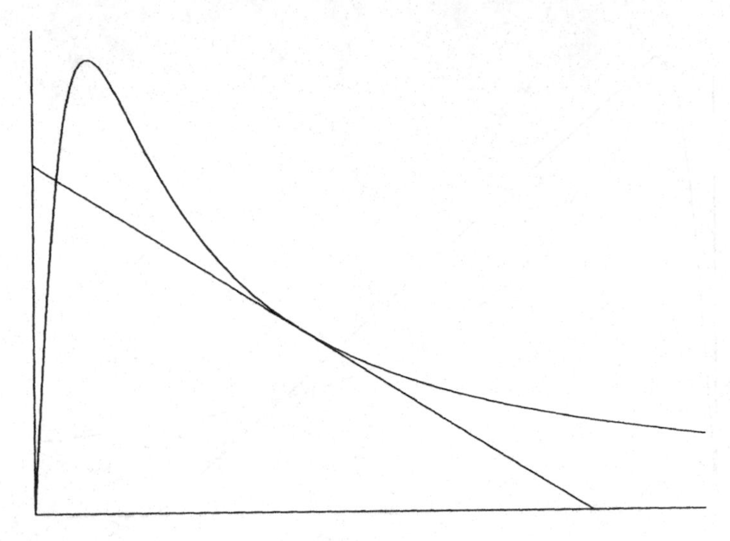

FIGURE 5.30.

The equilibrium conditions either $S = 0$ or $S = \frac{K_S}{k_E} E$ together with

$$S = \frac{pBK_E}{r_E} \frac{1}{E(K_E - E)}. \tag{5.16}$$

The curve (5.16) has vertical asymptotes at $E = 0$ and $E = K_E$ and has a minimum when $E = K_E/2$. Equilibria are intersections of this curve and the line $S = EK_S/K_E$ through the origin with positive slope. If B is small there are two equilibria (Figure 5.31), while if B is large, as during a budworm outbreak, the curve (5.16) is elevated and there are no equilibria (Figure 5.32).

We may see from the flow arrows that in Figure 5.31 the equilibrium on the left is a saddle point and the equilibrium on the right is asymptotically stable. The stable separatrices at the saddle point divide the plane into the domain of attraction of the stable equilibrium (to the right of the separatrices) and the region for which $E \to -\infty$ (to the left of the separatrices). (In our model, $E' < 0$ when $E = 0$ and this is unrealistic; later we shall describe how to modify the model to make $E' = 0$ when $E = 0$ and allow the forest to recycle).

When the budworm population is small enough that Figure 5.31 describes the slow time situation, S and E increase slowly toward the stable equilibrium. As B increases the curve (5.16) rises, causing the equilibrium values of S and E to decrease somewhat, indicating a deterioration of the health of the forest. When there is a catastrophic outbreak of budworms, the equilibria coalesce and disappear, shifting the situation to that of Figure 5.32. Then the forest collapses. There is another possibility: The budworm population level, though high, may be held down enough to avoid collapse

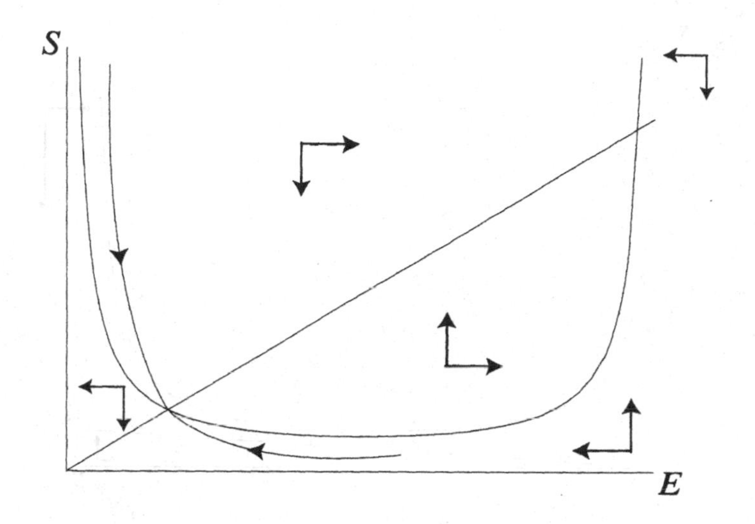

FIGURE 5.31.

of the forest. This is called "perpetual outbreak" and has been observed as a consequence of the spraying done to try to control the budworm.

We now have two separate models for budworm and forest. In order to combine them into one model we need to reconsider the two models by coupling the parameters appropriately. We would expect r_B, the intrinsic growth rate of the budworm population, to be independent of the forest parameters, but the carrying capacity should be proportional to the foliage surface area S. Thus, we replace K_B by a new parameter K with

$$K_B = KS.$$

The maximum predation rate β is not affected by the forest state, but the budworm population level α at which predation is half its maximum rate will also increase proportionally to S; as S increases predators will have to search more foliage and a higher budworm population can be supported at the same predation rate. Thus, we replace α by a new parameter α' with $\alpha = \alpha' S$. We now have a coupled model

$$\begin{aligned} B' &= r_B B\left(1 - \frac{B}{KS}\right) - \beta\frac{B^2}{(\alpha' S)^2 + B^2} \\ S' &= r_S S\left(1 - \frac{S}{K_S}\frac{K_E}{E}\right) \\ E' &= r_E E\left(1 - \frac{E}{K_E}\right) - P\frac{B}{S}. \end{aligned} \tag{5.17}$$

The parameters R and Q introduced in the analysis of 5.12 now become

$$R = \frac{\alpha' r_B}{\beta}S, \qquad Q = \frac{K}{\alpha}.$$

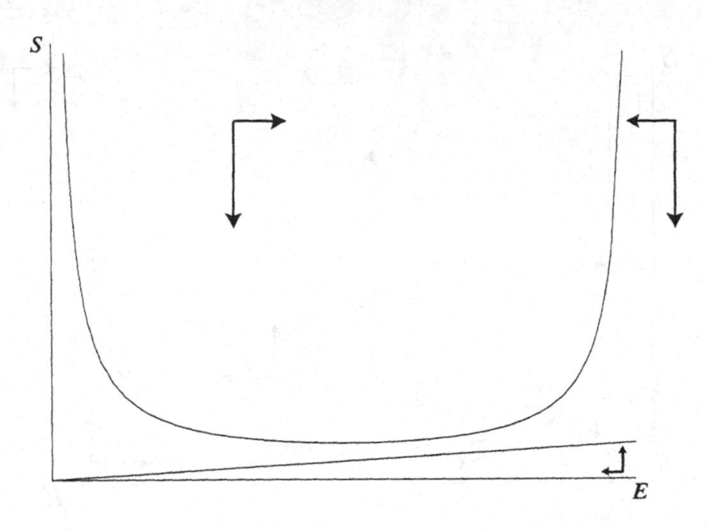

FIGURE 5.32.

Q remains constant but R increases proportionally to S as S increases. As a young forest matures, S increases and the consequent increase in R in the analysis of the fast variable will lead to a budworm outbreak. Whether there will be perpetual outbreak or forest collapse depends on parameter values.

Appropriate values for the parameters of the model (5.17) have been estimated [Ludwig, Jones, & Holling (1978)], first on the basis of general knowledge of the biology of the situation and then in more refinement from extensive field study of the forests of New Brunswick. These studies lead to the following choices for parameter values:

$$\begin{array}{lll} r_B = 1.52 & K = 335 & \beta = 43,200 \\ \alpha' = 1.11 & r_S = 0.095 & K_S = 25,440 \\ K_E = 1.0 & r_E = 0.92 & p = 0.00195. \end{array}$$

These lead to the values $R = 3.91 \times 10^{-5}S$, $Q = 302$. We could then use these parameter values in (5.17) to simulate the behavior of the spruce budworm forest system, and we would find outbreak followed by collapse of the forest. This conforms to observed behavior in part, but does not exhibit the cycles of budworm outbreak, forest collapse, and budworm collapse, followed by forest regeneration.

This shortcoming of our model is reflected in the fact that E, a variable meant to describe a general status and considered to vary from zero to one, may become negative. We may improve the model in an ecological sense by observing that budworm carrying capacity should depend on E as well as S and by incorporating dependence on E in the stress due to defoliation by budworms. We replace $K_B = KS$ by

$$K_B = KS\frac{E^2}{E^2 + T_E^2},$$

where T_E is a constant threshold value. The quantity $E^2/(E^2 + T_E^2)$ is near one in a healthy forest but decreases sharply as $E \to 0$. We also replace the stress term pB/S by

$$p\frac{B}{S}\frac{E^2}{T_E^2 + E^2}.$$

This means that the stress term will decrease rapidly as the forest collapses. The refined model would be

$$\begin{aligned}
B' &= r_B B\left(1 - \frac{B}{KS}\frac{E^2 + T_E^2}{E^2}\right) - \beta\frac{B^2}{(\alpha' S)^2 + B^2} \\
S' &= r_S S\left(1 - \frac{S}{K_S}\frac{K_E}{E}\right) \qquad (5.18) \\
E' &= r_E E\left(1 - \frac{E}{K_E}\right) - P\frac{B}{S}\frac{E^2}{E^2 + T_E^2}.
\end{aligned}$$

With suitably chosen T_E, this model will behave like the model (5.17) except that when B, S, and E are small it will allow regeneration of the forest and a new cycle.

The models (5.17) and (5.18) give a qualitative picture similar to what has been observed. For short term predictions and actual forest management it may be preferable to use a more detailed model that does not attempt to describe a whole forest by two variables. Such a model would not be analytically tractable but could be used for numerical simulations. However, there are, necessarily, approximations in the model and errors in the parameters, which could give accumulating errors in a simulation. Thus, such a model is best suited for short term predictions. Our less detailed model is better suited for explaining *why things happen.*

References: Arreola, Mijares-Bernal, Ortiz-Navarro, and Saenz (2000) have looked at the effects of insecticides on a two-dimensional version of the above model that supports oscillations. They show that insecticides can increase the frequency of these oscillations, a situation that, in the long run may be not be good for the trees.

5.7 The Community Matrix

In this chapter, we are concerned primarily with models for two interacting species. However, in most real life situations there are more than two species involved. For example, the spruce budworm model considered in Section 5.5 involves three species. In the remainder of this chapter, we shall give a very brief sketch of how one may use two-species models (competition, predator-prey, and mutualism) as building blocks to study some larger systems. In general, the study of multispecies models is very complicated and we shall confine ourselves to some remarks about their general structure and two three-species examples.

If $x_1, x_2, \ldots, x_n$ represent the population sizes of n different interacting species in a system for which it is assumed that the growth rate of each population size at any time depends only on the various population sizes at that time, then the system is modeled by an autonomous system of n first order differential equations

$$\begin{aligned} x_1' &= F_1(x_1, x_2, \ldots, x_n) \\ x_2' &= F_2(x_1, x_2, \ldots, x_n) \\ &\vdots \\ x_n' &= F_n(x_1, x_2, \ldots, x_n). \end{aligned} \tag{5.19}$$

We are interested only in nonnegative population sizes. As it is biologically reasonable to require that no species can generate spontaneously, it is reasonable to require that a solution for which some x_j is zero at any time should continue to have $x_j = 0$ for all time. Thus, it is natural to write the system (5.19) in the Kolmogorov form:

$$\begin{aligned} x_1' &= x_1 r_1(x_1, x_2, \ldots, x_n) \\ x_2' &= x_2 r_2(x_1, x_2, \ldots, x_n) \\ &\vdots \\ x_n' &= x_n r_n(x_1, x_2, \ldots, x_n). \end{aligned} \tag{5.20}$$

Exactly as for single-species models (Chapter 1) and two-species models (Chapter 4) we define an *equilibrium* of the system (5.19) to be a solution $(\xi_1, \xi_2, \ldots, \xi_n)$ of the system of equations

$$\begin{aligned} F_1(\xi_1, \xi_2, \ldots, \xi_n) &= 0 \\ F_2(\xi_1, \xi_2, \ldots, \xi_n) &= 0 \\ &\vdots \\ F_n(\xi_1, \xi_2, \ldots, \xi_n) &= 0. \end{aligned} \tag{5.21}$$

An equilibrium is a constant solution of the system of differential equations (5.19). The *linearization* of the system (5.19) at an equilibrium $(\xi_1, \xi_2, \ldots, \xi_n)$

is defined to be the *linear* system of differential equations

$$\begin{aligned}
u_1' &= \frac{\partial F_1}{\partial x_1}(\xi_1,\dots,\xi_n)u_1 + \frac{\partial F_1}{\partial x_2}(\xi_1,\dots,\xi_n)u_2 + \dots + \frac{\partial F_1}{\partial x_n}(\xi_1,\dots,\xi_n)u_n \\
u_2' &= \frac{\partial F_2}{\partial x_1}(\xi_1,\dots,\xi_n)u_1 + \frac{\partial F_2}{\partial x_2}(\xi_1,\dots,\xi_n)u_2 + \dots + \frac{\partial F_2}{\partial x_n}(\xi_1,\dots,\xi_n)u_n \\
\vdots &\quad \vdots \\
u_n' &= \frac{\partial F_n}{\partial x_1}(\xi_1,\dots,\xi_n)u_1 + \frac{\partial F_n}{\partial x_2}(\xi_1,\dots,\xi_n)u_2 + \dots + \frac{\partial F_n}{\partial x_n}(\xi_1,\dots,\xi_n)u_n.
\end{aligned}$$

It is convenient to use vector-matrix notations defining $\mathbf{x}$ to be the *column vector* with components $(x_1,\dots,x_n)$, ξ to be the column vector with components $(\xi_1,\dots,\xi_n)$, $\mathbf{u}$ to be the column vector with components $(u_1,\dots,u_n)$, $\mathbf{F}(\mathbf{x})$ to be the column vector function with components $(f_1(x_1,\dots,x_n),\dots,f_n(x_1,\dots,x_n))$, and $\mathbf{A}$ to be the matrix with element

$$\frac{\partial F_i}{\partial x_j}(\xi_1,\dots,\xi_n)$$

in the *ith* row, *jth* column. Then the system (5.19) can be written in the vector form

$$\mathbf{x}' = \mathbf{F}(\mathbf{x}), \tag{5.22}$$

an equilibrium is a vector ξ satisfying

$$\mathbf{F}(\xi) = \mathbf{0}, \tag{5.23}$$

and the linearization of the system (5.19) or (5.22) at an equilibrium is the linear system

$$\mathbf{u}' = \mathbf{A}\mathbf{u}. \tag{5.24}$$

The matrix

$$\mathbf{A} = \left(\frac{\partial F_i}{\partial x_j}(\xi_1,\dots,\xi_n)\right)$$

is called the *community matrix* of the system (5.19) or (5.22) at the equilibrium ξ.

There is a general theorem, analogous to Theorem 4.1, Section 4.3, which says that an equilibrium of the system (5.22) is asymptotically stable if all solutions of the linearization at this equilibrium tend to zero as $t \to \infty$, while an equilibrium ξ is unstable if the linearization has any solution that grows unbounded exponentially. It is also true that all solutions of the linearization tend to zero if all roots of the *characteristic equation*

$$\det(\mathbf{A} - \lambda\mathbf{I}) = \mathbf{0} \tag{5.25}$$

have negative real part and that there are solutions of the linearization that grow exponentially unbounded if the characteristic equation (5.25) has any roots with positive real part. The roots of the characteristic equation (5.25) are the eigenvalues of the matrix **A**. The characteristic equation has the property that its roots are the values of λ such that the linearization at the equilibrium has a solution $e^{\lambda t}\mathbf{c}$ for some constant column vector **c**.

Thus, the stability of an equilibrium ξ can be determined from the eigenvalues of the community matrix at the equilibrium.

We have:

Theorem 5.2. *If all eigenvalues of the community matrix of a system* (5.19) *or* (5.22) *at an equilibrium* ξ *have negative real part then the equilibrium is asymptotically stable.*

This is the generalization to n dimensions of Theorem 4.4 for two dimensional systems, but we no longer have such a simple set of conditions on the trace and determinant of the community matrix to determine the stability of an equilibrium.

The characteristic equation for an n-dimensional system is a polynomial equation of degree n for which it may be difficult or impossible to find all roots explicitly. There is, however, a general criterion for determining whether all roots of a polynomial equation have negative real part known as the *Routh-Hurwitz criterion.* This gives conditions on the coefficients of a polynomial equation

$$\lambda^n + a_1\lambda^{n-1} + a_2\lambda^{n-2} + \ldots + a_{n-1}\lambda + a_n = 0$$

under which all roots have negative real part. For $n = 2$, the Routh-Hurwitz conditions are

$$a_1 > 0, \ a_2 > 0$$

(equivalent to the conditions that the trace of the matrix **A** be negative and the determinant of the matrix **A** be positive). For $n = 3$, the Routh-Hurwitz conditions are

$$a_3 > 0, \ a_1 > 0, \ a_1a_2 > a_3.$$

For $n = 4$, the Routh-Hurwitz conditions are

$$a_4 > 0, \ a_2 > 0, \ a_1 > 0, \ a_3(a_1a_2 - a_4) > a_1^2a_4.$$

For a polynomial of degree n, there are n conditions. While the Routh-Hurwitz criterion may be useful on occasion, it is complicated to apply in problems of many dimensions.

For autonomous systems of two differential equations, the Poincaré-Bendixson theorem (Section 4.5) makes it possible to analyze the qualitative behavior of a system under very general conditions. Essentially, we know that a bounded orbit must approach either an equilibrium point or a limit cycle. For autonomous systems of more than two differential equations a much greater range of behavior is possible. An example is the *Lorenz equation*

$$\begin{aligned} x' &= \sigma(y-x) \\ y' &= rx - y - xz \\ z' &= xy - bz, \end{aligned}$$

with σ, r, and b three positive parameters. This equation, which arose originally as a highly simplified model of fluid convection in a meteorological system, can exhibit chaotic behavior for some ranges of the parameters [Lorenz (1963)].

There is some biological evidence to suggest that complicated population systems have a tendency to be more stable than simple systems. For example, a predator species that can switch between different prey species for its food supply may be less sensitive to disturbances than if it were dependent on a single food supply. On the other hand, removal of one species can lead to a collapse of population systems; an example was observed in which the removal of one species from a 15-species system led to its collapse to an 8-species system in less than two years [Paine (1966)]. The relationship between stability and complexity in population systems, a question raised by May (1974), is not well understood. One aspect of this question which can provide some useful information is the vulnerability of a system to invasion by a new species (Section 5.8).

5.8 The Nature of Interactions Between Species

If we model a two-species interaction by a system of Kolmogorov type

$$\begin{aligned} x' &= xf(x,y) \\ y' &= yg(x,y), \end{aligned} \tag{5.26}$$

it may appear reasonable to require

$$f_x(x,y) < 0, \quad g_y(x,y) < 0 \tag{5.27}$$

to express the idea that each species tends to limit its own growth. However, the condition (5.27) is not satisfied for all $x > 0, y > 0$ in the standard predator-prey models; we should only require that each species tend to limit its own growth if the other species is not present. The assumption $f_x(x,y) < 0$ is sometimes replaced by the requirement

$$xf_x(x,y) + yf_y(x,y) < 0$$

for $x > 0, y > 0$ [Kolmogorov (1936)], which is satisfied by models of Rosenzweig-MacArthur type. Another approach [Bulmer(1976)] is to require $f_x(x,0) < 0$ together with a condition modeling the idea that predators increase only through the consumption and conversion of prey, such as

$$yg(x,y) \leq \alpha x\big(f(x,0) - f(x,y)\big) - \mu y$$

for some positive constants α, μ.

In order to separate the growth rates in the system (5.26) into self-limiting and interaction terms, we write (5.26) in the form

$$\begin{aligned} x' &= xf(x,0) + x\big(f(x,y) - f(x,0)\big) \\ y' &= yg(0,y) + y\big(g(x,y) - g(0,y)\big). \end{aligned} \tag{5.28}$$

We then require that the respective single-species dynamics be self-limiting

$$f_x(x,0) < 0, \; g_y(0,y) < 0 \tag{5.29}$$

except that we will permit a species that is dependent on the interaction for survival (predator or obligate mutualist) to have, for example, $g_y(0,y) = 0$. This would mean that $g(0,y)$ is a constant, and we would require this constant to be negative so that $y(t) \to 0$ as $t \to \infty$ in the absence of the x species. We may then characterize the interaction as predator-prey, competitive, or mutualistic, depending on the signs of

$$\begin{aligned} \frac{\partial}{\partial y}\big(f(x,y) - f(x,0)\big) &= f_y(x,y) \\ \frac{\partial}{\partial x}\big(g(x,y) - g(0,y)\big) &= g_x(x,y). \end{aligned}$$

The terms $x\big(f(x,y) - f(x,0)\big)$ and $y\big(g(x,y) - g(0,y)\big)$ represent the interaction terms in (5.28). Thus, the system (5.26) or (5.28) is a predator-prey system if

$$f_y(x,y) < 0, \; g_x(x,y) > 0,$$

a competitive system if

$$f_y(x,y) < 0, \; g_x(x,y) < 0,$$

and a mutualistic system if

$$f_y(x,y) > 0, \; g_x(x,y) > 0.$$

If we attempt to decompose a three-species Kolmogorov model

$$\begin{aligned} x' &= xf(x,y,x) \\ y' &= yg(x,y,x) \\ z' &= zh(x,y,x) \end{aligned} \tag{5.30}$$

into self-limiting terms and two-species interaction terms we may write the first equation of (5.30) as either

$$x' = xf(x,0,0) + x\big(f(x,y,0) - f(x,0,0)\big) + x\big(f(x,y,z) - f(x,y,0)\big)$$

or

$$x' = xf(x,0,0) + x\big(f(x,y,z) - f(x,0,z)\big) + x\big(f(x,0,z,) - f(x,0,0)\big).$$

In characterizing the x-y interaction we would consider the sign of the partial derivative with respect to y of the effect of y on the per capita growth rate of x, which would be either

$$\frac{\partial}{\partial y}\big(f(x,y,0) - f(x,0,0)\big) = f_y(x,y,0)$$

or

$$\frac{\partial}{\partial y}\big(f(x,y,z) - f(x,0,z)\big) = f_y(x,y,z),$$

and these do not necessarily have the same sign.

Consider, for example, a population system consisting of two competing species x and y and a third species z acting as a predator on both. We may model such a system in more specific form by a system of Rosenzweig-MacArthur type:

$$\begin{aligned} x' &= rxf(x,y) - xz\phi(x,y) \\ y' &= syg(x,y) - yz\psi(x,y) \\ z' &= z\big(\lambda x\phi(x,y) - \mu y\psi(x,y) - c\big). \end{aligned} \tag{5.31}$$

This is of the form (5.30) with

$$\begin{aligned} f(x,y,z) &= rf(x,y) - z\psi(x,y) \\ g(x,y,z) &= sg(x,y) - z\psi(x,y) \\ h(x,y,z) &= \lambda x\phi(x,y) + \mu y\psi(x,y) - c, \end{aligned}$$

so that

$$f_y(x,y,0) = rf_y(x,y),\; f_y(x,y,z) = rf_y(x,y) - z\phi_y(x,y).$$

We model the competition between x and y by the assumptions

$$f_y(x,y) < 0,\; g_x(x,y) < 0.$$

If $\phi(x,y)$ and $\psi(x,y)$ are increasing functions of the ratios $x/(x+y)$ and $y/(x+y)$, respectively, then z is a predator that switches between x and y and $\phi_y(x,y) < 0, \psi_x(x,y) < 0$. This would imply that $f_y(x,y,0) < 0$ but $f_y(x,y,z) > 0$ for sufficiently large z. The same kind of sign change would occur for the partial derivative with respect to x of the effect of x on the per capita growth rate of y. If we try to describe the x-y interaction in this three-species model in terms of the signs of $f_y(x, y, z)$ and $g_x(x, y, z)$ in analogy to the two-species model, we would be led to say that this interaction is competitive when few predators are present, and mutualistic in the presence of many predators, with an intermediate stage in which the interaction is predator-prey. Such an interpretation suggests that the interpretation of the interaction between two species in a three-species model is more complicated. Presumably, a proper interpretation should be in terms of the signs of $f_y(x, y, 0)$ and $g_x(x, y, 0)$ together with some additional conditions whose form is not clear. For this reason, systems of more than two species are usually modeled by systems of differential equations less general than a Kolmogorov model. Frequently, a system of Lotka-Volterra type is used to give some indication of the type of behavior one might expect, even though the Lotka-Volterra model is a specific form not necessarily representative of general behavior, as we have seen for two-species models.

5.9 Invading Species and Coexistence

In analyzing population systems with more than two species there are many different cases that must be considered. For three species, even if we eliminate mutualism from consideration, there are six different cases: (i) three species in competition, (ii) one predator and two competing prey, (iii) two predators with a common prey, (iv) a prey whose predator is the prey of a higher level predator, (v) two species in competition and a third species that is the prey of one of them, (vi) two species in competition and a third species that is a predator of one of them. As general Kolmogorov-type models are not suitable for distinguishing these types, as we have seen in Section 5.7, each case must be analyzed separately.

A simplification which may be helpful is to ask for less information. Rather than trying to describe the qualitative behavior of a system completely, we might try to determine only the survival or extinction of each species. From a biological point of view, this may be the most significant aspect of the problem. A species with population size $x(t)$ is said to be persistent [Freedman and Waltman (1984)] if

$$\liminf_{t\to\infty} x(t) > 0$$

provided $x(0) > 0$. Persistence, however, does not assure survival of a species in a biological sense. Examples have been given of three-species systems that oscillate wildly so that each population size repeatedly comes arbitrarily close to zero (at which time a small perturbation could drive the population size to zero) and then recovers [Armstrong and McGehee (1980); May and Leonard (1975)]. A more realistic requirement would be uniform persistence [Butler and Waltman (1986)] or permanent coexistence [Hofbauer (1981)], a positive lower bound for population size. Conditions have been given, usually involving the behavior of lower-dimensional systems, under which persistence implies uniform persistence [Butler, Freedman, and Waltman (1986); Butler and Waltman (1986)]. We shall examine only persistence, avoiding the more delicate problem of uniform persistence.

In order to obtain some insight into the behavior of multi-species systems it may be helpful to consider the effect on an n-species system of introducing another species to give an $(n+1)$-species system. This additional species may be viewed biologically as an invading species disrupting an existing population system. Such situations have been observed in nature. For example, in systems consisting of two species in competition that are the prey of a common predator (to be examined in more detail later) the introduction of an alternate prey may lead to the extinction of the original prey [Holt (1977)]. Such a situation may also be viewed as the introduction of a predator into a system of coexisting competing species, leading to the exclusion of one of the prey species [Poole (1974); Dodson (1974); Holt (1977)].

We consider an n-species model

$$\begin{aligned} x_1' &= x_1F_1(x_1, x_2, \ldots, x_n, 0) \\ x_2' &= x_2F_2(x_1, x_2, \ldots, x_n, 0) \\ \vdots \quad & \quad \vdots \\ x_n' &= x_nF_n(x_1, x_2, \ldots, x_n, 0) \end{aligned} \tag{5.32}$$

with an equilibrium $(\bar{x}_1, \bar{x}_2, \ldots, \bar{x}_n)$ and having community matrix $\mathbf{A}$ at this equilibrium. We also consider the $(n+1)$-species model

$$\begin{aligned} x_1' &= x_1F_1(x_1, x_2, \ldots, x_n, y) \\ x_2' &= x_2F_2(x_1, x_2, \ldots, x_n, y) \\ \vdots \quad & \quad \vdots \\ x_n' &= x_nF_n(x_1, x_2, \ldots, x_n, y) \\ y' &= yG(x_1, x_2, \ldots, x_n, y) \end{aligned} \tag{5.33}$$

representing the introduction of an invading species y into the n-species system. The $(n+1)$-species system (5.33) has an equilibrium $(\bar{x}_1, \bar{x}_2, \ldots, \bar{x}_n, 0)$ and the community matrix at this equilibrium is

$$\begin{pmatrix} & \bar{x}_1\frac{\partial F_1}{\partial y}(\bar{x}_1, \bar{x}_2, \ldots, \bar{x}_n, 0) \\ & \bar{x}_2\frac{\partial F_2}{\partial y}(\bar{x}_1, \bar{x}_2, \ldots, \bar{x}_n, 0) \\ \mathbf{A} & \vdots \\ & \bar{x}_n\frac{\partial F_n}{\partial y}(\bar{x}_1, \bar{x}_2, \ldots, \bar{x}_n, 0) \\ \mathbf{0} & G(\bar{x}_1, \bar{x}_2, \ldots, \bar{x}_n, 0) \end{pmatrix}.$$

The eigenvalues of this matrix are the eigenvalues of $\mathbf{A}$ together with $G(\bar{x}_1, \bar{x}_2, \ldots, \bar{x}_n, 0)$.

We assume that the equilibrium $(\bar{x}_1, \bar{x}_2, \ldots, \bar{x}_n, 0)$ of the n-species model (5.32) is asymptotically stable, and thus that the eigenvalues of $\mathbf{A}$ have negative real part. Then if $G(\bar{x}_1, \bar{x}_2, \ldots, \bar{x}_n, 0) < 0$ the equilibrium $(\bar{x}_1, \bar{x}_2, \ldots, \bar{x}_n, 0)$ of the $(n+1)$-species system (5.33) is also asymptotically stable, and orbits starting near this equilibrium, corresponding to the introduction of a small number of members of the invading species, tend to this equilibrium. Thus, the invading species fails to survive.

On the other hand, if $G(\bar{x}_1, \bar{x}_2, \ldots, \bar{x}_n, 0) > 0$ the equilibrium $(\bar{x}_1, \bar{x}_2, \ldots, \bar{x}_n, 0)$ of (5.33) is unstable because the community matrix at this equilibrium has one positive eigenvalue. The hyperplane $y = 0$ is the stable manifold, but orbits starting near the equilibrium with $y(0) > 0$ tend away from the stable manifold. This does not assure survival of the invading species because the orbit could tend to another equilibrium on $y = 0$. However, if the per capita growth rate $G(\bar{x}_1, \bar{x}_2, \ldots, \bar{x}_n, 0)$ of the invading species is positive at every asymptotically stable equilibrium $(\bar{x}_1, \bar{x}_2, \ldots, \bar{x}_n, 0)$ of the n-species system (5.32) then the invading species survives. In this case, we will say that y is a survivor species. A survivor species is persistent, but as we have pointed out may not survive in the biological sense unless it is actually uniformly persistent.

In addition to using the idea of survivor species to model the effect of introducing an invading species into an existing population system, we may also try to analyze the behavior of an $(n+1)$-species system by considering each species in turn as an invader of the system consisting of the remaining n species. The basic result is that there is coexistence of the $(n+1)$-species system for all initial states if and only if every species is a survivor species. Two examples for $n = 1$ may be illuminating.

(i) Consider two species in competition with an unstable equilibrium, so that either species may win the competition depending on the initial state (Figure 5.33). Neither species is a survivor; the existence of an equilibrium with both species present is not sufficient to assure coexistence.

(ii) Consider a predator-prey system with an Allee effect for the prey. There is coexistence for some but not all initial states, but neither species is a survivor species (the predator species is a survivor at one prey equilibrium but not at the other) (Figure 5.34).

5.10 Example: A Predator and Two Competing Prey

As an example we shall now apply the ideas of Section 5.8 to a three species system consisting of a predator species z and two prey species x and y in competition with each other. Phenomena that have been observed for

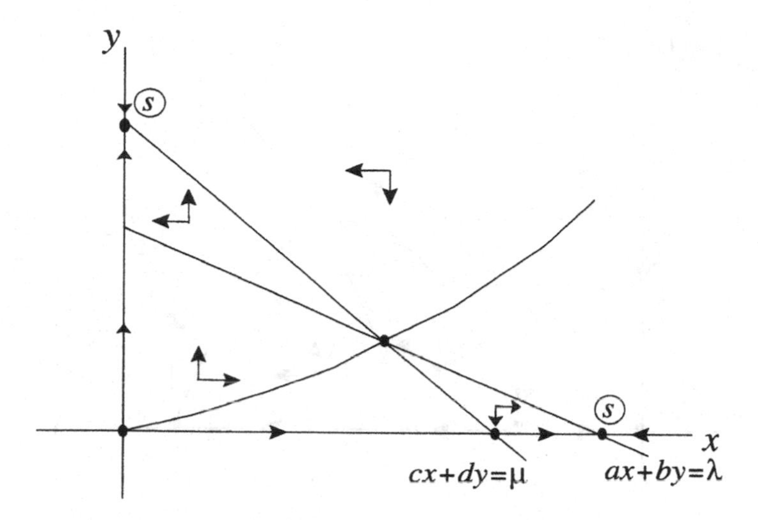

FIGURE 5.33.

such systems include predator-induced exclusion of one of the prey species [Poole (1974); Dodson (1974); Holt (1977)], exclusion of prey by alternate prey [Holt (1977)], and predator-induced coexistence [Oaten and Murdoch (1975); Paine (1966)].

In the Kolmogorov model

$$\begin{aligned} x' &= xf(x,y,z) \\ y' &= yg(x,y,z) \\ z' &= zh(x,y,z) \end{aligned} \tag{5.34}$$

we define the numbers $J_1, K_1, L_1, J_2, K_2, L_2$ by

$$\begin{aligned} f(K_1,0,0) &= 0 \qquad & f(0,L_2,0) &= 0 \\ g(0,K_2,0) &= 0 \qquad & f(L_1,0,0) &= 0 \\ h(J_1,0,0) &= 0 \qquad & h(0,J_2,0) &= 0. \end{aligned}$$

Thus, K_1 is the natural x population size in the absence of the other two species, J_1 is the minimum x population size that allows z to survive in the absence of y, and L_1 is the maximum x population size that allows y to survive unconditionally (for all initial states) in the absence of z. The numbers K_2, J_2, L_2 have analogous interpretations for the y population size. We may also define these quantities for the more explicit Rosenzweig-MacArthur model.

$$\begin{aligned} x' &= rxf(x,y) - xz\phi(x,y) \\ y' &= syg(x,y) - yz\psi(x,y) \\ z' &= z\big(\lambda x\phi(x,y) + \mu y\psi(x,y) - c\big). \end{aligned} \tag{5.35}$$

Before looking at the conditions for each species to be a survivor we must recall the properties of the three two-species subsystems.

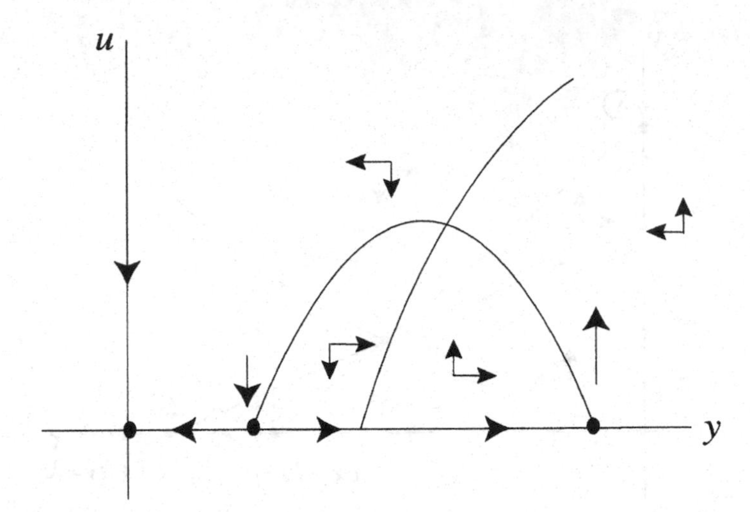

FIGURE 5.34.

(i) If $K_1 < L_1$ and $K_2 < L_2$ then x and y coexist unconditionally (written $x \sim y$) with equilibrium $(\bar{x}, \bar{y}, 0)$. If $K_1 > L_1$ and $K_2 < L_2$ we have the bi-stable case $(x \oplus y)$ with two possible equilibria, $(K_1, 0, 0)$ and $(0, K_2, 0)$. If $K_1 > L_1$ and $K_2 < L_2$ then x always wins the competition $(x \gg y)$ with equilibrium $(K_1, 0, 0)$.

(ii) If $J_1 < K_1$ then x and z coexist $(x \sim y)$ with equilibrium $(x^*, 0, z^*)$. If $J_1 > K_1$ then only x survives $(x \gg z)$ with equilibrium $(K_1, 0, 0)$.

(iii) If $J_2 < K_2$ then $y \sim z$, with equilibrium $(0, y^*, z^*)$. If $J_2 > K_2$ then $y \gg z$, with equilibrium $(0, K_2, 0)$.

The survivor conditions for (5.34) are as follows:

$$x \text{ is a survivor if } \begin{cases} f(0, y^*, z^*) > 0 & (y \sim z) \\ f(0, K_2, 0) > 0 \text{ or } K_2 < L_2 & (y \gg z) \end{cases}$$

$$y \text{ is a survivor if } \begin{cases} g(x^*, 0, z_1^*) > 0 & (x \sim z) \\ g(K_1, 0, 0) > 0 \text{ or } K_1 < L_1 & (x \gg z) \end{cases}$$

$$z \text{ is a survivor if } \begin{cases} h(\bar{x}, \bar{y}, 0) > 0 & (x \sim z) \\ h(K_1, 0, 0) > 0, h(0, K_2, 0) > 0 & \\ K_1 < J_1, K_2 > J_2 & (x \oplus y) \\ h(K_1, 0, 0) > 0 \text{ or } K_1 > J_1 & (x \gg y). \end{cases}$$

For the Rosenzweig-MacArthur model (5.35) some of these conditions may be described more explicitly. For example, if $y \sim z$ we have $y^* = J_2$, $z_2^* = sg(0, J_2)/\psi(0, J_2)$ and thus x is a survivor if $rf(0, J_2)\psi(0, J_2) - sg(0, J_2)\phi(0, J_2) > 0$. Similarly, if $x \sim z$, y is a survivor if $-rf(J_1, 0)\psi(J_1, 0) + sg(J_1, 0)\phi(J_1, 0) > 0$.

Next we use these conditions to determine whether our model will support the possibility of predator-induced exclusion, which in our terminology would be $x \sim y$ but x not a survivor-species. The assumption that $x \sim y$ is equivalent to $K_1 < L_1$ and $K_2 < L_2$. If $y \sim y$ so that $J_2 < K_2$, the condition that x not be a survivor is $f(0, y^*, z^*) < 0$. In the Rosenzweig-MacArthur model $y^* = J_2$ and $z_2^* = sg(0, J_2)/\psi(0, J_2)$, so that this condition becomes $rf(0, J_2)\psi(0, J_2) - sg(0, J_2)\psi(0, J_2) < 0$. Since $J_2 < L_2 < K_2$, $f(0, J_2) > 0$ and $g(0, J_2) > 0$. Thus if s/r is sufficiently large, x is not a survivor. If $y \gg z$, so that $J_2 > K_2$, the condition for x to be a survivor is $K_2 < L_2$, which is satisfied. Thus, it is possible to have $x \sim y, y \sim z$ with x not a survivor, which can be viewed as predator-induced exclusion. It may also be viewed as exclusion of the prey x by the alternate prey.

A more complicated analysis is needed to examine the possibility of predator-induced coexistence. In our terminology this means that x and y do not coexist unconditionally, either $x \gg y$ or $x \oplus y$, but all three species are survivors. Let us begin with the possibilities when $x \gg y$ or $x \oplus y$, but all three species are survivors. Let us begin with the possibilities when $x \gg y$ or $L_1 < K_1$ and $K_2 < L_2$.

(i) If $x \sim z$ and $y \gg z$, or $J_1 < K_1$ and $J_2 > K_2$, then x and z are survivors, and y is a survivor if $g(x^*, 0, z_1^*) > 0$, which is possible if $J_1 < L_1$. In the Rosenzweig-MacArthur case y will be a survivor if $J_1 < L_1$ and s/r are large.

(ii) If $x \sim y, y \sim z$, or $J_1 < K_1$ and $J_2 < K_2$, z is a survivor, while x is a survivor if $f(0, y^*, z_2^*) > 0$, and y is a survivor if $g(x^*, 0, z_1^*) > 0$. In the Rosenzweig-MacArthur case x is a survivor if s/r is small enough and y is a survivor if s/r is large enough. Thus, either or both competitors may survive, but for both to be survivors two conditions that may be incompatible must be satisfied.

(iii) If $x \gg z$, $y \gg z$, or $J_1 > K_1$ and $J_2 > K_2$, x is a survivor but y and z are not survivors.

(iv) If $x \gg z$, $y \sim z$, or $J_1 > K_1$ and $J_2 > K_2$, x is a survivor if $f(0, y^* z_2^*) > 0$ but y and z are not survivors.

From this we conclude that predator-induced coexistence is impossible if $x \gg y$ and $x \gg z$, but is possible if $x \gg y, x \sim z, y \gg z$, and may be possible if $x \gg y, x \sim z, y \sim z$. Further, if $x \sim z, y \sim z, x \gg y$, it is possible for the predator to reverse the outcome of the competition.

If $x \oplus y$, we have $K_2 > L_2$ in place of $K_2 < L_2$. In this case x cannot be a survivor if $x \sim z, y \gg z$, or if $x \gg z, y \gg z$ but otherwise the results are the same. The only case in which both x and y can be survivors is the case $x \oplus y, x \sim z$, and the survivor conditions for x and y could be incompatible. In the Lotka-Volterra model, $f(x, y) = 1 - x/K_1 - y/L_2, g(x, y) = 1 - xL_1 -$

$y/K_2, \phi(x,y) = c/\lambda J_1, \psi(x,y) = c/\mu J_2$, it is not difficult to show that the two survivor conditions are indeed incompatible.

Three-species coexistence is not possible in the bi-stable case [Hutson and Vickers (1983)]. However, with more complicated predator response terms it is possible to give examples in which three-species coexistence may or may not be possible, depending on the parameters of the model.

The possibilities of predator-induced exclusion of one of the prey species and predator-induced coexistence are both supported by the Rosenzweig-MacArthur model (5.35). This both vindicates the Rosenzweig-MacArthur model as an approximation to biological reality and illustrates the use of the survivor species idea to study three-dimensional systems.

5.11 Example: Two Predators Competing for Prey

We discussed the principle of competitive exclusion of Gause (1934) in Section 5.1. There we analyzed a two-dimensional model for species in competition, without going into detail as to the nature of the competition. The principle of competitive exclusion is sometimes formulated as saying that it is not possible for n species to coexist when competing for fewer than n different resources as food supply. If we consider the resources as prey species, the simplest case would be two species each acting as predator on a third species. Let us consider a three-species Rosenzweig-MacArthur model

$$\begin{aligned} x' &= sf(x) - xy_1\phi_1(x) - xy_2\phi_2(x) \\ y_1' &= y_1\bigl(c_1 x\phi_1(x) - e_1\bigr) \\ y_2' &= y_2\bigl(c_1 x\phi_2(x) - e_2\bigr) \end{aligned} \tag{5.36}$$

to describe such a situation. Here x is the population size of the common prey species and y_1 and y_2 are the population sizes of the competing predator species. For three-species coexistence to be possible, the two-species systems describing x and y_1 in the absence of y_2 and x and y_2 in the absence of y_1 must both admit coexistence. The x-y_1 system model is

$$x' = xf(x) - xy_1\phi_1(x) \quad \text{and} \quad y_1' = y_1\bigl(c_1 x\phi_1(x) - e_1\bigr) \tag{5.37}$$

and coexistence requires an equilibrium with

$$x = J_1, \text{ where } c_1 J_1\phi_1(J_1) = e_1 \tag{5.38}$$

and $y_1 = f(J_1)/\phi_1(J_1)$. Then y_2, considered as an invading species, is a survivor species if

$$c_2 J_1\phi_2(J_1) > e_2. \tag{5.39}$$

Similarly, the x-y_2 system

$$x' = xf(x) - xy_2\phi_2(x) \quad \text{and} \quad y_2' = y_2\big(c_2x\phi_2(x) - e_2\big) \tag{5.40}$$

has an equilibrium with $x = J_2$, where

$$c_2J_2\phi_2(J_2) > e_2 \tag{5.41}$$

and $y_2 = f(J_2)/\phi_2(J_2)$. Then y_1 is a survivor species if

$$c_1J_2\phi_1(J_2) > e_1. \tag{5.42}$$

Because of the standard assumption that the predator functional responses $x\phi_1(x)$ and $x\phi_2(x)$ are increasing functions, the relationships (5.38) and (5.42) imply $J_2 > J_1$, while the relationships (5.39) and (5.41) imply $J_1 > J_2$. Thus, y_1 and y_2 cannot both be survivor species, and this at least suggests the possibility that y_1 and y_2 cannot coexist.

However, an example has been constructed by R.A. Armstrong and R. McGehee (1980) of a system of the form (5.36) that does admit coexistence. The construction is indirect and quite complicated, and it is not known whether the asymptotically stable orbit of the example is periodic or a non-periodic recurrent orbit. In any case, the example violates the competitive exclusion principle.

5.12 Project 5.1: A Simple Neuron Model

Neurons are cells in the body that transmit information to the brain and the body by amplifying an incoming stimulus (electrical charge input) and transmitting it to neighboring neurons, then turning off to be ready for the next stimulus. Neurons have fast and slow mechanisms to open ion channels in response to electrical charges. The key quantities are the concentrations of sodium ions and potassium ions (both positively charged). A resting neuron has an excess of potassium and a deficit of sodium and a negative resting potential (an excess of negative ions). Neurons use changes of sodium and potassium ions across the cell membrane to amplify and transmit information. *Voltage-gated channels* exist for each kind of ion, which open and close in response to voltage differences, which are closed in a resting neuron. When a burst of positive charge enters the cell, making the potential less negative, the voltage-gated sodium channels open. Since there is an excess of sodium ions outside the cell, more sodium ions enter, increasing the potential until it eventually becomes positive. Next a slow mechanism acts to block the voltage-gated sodium channels, and another slow process begins to open voltage-gated potassium channels. Both of these diminish the buildup of positive charge by blocking sodium ions from entering and by allowing excess potassium ions to leave. When the potential decreases to or below the resting potential these slow mechanisms turn off, and then the process can start over. If the electrical excitation reaches a sufficiently high level, called an *action potential*, the neuron fires and transmits the excitations to other neurons.

In order to describe a simple model for this process, we let the potential be v, scaled so that $v = 0$ is the resting potential. We let $v = a$ be the potential above which the neuron fires and $v = 1$ the potential at which sodium channels open $(0 < a < 1)$. A model of the form

$$v' = -v(v-a)(v-1)$$

with asymptotically stable equilibria at $v = 0$ and $v = 1$ and unstable equilibrium at $v = a$ would explain part of the observed behavior. If the initial potential is above a, the potential increases to one and if the initial potential is below a, it decreases to zero. Thus, the model allows the signal amplification of the neuron but stops at $v = 1$. We must also build in a blocking mechanism.

Let w denote the strength of the blocking mechanism with $w = 0$ (turned off) when $v = 0$. As v approaches one, the blocking mechanism becomes stronger but remains bounded, and we assume an equation of the form

$$w' = \epsilon(v - \xi w)$$

with a limiting value v/ξ for w if v is fixed. If $v = 0$, $w \to 0$ and if $v = 1$, $w \to 1/\xi$ (the maximum strength of the blocking mechanism). The parameter ϵ influences the rate of approach to equilibrium but does not affect the equilibrium value. We use a small value of ϵ to indicate a slow-acting mechanism.

In order to formulate a model that includes both v and w, we must also take account of the effect of the blocking mechanism on v. The model we shall examine is the two-dimensional system

$$\begin{aligned} v' &= -v(v-a)(v-1) - w \\ w' &= \epsilon(v - \xi w) \end{aligned} \tag{5.43}$$

known as the *Fitzhugh-Nagumo* system. [Fitzhugh (1961); Nagumo, Arimato & Yoshizawa (1962)] It is a simplification of the four-dimensional *Hodgkin-Huxley* model proposed in the early 1950s by Sir Alan Hodgkin and Sir Andrew Huxley [Hodgkin and Huxley (1952)], for which they received the 1963 Nobel Prize in Physiology and Medicine, and which is still being used in the study of neurons and other kinds of cells. Excitable systems occur in a large variety of biological systems, and the Fitzhugh-Nagumo and Hodgkin-Huxley models are prototypical models of excitable systems.

Question 1
Show that the only equilibrium of the Fitzhugh-Nagumo system (5.43) is $v = 0$, $w = 0$, and that this equilibrium is asymptotically stable.

Question 2
Use a computer algebra system to draw the orbit of the system (5.43) with $a = 0.3$, $\xi = 1$, $\epsilon = 0.01$ and a starting point $(v_0, 0)$ with $v_0 > a$, say $v_0 = 0.4$. You should observe that v amplifies quickly and then returns to zero. You should also note that after the neuron fires and the potential drops, it overshoots zero (this can also be observed experimentally).

Another experiment gives the cell a constant input of positive ions instead of a single pulse. If we apply a constant electrical current J, we add J to the rate of change of potential (taking units of current so that one unit of current raises the potential by one unit in unit time). Thus, we replace the system (5.43) by

$$\begin{aligned} v' &= -v(v-a)(v-1) - w + J \\ w' &= \epsilon(v - \xi w). \end{aligned} \tag{5.44}$$

Question 3
Show that increasing J moves the equilibrium from $(0, 0)$ into the first quadrant of the (v, w)-plane. Show that this equilibrium is asymptotically stable for small values of J but becomes unstable for larger values of J.

Question 4
Use a computer algebra system to experiment with different values of J, taking $a = 0.3$, $\xi = 1$, $\epsilon = 0.01$, and find a value for which the system (5.44) has a periodic orbit.

Question 5
Use a computer algebra system to graph v as a function of t for the periodic orbit found in Question 4. You should observe a "bursting" behavior, with potential rising close to one and dropping below zero (similar to observations in real neurons).

Question 6
Add a periodic term to the equation for u in the model and experiment using a computer algebra system with different amplitudes and periods to see the effects on the behavior of the model.

6

Harvesting in two-species models

6.1 Harvesting of species in competition

The topics in this chapter are part of the subject of natural resource management and bioeconomics. This is an important and rapidly-developing subject. The classical reference is the book by Clark (1990), where additional references may be found.

In Section 1.5 we examined the effects of both constant-yield harvesting and constant-effort harvesting on single species populations modelled by an ordinary differential equation. In this section we shall study the effects of harvesting one of a pair of interacting species in competition.

In studying models for competition between two species (Section 5.1), we began by assuming linear per capita growth rates, obtaining a system

$$\begin{aligned} x' &= x(\lambda - ax - by) \\ y' &= y(\mu - cx - dy) \end{aligned} \tag{6.1}$$

We observed that there are four cases depending on the relation between the x-isocline $ax + by = \lambda$ on which $x' = 0$ and the y-isocline $cx + dy = \mu$ on which $y' = 0$. These cases may be depicted graphically in Figure 6.1, with equilibria marked $\times$ and asymptotically stable equilibria marked $\bullet$.

We will consider only the harvesting of one of the two species, say the x-species. For constant-effort harvesting, the model would be

$$\begin{aligned} x' &= x(\lambda - ax - by) - Ex \\ y' &= y(\mu - cx - dy). \end{aligned}$$

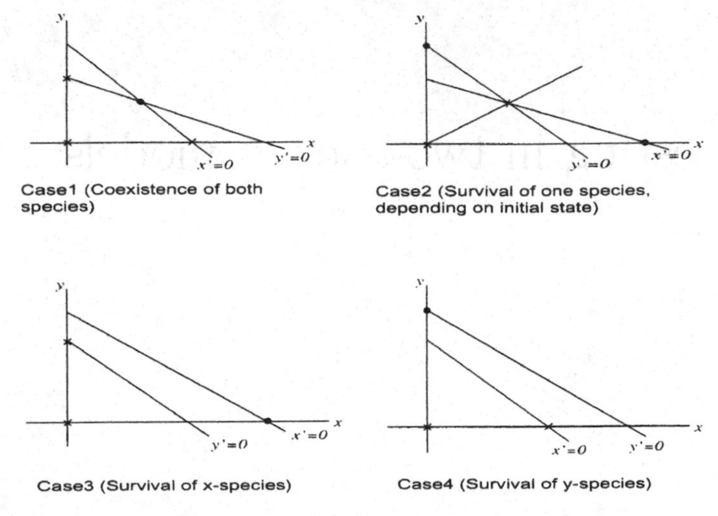

FIGURE 6.1.

This is the same as the unharvested model but with λ replaced by $\lambda - E$, or with the x-isocline moved parallel to itself towards the origin. It has boundary equilibria $(K(E), 0)$ and $(0, M)$ with $M = \frac{\mu}{d}$ and $K(E) = \frac{\lambda}{a+E}$. Examination of the four cases shows that it is possible by harvesting to change Case 1 to Case 4 (coexistence to x-extinction), Case 2 to Case 4 (survival of either species alone to x-extinction), or Case 3 to Case 1 or Case 2 and then to Case 4 (y-extinction to x-extinction via coexistence). If the growth rates are not linear, the same type of argument can be used to see how the isoclines move under harvesting and obtain the same results, all of which are in accordance with our intuition.

With constant-yield harvesting, the model is

$$x' = x(\lambda - ax - by) - H, y' = y(\mu - cx - dy).$$

The x-isocline, instead of being the pair of lines $x = 0$, $ax + by = \lambda$, is now the curve $x(\lambda - ax - by) = H$, which is a hyperbola having the lines $x = 0$ and $ax + by = \lambda$ as asymptotes and which moves away from these asymptotes as H increases. The effect of harvesting on the system can be depicted graphically by sketching these hyperbolas; the results are somewhat different in the four cases.

Case 1, the asymptotically stable (coexistence) equilibrium moves up and to the left and the saddle point moves from $(0, M)$ down and to the right (see Figure 6.2). The stable separatrices at this saddle point move from the y-axis into the first quadrant, and divide this quadrant into a coexistence region (initial states which tend to the asymptotically stable equilibrium) and a region for which the x-species becomes extinct in finite time. Since $x = 0$ is not a solution of the harvested system, it is possible for orbits to cross the y-axis, corresponding to extinction of the x-species in finite time. As the hyperbola moves down, the saddle point and asymptotically stable equilibrium coalesce and disappear, producing a catastrophe. For harvest rates larger than the critical rate at which this coalescence occurs, the x-species becomes extinct in finite time for all initial states.

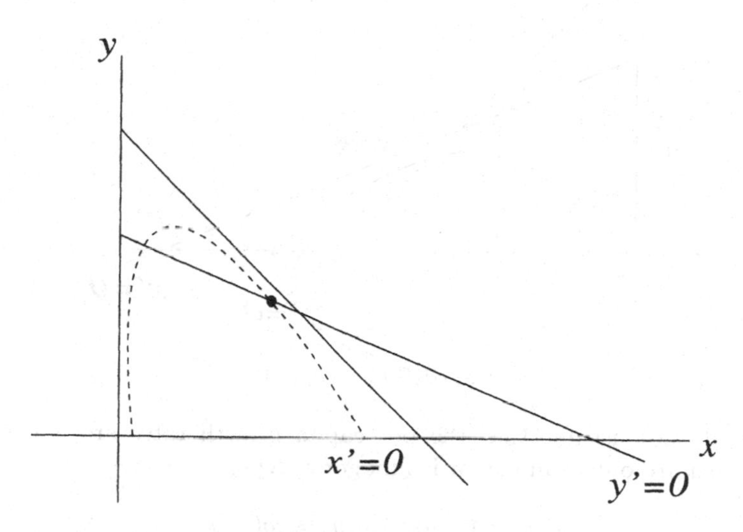

FIGURE 6.2. *Case 1*

Case 2, the asymptotically stable equilibrium at $(K, 0)$ moves to the left along the x-axis and the asymptotically stable equilibrium at $(0, M)$ moves into the second quadrant, corresponding to extinction of the x-species in finite time (see Figure 6.3). The saddle point at (x_∞, y_∞) moves down and to the right, remaining a saddle point, and its stable separatrices divide the first quadrant into a region of x-extinction in finite time and a region of ultimate y-extinction. As the hyperbola moves down with increasing H, the equilibrium at (x_∞, y_∞) coalesces either with the equilibrium at $(0, 0)$ or the equilibrium at $(K, 0)$ and disappears; then the two remaining equilibria on the x-axis coalesce and disappear. The second coalescence signifies x-extinction for all initial states.

Case 3, the asymptotically stable equilibrium moves from $(K, 0)$ along the x-axis. If $ad - bc > 0$, when the hyperbola meets the line $cx + dy = \mu$ this equilibrium moves into the interior of the first quadrant, giving coexistence (see Figure 6.4(A)). The other equilibrium is a saddle point whose stable separatrices give a division into a coexistence region and a region of extinction of the x-species in finite time (see Figure 6.4(B)). When the two equilibria coalesce and disappear, we have extinction of the x-species in finite time for all initial states.

Case 4, the asymptotically stable equilibrium moves from $(0, M)$ into the second quadrant, giving extinction of the x-species in finite time for all initial states (see Figure 6.5).

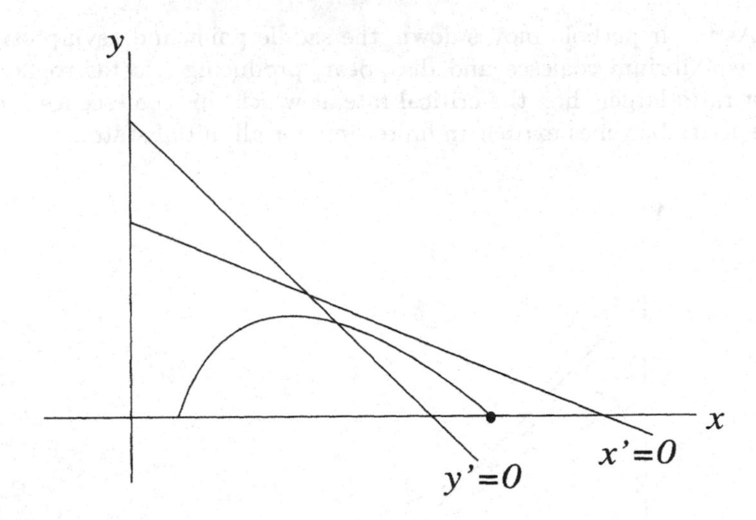

FIGURE 6.3. *Case 2*

If we do not assume that the per capita growth rates are linear, but instead require only a model of Kolmogorov type,

$$x' = xf(x, y), \qquad y' = yg(x, y),$$

with $f_x(x, y) < 0, f_y(x, y) < 0, g_x(x, y) < 0, g_y(x, y) < 0$ for $x > 0, y > 0$, then the isoclines are curves with negative slope but are not necessarily straight lines. The x-isocline under constant-yield harvesting is no longer necessarily a hyperbola, but it does approach the line $x = 0$ and the curve $f(x, y) = 0$ asymptotically. The same cases as those described for linear growth rates may occur, and the qualitative responses to harvesting are the same.

Example 1: Determine the response of the system

$$\begin{aligned} x' &= x(100 - 4x - y) \\ y' &= y(60 - x - 2y) \end{aligned}$$

to constant-effort harvesting of the x-species.

Solution: With no harvesting, there is an asymptotically stable equilibrium at $(20, 20)$ [Example 1, Section 5.1]. A harvested system has the form

$$\begin{aligned} x' &= x\big[(100 - E) - 4x - y\big] \\ y' &= y(60 - x - 2y). \end{aligned}$$

Equilibria are given by the pair of equations

$$4x + y = 100 - E, \quad x + 2y = 60$$

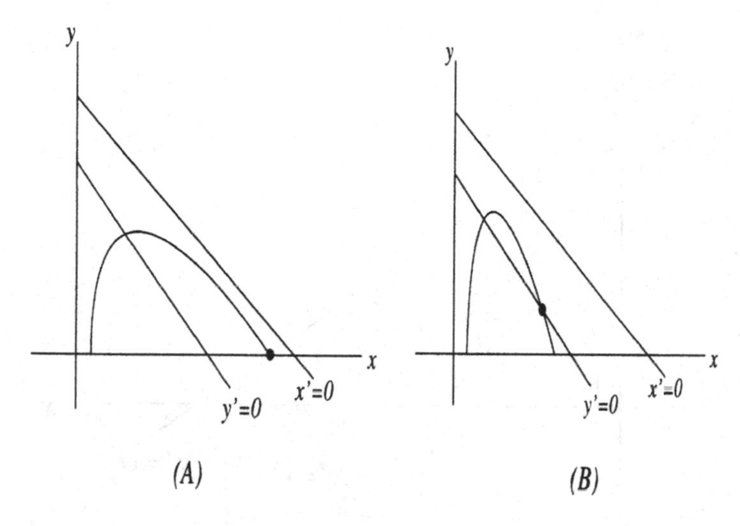

FIGURE 6.4. *Case 3*

and elimination gives the solution

$$x = 20 - \frac{2}{7}E, \quad y = 20 + \frac{E}{7}.$$

When $E = 70$, x reaches zero, and $y = 30$. Thus harvesting decreases the x-population size and increases the y-population size, eventually moving coexistence to x-extinction.

Example 2: Determine the response of the system

$$\begin{aligned} x' &= x(100 - 4x - y) \\ y' &= y(60 - x - 2y) \end{aligned}$$

to constant-yield harvesting of the x-species.

Solution: With no harvesting, there is an asymptotically stable equilibrium at $(20, 20)$ [Example 1, Section 5.1]. A harvested system has the form

$$\begin{aligned} x' &= x(100 - 4x - y) - H \\ y' &= y(60 - x - 2y). \end{aligned}$$

Equilibria are given by the pair of equations

$$x(100 - 4x - y) = H, \quad x + 2y = 60.$$

Replacing x by $60 - 2y$ in the first of these equations, we obtain a quadratic equation for y

$$\begin{aligned} (60 - 2y)\left[100 - 4(60 - 2y) - 2y\right] &= H, \\ 14(y - 30)(y - 20) + H &= 0, \\ 14y^2 - 700y + (8400 + H) &= 0. \end{aligned}$$

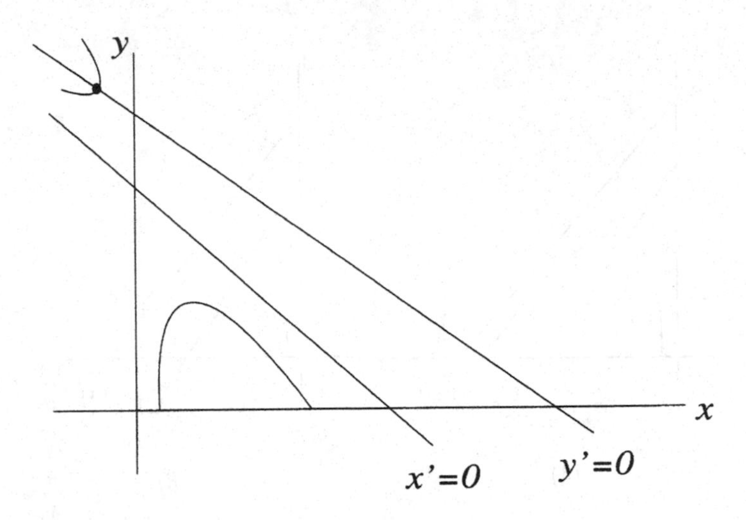

FIGURE 6.5. *Case 4*

The roots of this equation may be found using the quadratic formula:

$$y = 25 \pm \tfrac{1}{2}\sqrt{100 - \tfrac{2H}{7}}.$$

For $H = 0$, these roots are $y = 20$ (which gives $x = 20$) and $y = 30$ (which gives $x = 0$). There is an asymptotically stable equilibrium at $(20, 20)$ and a saddle point at $(0, 30)$. As H increases, these equilibria move along the line $x + 2y = 60$ until they coalesce at $(10, 25)$ for $H = 350$.

Exercises

In each of Exercises 1-4, determine the response of the system to constant-effort harvesting of the x-species.

1. $x' = x(80 - x - y), \quad y' = y(120 - x - 3y)$ [cf. Exercise 1, Section 5.1]
2. $x' = x(60 - 3x - y), \quad y' = y(75 - 4x - y)$ [cf. Exercise 2, Section 5.1]
3. $x' = x(40 - x - y), \quad y' = y(90 - x - 2y)$ [cf. Exercise 3, Section 5.1]
4. $x' = x(80 - 3x - 2y), \quad y' = y(80 - x - y)$ [cf. Exercise 4, Section 5.1]

   In each of Exercises 5-8, determine the response of the system to constant-yield harvesting of the x-species.

5. $x' = x(80 - x - y), \quad y' = y(120 - x - 3y)$ [cf. Exercise 1, Section 5.1]
6. $x' = x(60 - 3x - y), \quad y' = y(75 - 4x - y)$ [cf. Exercise 2, Section 5.1]
7. $x' = x(40 - x - y), \quad y' = y(90 - x - 2y)$ [cf. Exercise 3, Section 5.1]
8. $x' = x(80 - 3x - 2y), \quad y' = y(80 - x - y)$ [cf. Exercise 4, Section 5.1]

6.2 Harvesting of Predator-Prey Systems

Predator-prey systems are most commonly used to describe the interaction of a species with its food supply. In other words, the primary interest is in the predator population size and the prey population size is of interest only because of its effect on the predator growth rate. A predator-prey model may be viewed as a refinement of a single-species model through more realistic description of the limitations on the growth rate imposed by resource limitations. Thus we will study only the harvesting of predators. It would be possible, however, to use an analogous approach to study the harvesting of prey and this would be appropriate in an examination of the extent to which one can control a population by tampering with its food supply.

Constant-effort harvesting: For a system modelled by

$$x' = xf(x) - xy\phi(x), \qquad y' = y(cx\phi(x) - e)$$

(Rosenzweig-MacArthur model), the harvested system is modelled by

$$x' = xf(x) - yx\phi(x), \qquad y' = y(cx\phi(x) - e - E).$$

The effect of harvesting is to move the predator isocline $x = J$, where $cJ\phi(J) = e$, to the right. The same applies the more general (Kolmogorov type) model

$$x' = xf(x,y), \qquad y' = y(g(x) - E)$$

with the predator per capita growth rate independent of predator population size. If the prey isocline $y = \frac{f(x)}{\phi(x)}$ (as in the Rosenzweig-MacArthur model) or $f(x,y) = 0$ (as in the Kolmogorov model) is concave downwards and has a maximum, an equilibrium which is unstable for $E = 0$ is stabilized by harvesting as the harvested predator isocline moves past the maximum of the prey isocline (see Figure 6.6).

For a given model, the effort which maximizes yield may be calculated by solving

$$\frac{d}{dE}(Ey_\infty) = E\frac{dy_\infty}{dE} + y_\infty = 0$$

for E and then calculating $Ey_\infty(E)$ for this value of E. Alternately, one would proceed as follows for a Kolmogorov model. For a given effort E, the equilibrium conditions are $f(x_\infty(E), y_\infty(E)) = 0$, $g(x_\infty(E)) = E$ and the yield is $Ey_\infty(E) = y_\infty(E)g(x_\infty(E))$. The maximum yield is thus the maximum of the function $yg(x)$ subject to the constraint $f(x,y) = 0$. So long as $E < cK\phi(K) - e$ where $f(K) = 0$ (Rosenzweig-MacArthur model) or $E < g(K)$ where $f(K,0) = 0$ (Kolmogorov model), $y_\infty(E) > 0$ and the equilibrium depends continuously on E. For $E \geq cK\phi(K) - e$ or $E \geq g(K)$

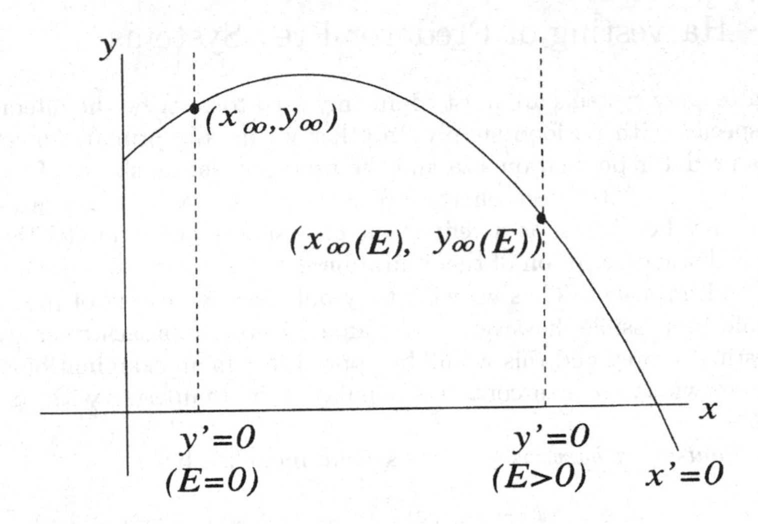

FIGURE 6.6.

in the respective models, we have $y_\infty(E) = 0$, corresponding to the ultimate extinction of predators.

Constant-yield Harvesting: For a Rosenzweig-MacArthur model the harvested system is

$$x' = xf(x) - xy\phi(x), \qquad y' = y(cx\phi(x) - e) - H.$$

The predator isocline $cx\phi(x) = e$ is replaced by the curve $y(cx\phi(x) - e) = H$ to the right of $cx\phi(x) = e$ and having $cx\phi(x) = e$ as a vertical asymptote (since $cx\phi(x) - e \to 0$ as $y \to \infty$) and a horizontal asymptote $y = \frac{H}{c\lim_{x\to\infty} x\phi(x) - e}$ $\left(\text{since } y \to \frac{H}{c\lim_{x\to\infty} x\phi(x) - e} \text{ as } x \to \infty\right)$. It is no more difficult to study the more general system

$$x' = xf(x, y), \qquad y' = yg(x),$$

or indeed the general Kolmogorov model

$$x' = xf(x, y), \qquad y' = yg(x, y),$$

with harvested system

$$x' = xf(x, y), \qquad y' = yg(x, y) - H,$$

where $f_y(x, y) < 0, g_x(x, y) > 0, g_y(x, y) \leq 0, f(K, 0) = 0, g(J, 0) = 0$; we will usually think in terms of the special case where g is independent of y. As H increases, the "hyperbola-like" curve $y(cx\phi(x) - e) = H$ or $yg(x, y) = H$ moves away from the line $x = J$ or the curve $g(x, y) = 0$ and upwards to the right.

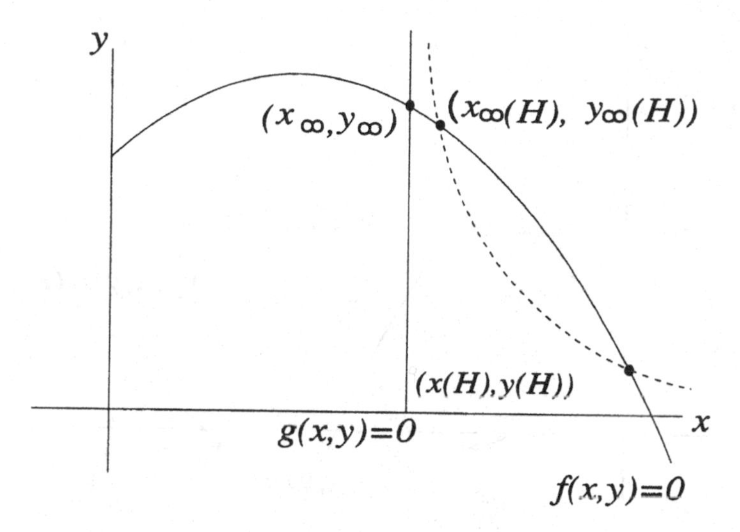

FIGURE 6.7.

It intersects the prey isocline $f(x,y) = 0$ at two points–$(x_\infty(H), y_\infty(H)$ coming from (x_∞, y_∞) when $H = 0$ and $\big(\xi(H), y(H)\big)$ coming from $(K, 0)$ when $H = 0$ (Figure 6.7). As H increases, these two equilibria move together, coalesce, and disappear. The value of H for which they coalesce is the value of for which the level curve $yg(x,y) = H$ of the function $yg(x,y)$ is tangent to the prey isocline. This is the largest value of the function $yg(x,y)$ attained on the curve $f(x,y) = 0$; we define

$$H_c = \max_{f(x,y)=0} yg(x,y).$$

Then for $H > H_c$ there are no equilibria in the first quadrant (Figure 6.8).

If $H > 0$, then $y' < 0$ on the x-axis. This means that solutions can reach and cross the x-axis in finite time (predator extinction). Indeed, if $H > H_c$ this occurs for all solutions, independent of initial values.

It is possible to prove that no solution can be unbounded; the proof is similar to the proof in the case $H = 0$ but the details are more complicated. Thus a solution either tends to an equilibrium, or tends to a limit cycle, or reaches the x-axis in finite time. If a solution reaches the x-axis in finite time, we consider the system to have collapsed.

To study the stability of any equilibrium (x_0, y_0) we look at the community matrix

$$A(x,y) = \begin{pmatrix} x_0 f_x(x_0,y_0) & x_0 f_y(x_0,y_0) \\ y_0 g_x(x_0,y_0) & g(x_0,y_0) + y_0 g_y(x_0,y_0) \end{pmatrix}.$$

If $\det A(x_0,y_0) < 0$ then (x_0,y_0) is a saddle point and if $\det A(x_0,y_0) > 0$ then (x_0,y_0) is a node or spiral point which is unstable if $\operatorname{tr} A(x_0,y_0) > 0$ and asymptotically stable if $\operatorname{tr} A(x_0,y_0) < 0$.

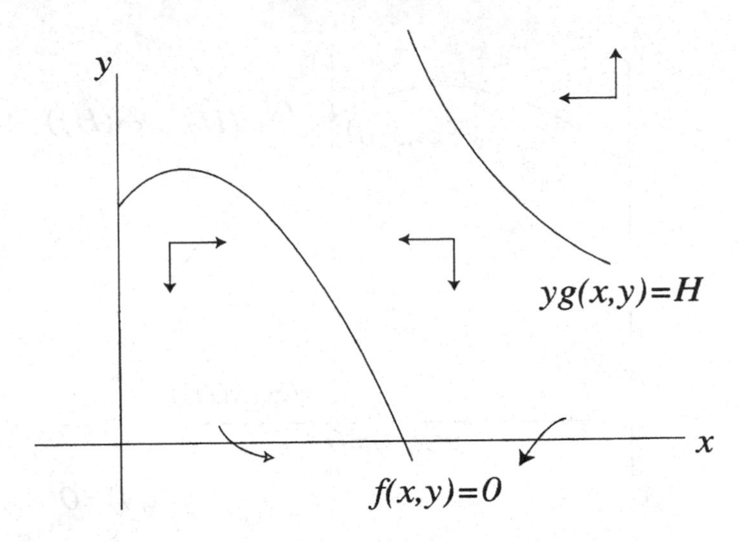

FIGURE 6.8.

The slope of the prey isocline $f(x,y) = 0$ at (x_0, y_0) is $-\frac{f_x(x_0,y_0)}{f_y(x_0,y_0)}$, and the slope of the (harvested) predator isocline $yg(x,y) = H$ at (x_0, y_0) is $-\frac{y_0 g_x(x_0,y_0)}{y_0 g_y(x_0,y_0)+g(x_0,y_0)}$. If we assume $g(x_0, y_0) + y_0 g_y(x_0, y_0) \geq 0$, which is certainly true if $g(x, y)$ is independent of y, and use the hypothesis $f_y(x_0, y_0) < 0$, we see that the slope of the prey isocline is less than the slope of the predator isocline if and only if $f_x(yg_y + g) < yg_x f_y$, or

$$y_0\big(f_x(x_0, y_0)g_y(x_0, y_0) - f_y(x_0, y_0)g_x(x_0, y_0)\big) + g(x_0, y_0)f_x(x_0, y_0) < 0.$$

Because

$$\begin{aligned} \det A(x_0, y_0) &= x_0 y_0\big(f_x(x_0, y_0)g_y(x_0, y_0) - f_y(x_0, y_0)g_x(x_0, y_0)\big) \\ &\quad + g(x_0, y_0)f_x(x_0, y_0), \end{aligned}$$

(x_0, y_0) is a saddle point if and only if the slope of the prey isocline at (x_0, y_0) is less than the slope of the predator isocline at (x_0, y_0). Thus $(\xi(H), y(H))$ is a saddle point and $(x_\infty(H), y_\infty(H))$ is not a saddle point. If $(x_\infty(H), y_\infty(H))$ is asymptotically stable, some orbits tend to it, and if $(x_\infty(H), y_\infty(H))$ is unstable, then either there is a limit cycle to which some orbits tend or every orbit reaches the x-axis in finite time. We define the region of coexistence to be the set of initial population sizes $(x(0), y(0))$ for which the orbit tends either to the equilibrium $(x_\infty(H), y_\infty(H))$ or to a limit cycle around this equilibrium as $E \to \infty$. We know that if $H = 0$, the region of coexistence is the whole first quadrant of the x-y plane. If $H > 0$, the region of coexistence is smaller and depends on the separating orbits at the saddle point $(\xi(H), \eta(H))$. There is one stable separatrix in the region $x' < 0, y' > 0$, and there is one unstable separatrix in the region $x' > 0, y' < 0$ (Figure 6.9).

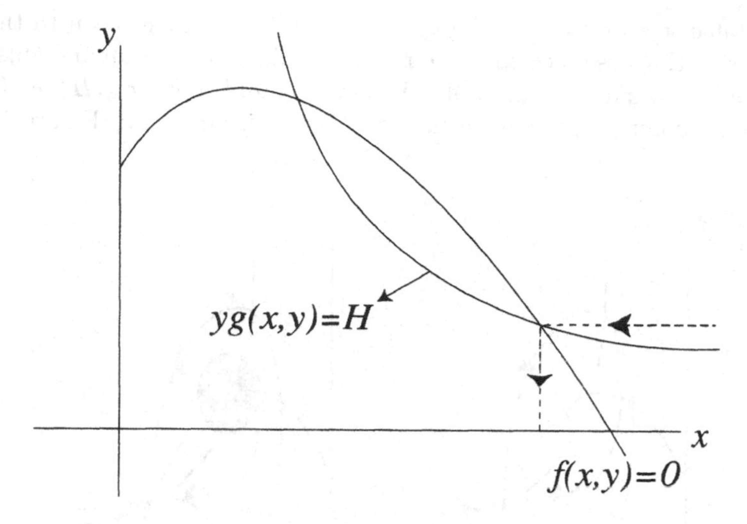

FIGURE 6.9.

There are two possible cases for the other two separatrices:
"Good case": There is a stable separatrix from $x \to +\infty$ around $\big(x_\infty(H), y_\infty(H)\big)$ to $\big(\xi(H), \eta(H)\big)$ and an unstable separatrix from $\big(\xi(H), \eta(H)\big)$ to $\big(x_\infty(H), y_\infty(H)\big)$ or a limit cycle around $\big(x_\infty(H), y_\infty(H)\big)$. The coexistence region is bounded by the two stable separatrices, and orbits starting in this region tend to $\big(x_\infty(H), y_\infty(H)\big)$ or to a limit cycle around $\big([x_\infty(H), y_\infty(H)\big)$ (Figure 6.10).

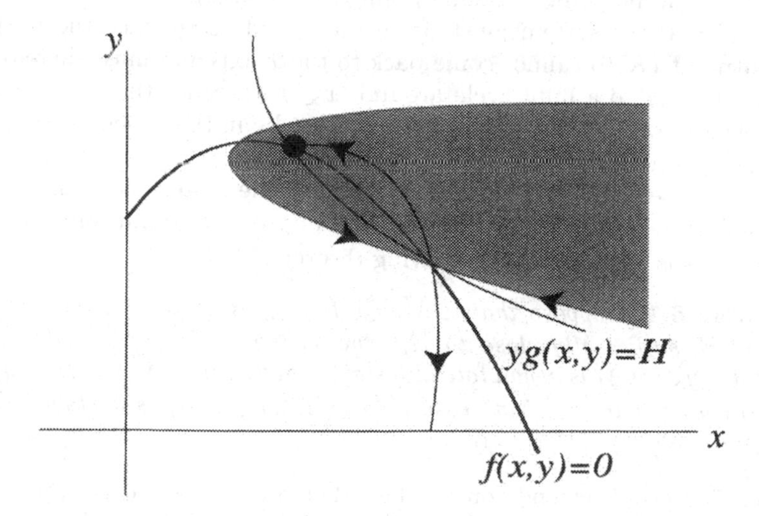

FIGURE 6.10.

"Bad case": There is a stable separatrix from the equilibrium $\big(x_\infty(H), y_\infty(H)\big)$ or from a periodic solution around $\big(x_\infty(H), y_\infty(H)\big)$ and

an unstable separatrix around $(x_\infty(H), y_\infty(H))$ and then down to the x-axis. Either there is no coexistence region at all, or there is an unstable periodic orbit around the asymptotically stable equilibrium $(x_\infty(H), y_\infty(H))$ and the coexistence region is the interior of this periodic orbit (Figure 6.11).

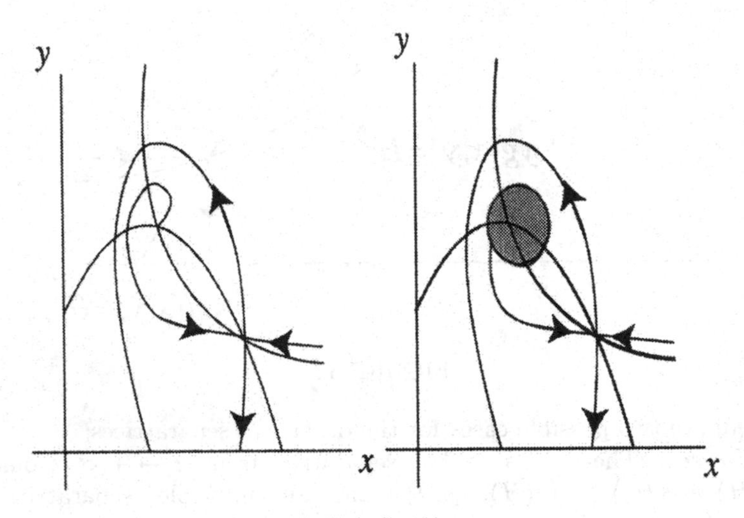

FIGURE 6.11.

It is possible to have a transition from "good" to "bad" case as H increases. If this occurs, then at the transition there must be a homoclinic orbit from saddle point to saddle point (Figure 6.12).

For $H = 0$ the system must be in the "good" case since the unstable separatrix at $(K, 0)$ cannot come back to the x-axis and must therefore go to (x_∞, y_∞) or to a limit cycle around (x_∞, y_∞). Since the orbits depend continuously on the system, the system must be in the "good" case if H is sufficiently small.

For H close to H_c we can decide whether the system is in the "good" or "bad" case by looking at the local stability of the equilibrium $(x_\infty(H), y_\infty(H))$, as is shown by the following theorem.

Theorem 6.1. *Suppose that* $\operatorname{tr} A(x_\infty(H_c), y_\infty(H_c)) \neq 0$. *Then for* $H < H_c$ *but* H *sufficiently close to* H_c, *the system is in the "good" case if* $(x_\infty(H_c), y_\infty(H_c))$ *is asymptotically stable, that is, if* $\operatorname{tr} A\,(x_\infty(H_c), y_\infty(H_c)) < 0$ *and it is in the "bad" case if* $(x_\infty(H_c), y_\infty(H_c))$ *is unstable, that is if* $\operatorname{tr} A(x_\infty(H_c), y_\infty(H_c)) > 0$.

Proof. The proof depends on the fact that for H close to H_c there must be a (full) orbit connecting $(x_\infty(H), y_\infty(H))$ and $(\xi(H), \eta(H))$ as t runs from $-\infty$ to $+\infty$. This implies that there cannot be a closed orbit around $(x_\infty(H), y_\infty(H))$ and thus rules out the "good" unstable and "bad" asymptotically stable cases.

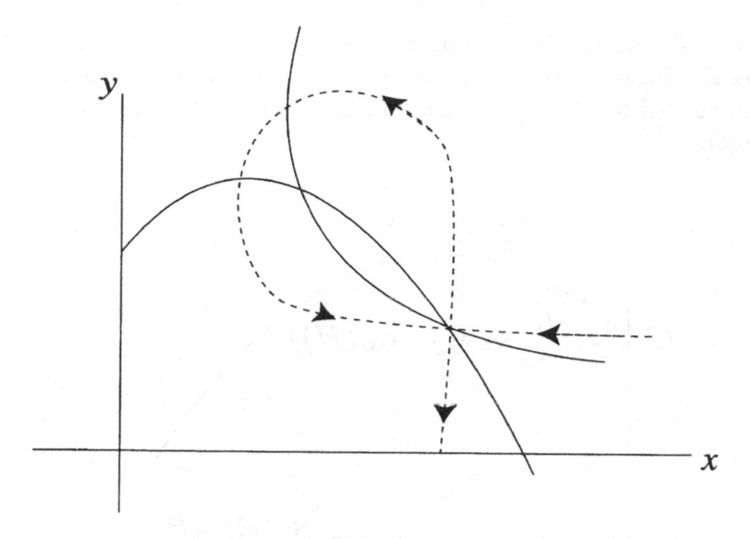

FIGURE 6.12.

The details of the proof of the existence of a connecting orbit may be omitted without serious loss. To prove the existence of a connecting orbit, we proceed as follows. For $H < H_c$ but H sufficiently close to H_c, the continuity of $\operatorname{tr} A(x_0, y_0)$ in (x_0, y_0) and of $\big(x_\infty(H), y_\infty(H)\big)$ in H imply that $\operatorname{tr} A\big(x_\infty(H), y_\infty(H)\big) \neq 0$. Again by continuity, there is a neighborhood of $\big(x_\infty(H), y_\infty(H)\big)$ in which $\operatorname{tr} A(x, y) \neq 0$. We take a point C on $f(x, y) = 0$ above $\big(x_\infty(H), y_\infty\infty(H)\big)$ but close to $\big(x_\infty(H), y_\infty(H)\big)$ and construct the orbit of $x' = xf(x, y), y' = yg(x, y) - H$ backwards in time until it meets $yg(x, y) = H$ in a point B. We may assume that B is close to $\big(x_\infty(H), y_\infty(H)\big)$ for if it is not we can take a point B' on $yg(x, y) = H$ close to $\big(x_\infty(H), y_\infty(H)\big)$ and construct the orbit forwards in time until it meets $f(x, y) = 0$ closer to $\big(x_\infty(H), y_\infty\infty(H)\big)$ than C, and use B' and C' in place of B and C. Now continue the orbit through B backwards in time to a point A below and to the right of the saddle point $\big(\xi(H), \eta(H)\big)$ and above the curve $yg(x, y) = H$. This is possible unless the orbit meets $f(x, y) = 0$ between $\big(x_\infty(H), y_\infty(H)\big)$ and $\big(\xi(H), \eta(H)\big)$, in which case we choose H larger so that $\big(\xi(H), \eta(H)\big)$ moves up to the point on $f(x, y) = 0$ where the orbit meets it; increasing H reduces y' and since $x' < 0$, this flattens the orbit and moves the intersection of the orbit with $f(x, y) = 0$ down below $\big(\xi(H), \eta(H)\big)$. Thus we may assume that A is below and to the right of the saddle point. By an argument similar to that used to show that B may be assumed close to $\big(x_\infty(H), y_\infty(H)\big)$, we may assume that A is close to $\big(\xi(H), \eta(H)\big)$. Now draw a vertical line through C and a horizontal line through A, meeting in D. For H sufficiently close to H_c, the points $\big(x_\infty(H), y_\infty(H)\big)$ and $\big(\xi(H), \eta(H)\big)$ are close together and we may assume that in the region $\mathcal{N}$ bounded by the arc ABC, the vertical line segment CD and the horizontal line segment DA we have $\operatorname{tr} A(x,y) \neq 0$. By the Bendixson criterion there can be no closed orbit contained in $\mathcal{N}$, and no homoclinic orbit from the saddle point $(\xi(H), \eta(H))$ contained in $\mathcal{N}$ (Figure 6.13).

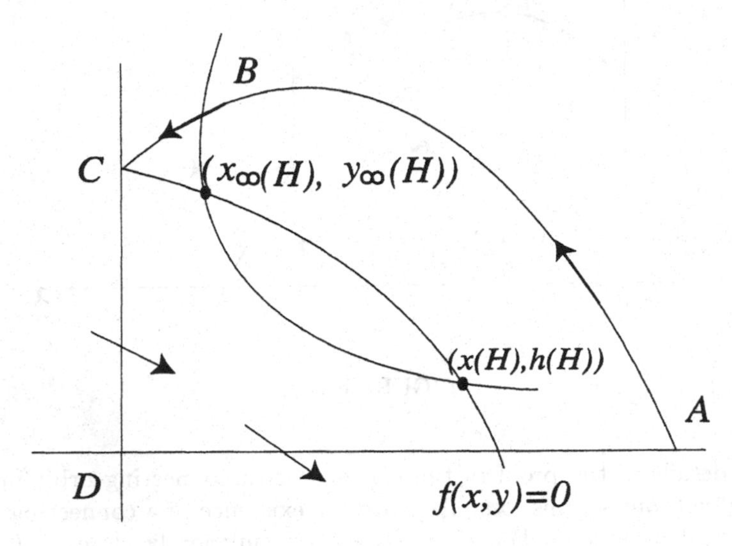

FIGURE 6.13.

By the Poincaré-Bendixson theorem, a nonconstant orbit which remains in $\mathcal{N}$ for $-\infty < t < \infty$ must have the equilibrium $(x_\infty(H), y_\infty(H))$ and the saddle point $(\xi(H), \eta(H))$ as its limiting sets for $t \to \pm\infty$ and must therefore be a connecting orbit. Thus if there is no connecting orbit, then every orbit starting in $\mathcal{N}$ must leave $\mathcal{N}$. We know that one of the unstable separatrices at the saddle point leaves $\mathcal{N}$ (downward). If the other one also leaves $\mathcal{N}$ as $t \to \infty$, it must do so across the line segment DA. But then there is an asymptotically stable separatrix at the saddle point between these two unstable separatrices which can not leave $\mathcal{N}$ as $t \to -\infty$ because it could leave $\mathcal{N}$ only across the line segment DA and such a crossing would be in the wrong direction. Thus there must be a separatrix which remains in $\mathcal{N}$ for $-\infty < t < \infty$, and this is the desired connecting orbit. We have now established Theorem 6.1. □

The value H_c is the maximum harvest rate for which coexistence is possible. However, if the system is in the "bad" case (unstable) at H_c, there must in fact be a value $H^* < H_c$ such that coexistence is possible only for $H \leq H^*$. Then H^* is the true maximum harvest rate. The value H_c may be calculated by looking for equilibria, but to find H^* it is necessary to compute orbits near the saddle point and determine whether as $t \to \infty$ they are unbounded ("good") or go to $(x_\infty(H), y_\infty(H))$ ("bad").

Example 1: Consider the model of Rosenzweig-MacArthur type

$$\begin{aligned} x' &= rx\Big(1-\frac{x}{K}\Big) - \frac{cxy}{x+A} \\ y' &= sy\Big(\frac{x}{x+A} - \frac{J}{J+A}\Big) - H = \frac{sA}{J+A}\cdot\frac{x-J}{x+A} - H, \end{aligned}$$

in which the prey satisfies a logistic equation in the absence of predators and $\frac{cx}{x+A}$ is the *predator functional response*-the rate at which predators remove prey. Here, A represents a prey population level at which the predators' attack capability begins to saturate, K is the equilibrium population size for prey in the absence of predators, and J is the prey population size in equilibrium with predators without harvesting. For this model we may calculate that

$$\begin{aligned} x_\infty(H) &= \frac{1}{2}\Big((K+J) - \sqrt{(K-J)^2 - \frac{4K(J+A)}{rsA}H}\Big) \\ y_\infty(H) &= \frac{r}{K}\big(x_\infty(H)+A\big)\big(K - x_\infty(H)\big). \end{aligned}$$

In particular, $x_\infty(0) = J$ and $x_\infty(H_c) = \frac{1}{2}(K+J)$, with $H_c = \frac{rsA(K-J)^2}{4K(J+A)}$. An equilibrium at $\big(x_\infty(H), y_\infty(H)\big)$ is asymptotically stable if and only if

$$sAK\big(x_\infty(H) - J\big) < rx_\infty(H)(J+A)(2x_\infty(H) + A - K).$$

For example, if $A = 10, J = 20, K = 40, c = 1, r = 1, s = 1$, we have $H_c = 0.833$. Calculation shows that the equilibrium is asymptotically stable for $0 \leq H < H_c$, and numerical simulation of orbits shows that the system is in the "good" case for $0 \leq H < H_c$.

If $A = 10, J = 20, K = 40, c = 1, r = 1, s = 7$, we have $H_c = 5.833$. Calculation shows that the equilibrium is asymptotically stable for $0 \leq H < 5.185$, and simulation shows a transition from "good" to "bad" case at $H = 5.185$. Thus $H^* = 5.185$, the effective maximum harvest.

If $A = 10, J = 20, K = 45, c = 1, r = 1, s = 5$, we have $H_c = 5.787$. Calculation shows that the equilibrium is asymptotically stable if $0 \leq H < 3.704$ or $5.556 < H < H_c$. Thus if $3.704 < H < 5.556$ there is an asymptotically stable limit cycle; otherwise there is an asymptotically stable equilibrium.

If $A = 10, J = 20, K = 60, c = 1, r = 2, s = 1$, we have $H_c = 4.444$. Calculation shows that the equilibrium is unstable for $0 \leq H < 2.322$ and asymptotically stable for $2.322 < H < H_c$. Thus there is an asymptotically stable limit cycle for small H which disappears when $H = 2.322$ and is replaced by an asymptotically stable equilibrium.

A final example is $A = 10, J = 20, K = 60, c = 1, r = 1, s = 4$, for which $H_c = 8.889$. Calculation shows that the equilibrium is unstable for all H and simulation indicates a transition from "good" to "bad" cases for

$H = 0.785$. Thus there is an asymptotically stable limit cycle for $0 \leq H < 0.785$ and predator extinction for all harvest rates greater than $H^* = 0.785$. Observe that the true maximum harvest rate is much less than the critical harvest rate H_c which can be calculated from equilibrium analysis.

Exercises

In each of Exercises 1-5, find the critical harvest rate of the system 6.2 with the given parameter values, find the range of harvest rates for which there is an asymptotically stable equilibrium, and carry out numerical simulations using a computer algebra system to describe the behavior of the system.

1. $A = 10$, $J = 20$, $K = 40$, $c = 1$, $r = 1$, $s = 1$
2. $A = 10$, $J = 20$, $K = 40$, $c = 1$, $r = 1$, $s = 7$
3. $A = 10$, $J = 20$, $K = 45$, $c = 1$, $r = 1$, $s = 5$
4. $A = 10$, $J = 20$, $K = 60$, $c = 1$, $r = 2$, $s = 1$
5. $A = 10$, $J = 20$, $K = 60$, $c = 1$, $r = 1$, $s = 4$

6.3 Intermittent Harvesting of Predator-Prey Systems

The harvesting of predators in a predator-prey system at a rate u which may depend on time is modelled by the system

$$x' = xf(x,y), \qquad y' = yg(x,y) - u(t).$$

Let us assume that the harvest rate is constrained by a condition $0 \leq u(t) \leq H$. Our goal is to determine a harvesting policy, or choice of u as a function of t, which will maximize the long-term average yield from harvesting,

$$Y = \lim_{t\to\infty} \frac{1}{t} \int_0^t u(\tau)d\tau,$$

but we will also require that u be chosen in such a way as to avoid the collapse of the system by demanding that $x(t) > 0$, $y(t) > 0$ for $0 \leq t < \infty$.

The problem which we have formulated is an *optimal control* problem with an objective function $Y[u]$ which is linear in the control variable. It is an immediate consequence of the general theory of optimal control, specifically the Pontrjagin maximum principle [Pontrjagrin, et al (1962)] that the optimal control must be a "bang-bang" control, that is, a control which switches between the values $u = 0$ and $u = H$, possibly combined with some time intervals on which the control is "singular."

If an optimal control makes only a finite number of switches between the values $u = 0$ and $u = H$, then we must have $u = H$ for all large t and this implies that the yield Y is equal to H. As we have seen in the preceding section, there is a value $H_c = \max_{f(x,y)=0} yg(x,y)$ with the property that if $H > H_c$ then every solution of $x' = xf(x,y), y' = yg(x,y) - H$ reaches the x-axis in finite time. Thus an optimal control can make a finite number of switches only if $H \leq H_c$. Even for $H < H_c$, the system may collapse if it is in the "bad" case, which occurs if its equilibrium is unstable as $H \to H_c$, and then there is a value $H^* < H_c$ such that the system actually collapses for $H > H^*$. If the equilibrium is asymptotically stable as $H \to H_c$ then for every $H < H_c$ it is possible to bring the solution of $x' = xf(x,y), y' = yg(x,y) - u(t)$ into the set of initial conditions for which the solution of $x' = xf(x,y), y' = yg(x,y) - H$ tends to the equilibrium by making a finite number of switches between $u = 0$ and $u = H$. Thus for such a system we can obtain a yield H for every $H \leq H_c$ with a finite number of switches.

In considering controls with an infinite number of switches between $u = 0$ and $u = H$, it would be desirable for practical applications to have a reasonably simple criterion for determining switching points. One very simple control strategy would be to assign upper and lower thresholds y_U and y_L, respectively, for predator population size, start the system at $t = 0$ with $y = y_L$ and $u = 0$, and switch from $u = 0$ to $u = H$ whenever y increases through y_U and switch from $u = H$ to $u = 0$ whenever y decreases through y_L. This procedure is safe, in the sense that it prevents predator extinction. Numerical simulations indicate that it is always possible to choose the parameters H, y_U, and y_L so as to make the yields Y arbitrarily close to H_c, even for systems which are unstable at H_c and therefore collapse for smaller yields under constant yield harvesting. For such systems there is a tendency for the orbit to approach a periodic orbit with two switch points per cycle (Figure 6.14).

There may be an interval of values of H for given y_U and y_L on which the behavior is more complicated, with period-doubling indicated by periodic orbits having four or eight switch points per cycle, and apparent chaotic behavior (Figure 6.15). Such behavior appears to occur only for $H < H_c$ and with yields which are not close to optimal.

It is possible to prove that if the functions $f(x,y)$ and $g(x,y)$ satisfy some additional concavity conditions (always satisfied by Rosenzweig-MacArthur models and by Kolmogorov models with $g(x,y)$ independent of y and $f(x,y)$ linear in y), then if the control which maximizes the yield gives either a periodic solution of the controlled system or a solution which tends to an equilibrium, then the yield cannot be greater than H_c for any H. If $H < H_c$ it is clear that the yield cannot be greater than H_c. For $H > H_c$ it appears plausible that optimal controls lead to periodic orbits and therefore that the yield cannot be greater than H_c.

If no bound $u(t) \leq H$ is imposed on the harvest rate, then the analog of a bang-bang control which switches between the value $u = 0$ and the

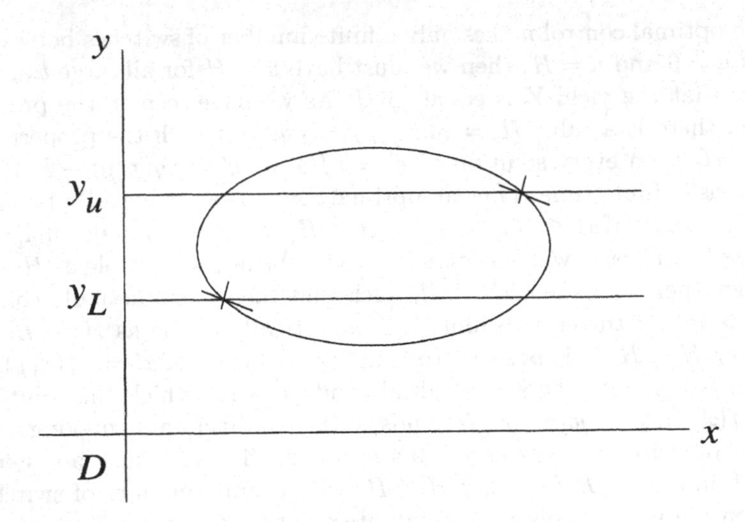

FIGURE 6.14.

maximum permissible value of u would be an *instantaneous harvest*. We may model this procedure by setting an upper threshold y_U and a lower threshold y_L and considering only the unharvested system $x' = xf(x,y)$, $y' = yg(x,y)$; whenever $y(t)$ reaches y_U we remove $(y_U - y_L)$ predators to make $y(t) = y_L$ and restart the system with the same x-value.

Another instantaneous harvesting policy which occurs more commonly in practice is to set a fixed "closed season" of length T at the end of which the predator population is harvested instantaneously down to a set escapement level y_L. These two policies are equivalent mathematically; the only difference is that in the first y_U and y_L are assigned and T is to be determined while in the second y_L and T are assigned and y_U is determined from the orbit. Another equivalent variant would assign T and $(y_U - y_L)$ with y_L or y_U to be determined from the orbit. In practice, there may be a very short hunting season and a relatively long closed season, and instantaneous harvesting approximates this procedure.

For instantaneous harvesting the following convergence result holds.

Theorem 6.2. *Let the value y_L be small enough that the line $y = y_L$ and the curve $f(x,y) = 0$ intersect. If $y_U > y_L$ is sufficiently close to y_L, then the instantaneous harvesting procedure converges to a periodic orbit consisting of a solution of $x' = xf(x,y), y' = yg(x,y)$ from a point (x_0, y_L) to a point (x_0, y_U) and a vertical line segment from (x_0, y_U) to (x_0, y_L) (see Figure 6.16).*

Proof. The unharvested system has a unique equilibrium P_∞ in the interior of the first quadrant which may be asymptotically stable or unstable, and a saddle point $S(K, 0)$. There is a full orbit from S to P_∞ or to a limit

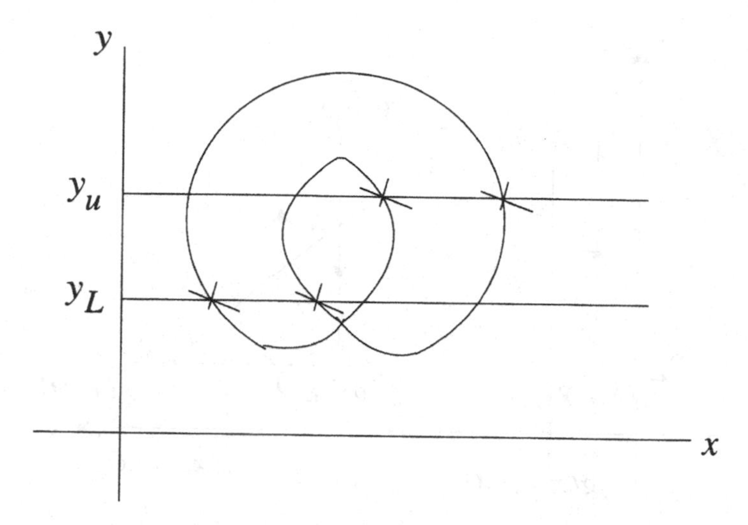

FIGURE 6.15.

cycle around P_∞ for $-\infty < t\infty$. Let $\hat{x}$ be the minimum value of x on this orbit and let Q be the point $(\hat{x}, y_L)$, R the point (K, y_L) (Figure 6.17).

For any solution $\big(x(t), y(t)\big)$, if $x(0) < x_L, y(0) = y_L$ then $x(t) > x(0)$ for $0 < t < \infty$, and if $x(0) > K, y(0) = y_L$ then $x(t) < x(0)$ for $0 < t < \infty$. The verification of these facts depends on the observation that no other orbit can cross the orbit from S to P_∞ or a limit cycle around P_∞. It now follows that for each $T > 0$ there is an x_0 with $\hat{x} < x_0 < K$ such that if $x(0) = x_0, y(0) = y_L$ then $x(T) = x(0) = x_0$. Further, the policy of starting the system at a point (x^*, y_L) at $t = 0$, letting it run to $\big(x(T), y(T)\big)$ and then harvesting $y(T) - y_L$ predators to bring the system to the point $\big(x(T), y_L\big)$ will bring $x(T)$ closer to x_0 than x^* is. Thus repetition of the process must give convergence to a periodic orbit starting at (x_0, y_L), following the orbit of the system to $\big(x_0, y(T)\big)$ at time T, and then going vertically to (x_0, y_L). The procedure of setting $y_U = y(T)$ rather than specifying T gives the same periodic orbit, which then determines T provided y_U is close enough to y_L that the orbit starting at (x_0, y_L) reaches the horizontal line $y = y_U$. This completes the proof of the theorem. □

The yield Y under instantaneous harvesting is given by $Y = \frac{y_U - y_L}{T}$. Since the solution of the controlled system is periodic, it follows from the results cited earlier in this section that this yield is at most H_c. Obviously, it is possible to make the yield arbitrarily close to H_c by making T very small and arranging that the orbit lie very close to $\big(x_\infty(H_c), y_\infty(H_c)\big)$ so that $y(t)g\big(x(t), y(t)\big)$ is very close to H_c for all points of the orbit. Since the yield is the average of $y'(t) = y(t)g\big(x(t), y(t)\big)$ over $0 \le t \le T$, this yield is close to H_c. Of more interest is the fact that numerical simulations indicate that yields arbitrarily close to H_c can always be obtained for a range of

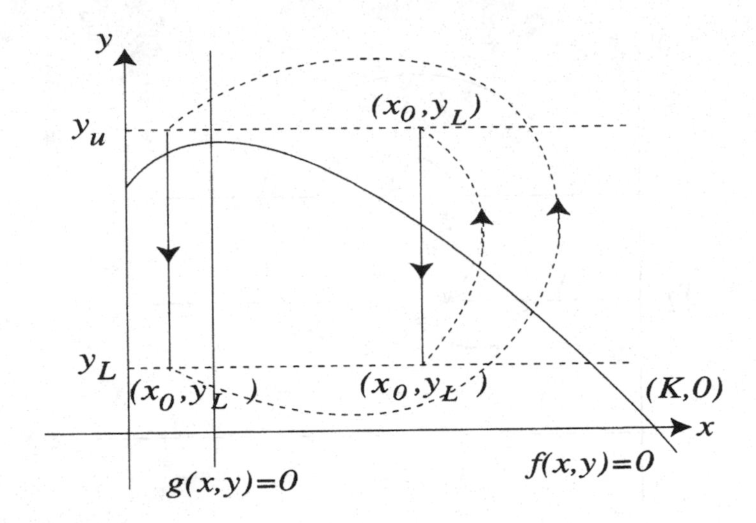

FIGURE 6.16.

values of T not confined to very small values. This is true in the "bad" case in which such yields cannot be achieved by constant-yield harvesting.

The material in this section is from [Brauer & Soudack (1979)], where additional discussion may be found.

Exercises

1. Consider the system

$$x' = x\left(1 - \frac{x}{60}\right) - \frac{xy}{x+10}$$
$$y' = 4y\left(\frac{x}{x+10} - \frac{2}{3}\right) = \frac{4y(x-20)}{3(x+10)}$$

We will choose $y_L = 5$, $T = 4$. Use a computer algebra system to simulate this system with $x(0) = 40$, $y(0) = 5$ for $0 \le t \le 4$. After recording the instantaneous harvest $y(4) - 5$, restart with initial values $(x(4), 5)$ at $t = 4$ and run for $4 \le t \le 8$. Repeat this process several times and estimate the average instantaneous harvest.

6.4 Some Economic Aspects of Harvesting

The incorporation of economic considerations into resource harvesting models leads to a rather new subject, bioeconomics. We shall only scratch the surface of this subject, considering some simple fisheries models and introducing the idea of an optimal control problem and illustrating the use of the maximum principle of Pontrjagin (1962). For further exploration of

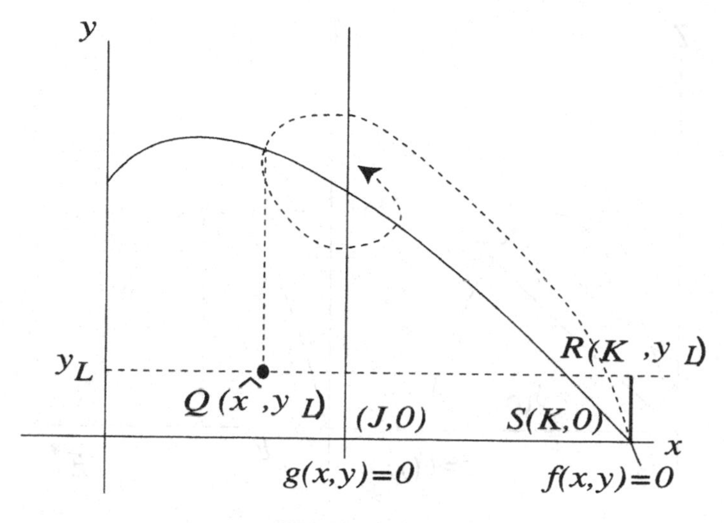

FIGURE 6.17.

bioeconomics, we refer the reader to the standard text reference by Clark (1990).

Let us consider a model for an open access fishery, that is, a fishery in which the exploitation of resources is completely uncontrolled. Our model will be an idealization, but in practice fisheries are usually relatively unregulated. However, the abuses of open access exploitation have led to the imposition of regulations in some instances.

We consider a fish population which is governed by a first-order differential equation,

$$y' = F(y),$$

and which is subjected to constant-effort harvesting, so that the model becomes

$$y' = F(y) - Ey. \tag{6.2}$$

The *yield*, or harvest in unit time, is

$$Y(E) = Ey.$$

Corresponding to an effort E there is an equilibrium population size $y_\infty = y_\infty(E)$ given by

$$F(y_\infty) = Ey_\infty$$

for $0 \leq E \leq E^* = F'(O)$. As we have seen in Section 1.5, we may plot $Y(E)$ as a function of E to obtain the *yield-effort curve* (Figure 6.18).

In Figure 6.18, the yield-effort curve is shown for a compensation model, rising continuously to a maximum, called the *maximum sustainable yield*

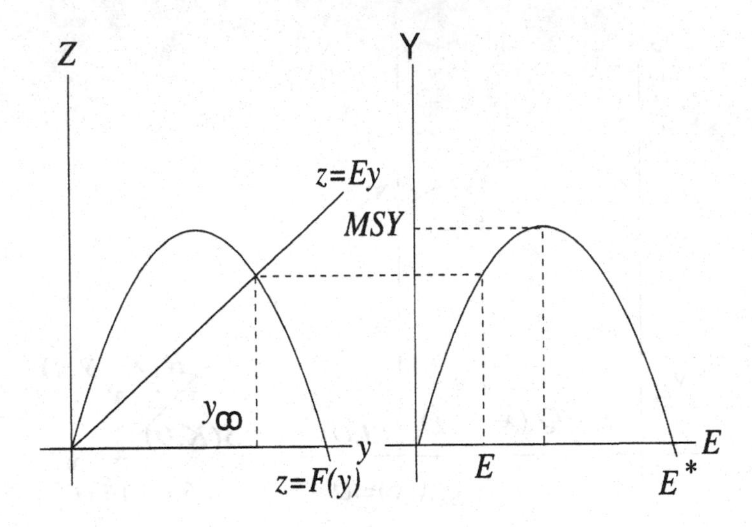

FIGURE 6.18.

(MSY) and then decreasing continuously to zero for $E = E^*$. Recall from Section 1.5, however, that for a depensation model there may be a critical effort E^* at which the yield drops to zero discontinuously.

Example 1: For logistic growth, $F(y) = ry\left(1 - \frac{y}{K}\right)$ we have

$$F\left(y_\infty(E)\right) = ry_\infty(E)\left[1 - \frac{y_\infty(E)}{K}\right] = Ey_\infty(E)$$

$$r\left[1 - \frac{y_\infty(E)}{K}\right] = E$$

$$y_\infty(E) = K\left(1 - \frac{E}{r}\right).$$

Thus

$$Y(E) = Ey_\infty(E) = KE\left(1 - \frac{E}{r}\right).$$

The maximum sustainable yield (MSY) is obtained by solving $\frac{dY}{dE} = 0$, giving $E_{MSY} = \frac{r}{2}$, $y(E_{MSY}) = \frac{K}{2}$, and $Y(E_{MSY}) = \frac{rK}{4}$.

Example 2: For the Ricker model $F(y) = rye^{-ay}$, we have

$$re^{-ay_\infty(E)} = E$$

which leads to

$$y_\infty(E) = \frac{1}{a}\log\frac{r}{E} \quad \text{and} \quad Y(E) = \frac{E}{a}\log\frac{r}{E}.$$

Then

$$Y'(E) = \frac{E}{a} \log \frac{r}{E} - \frac{1}{a}$$

and maximum sustainable yield is obtained when $E \log \frac{r}{E} = 1$, which gives $Y(E_{MSY}) = \frac{1}{a}$.

To incorporate the economic aspects of a harvesting model we must attach a revenue from the sale of this fish or other item being harvested as well as a cost to carry out the harvesting. The simplest assumptions would he that the *cost* C of fishing is proportional to the effort,

$$C = cE,$$

and that the price does not depend on the supply of fish available to sell, so that the *revenue* R is a constant multiple of the quantity harvested,

$$R = pY(E).$$

This would give a *sustainable economic rent*, meaning a profit per unit time which can be maintained indefinitely, of

$$R - C = rY(E) - cE.$$

More generally, economic models may assume a functional relation, known as a *demand curve*, between the unit selling price p and the quantity Y available for sale. It would be reasonable to assume that p is a non-increasing function of Y or, equivalently, that Y is a non-increasing function of p.

A basic (but considerably oversimplified) principle is that in an open access fishery the effort tends to approach an equilibrium effort E_∞, called the *bionomic equilibrium effort*, at which the sustainable economic rent is zero. The argument is that at the bionomic equilibrium $R = C$ and that $R > C$ if $E < E_\infty$, $R < C$ if $E > E_\infty$. If $E > E_\infty$, then $R < C$ and sustainable economic rent is negative. This means that some fisheries are losing money and therefore drop out of the market, thus decreasing the total effort. If $E < E_\infty$, then $R > C$ and fishing is profitable, encouraging new fishers to join the market and increase the total effort. Thus there should be a tendency to approach the bionomic equilibrium, although we should recognize that the delay which represents the time needed to respond to the state of the market could produce oscillations around the equilibrium (Figure 6.19).

There are many oversimplifications in this bionomic equilibrium principle. For example, the cost of fishing should also include the loss from not undertaking some alternate activity (opportunity cost), which in practice might be the most important component of fishing cost. If opportunity cost is included in the total cost, then the fishers who remain in an open access would tend to be those with the poorest alternate economic opportunities.

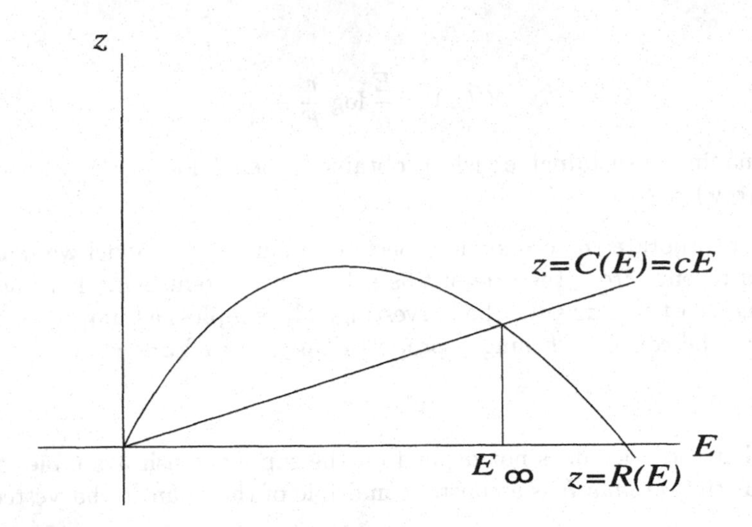

FIGURE 6.19.

Bionomic equilibrium is a situation in which there is *economic overfishing*, because a fishery which would produce positive economic rent if fishing effort were decreased is producing zero economic rent. Another possibility is *biological overfishing*, where $E_\infty > E_{MSY}$ and thus the sustainable yield and equilibrium population size are smaller than they would be if the effort were decreased.

Example 1: Consider the logistic model

$$y' = ry\left(l - \frac{y}{K}\right) - Ey$$

with $R = py_\infty E$ and $C = cE$. At equilibrium, $y_\infty = \frac{K}{r}(r - E)$, $R = \frac{pKE}{r}(r-E)$, and $C = cE$. Then the bionomic equilibrium E_∞ is determined from

$$\frac{pKE}{r}(r - E) = cE,$$

which gives

$$pK(r - E) = rc \quad \text{and} \quad r - E = \frac{rc}{pK},$$

so that

$$E_\infty = r - \frac{rc}{pK} = r\left(1, \frac{c}{pK}\right).$$

The corresponding equilibrium population size is

$$y_\infty = K\left(1 - \frac{E_\infty}{r}\right) = K\left[1 - \left(1 - \frac{c}{pK}\right)\right] = \frac{c}{p}.$$

Thus if $\frac{c}{p} > K$ there is no exploitation at the bionomic equilibrium. If $\frac{c}{p} < \frac{K}{2}$ there is biological over-fishing at bionomic equilibrium, but so long as $p > 0$ we do not have biological extinction at bionomic equilibrium. The bionomic equilibrium depends upon the cost-price ratio $\frac{c}{p}$ as well on the biological parameters r, K. In fact, under the assumptions of constant unit price p and constant cost c of unit effort, the equilibrium population size at bionomic equilibrium is equal to the cost-price ratio and the harvesting effort at bionomic equilibrium may he found by simply calculating the effort which leads to this equilibrium population size.

Exercises

1. The model for fisheries

$$\frac{dx}{dt} = a - bx$$

was proposed by Schoener in 1973. Determine the equation of the yield-effort curve and explain its peculiar feature.

2. Find the x_{MSY} (density of maximum sustainable yield) and $F(x_{MSY})$ (the maximum sustainable yield) for the Gompertz law of population growth:

$$\frac{dx}{dt} = rx \log\left(\frac{K}{x}\right).$$

Also sketch the yield-effort curve. What is E^*(effort at MSY)?

3. Find the $x_{\max}$ (density of maximum sustainable yield) and $F(x_{\max})$ (the maximum sustainable yield) for a population governed by the equation

$$\frac{dx}{dt} = rxe^{-x} - dx$$

with $r > d$.

4. What happens to the yield in a population governed by the equation

$$\frac{dx}{dt} = rxe^{-x} - dx$$

with $r < d$?

5. Show that if a population is governed by the equation

$$\frac{dx}{dt} = r\sqrt{x},$$

there is no maximum sustainable yield.

6.5 Optimization of Harvesting Returns

Consider an open access fishery which is in bionomic equilibrium with biological overfishing. Let us suppose that we try to regulate this fishery by reducing the harvesting effort in order to increase the yield. The immediate effect of reducing effort is to reduce the yield, but the long - term effect will be to increase the yield by increasing the equilibrium population level. We would like to determine the optimal trade-off between current and future harvesting revenue. In order to do this we must take into account the time value of money because of interest. As the quantity which we shall study is the present value of all future harvest rents, we shall do this by using a discount rate δ, and as we are studying a continuous model we shall assume that this discount rate is compounded continuously.

When we speak of a discount rate δ, we mean that the present value of an amount which would have value 1 at a time t units in the future is $(1-\delta)^t$. Normally the time unit is one year. If the discount is *compounded* K times per year, we use a discount rate of $\frac{\delta}{K}$ per discount period. Then t years would be tK discount periods and the present value of an amount 1 at a time t units in the future would be $\left(1-\frac{\delta}{K}\right)^{tK}$. By continuous compounding, we mean that the present value would be the limit of this quantity as $K \to \infty$. Because

$$\lim_{K\to\infty}\left(1-\frac{\delta}{K}\right)^{Kt} = e^{-\delta t},$$

the effect of a continuously compounded discount rate δ would be to introduce a factor $e^{-\delta t}$ for a rent at time t years in the future.

We now generalize the model (6.2) slightly by assuming that the harvest at time t is proportional to the effort E as before, but is not necessarily proportional to the population size $y(t)$. Thus we write the harvest at time t as

$$h(t) = EG\{y(t)\}, \tag{6.3}$$

where $G(y)$ is a non-negative, non-decreasing function of y. We continue to assume a constant unit price p and a constant cost c of unit effort. Then the unit harvest cost when the population level is y is a function $c(y)$ given by

$$c(y) = \frac{c}{G(y)}. \tag{6.4}$$

An input of effort $E\Delta t$ from time t to time $t+\Delta t$ produces a harvest $EG\{y(t)\}\Delta t$ which produces a revenue of $pEG\{y(t)\}\Delta t$ at a cost of $cE\Delta t$. Thus the net revenue is

$$pEG\{y(t)\}\Delta t - cE\Delta t = [p - c\{y(t)\}]h(t)\Delta t, \tag{6.5}$$

making use of (6.3) and (6.4). The present value of all future harvesting effort is thus

$$P = \int_0^\infty e^{-\delta t}[p - c\{y(t)\}]h(t)dt. \tag{6.6}$$

Note that we must use a positive discount rate δ as otherwise the infinite integral in (6.6) would diverge. If we choose to use a *finite time horizon* by considering only the harvesting effort until time T, then we would replace (6.6) by

$$P = \int_0^T e^{-\delta t}[p - c\{y(t)\}]h(t)dt. \tag{6.7}$$

With a finite time horizon we could choose to disregard the time value of money by taking $\delta = 0$ without introducing any problems with divergence of the integral.

The optimal control problem which we wish to solve is to choose the harvest rate function which will maximize the integral (6.6) subject to the *state equation*

$$y' = F(y) - h(t) = F\{y(t)\} - EG\{y(t)\}, \tag{6.8}$$

and the constraints $y(t) \geq 0$ (because a realistic model requires a non-negative population level) and $0 \leq h(t) \leq h_{max}$ (because a fishery always has a maximum harvesting capability).

The solution of this optimal control problem, which we shall justify later, is that we should first choose the equilibrium population level y^* as the solution of the equation

$$F'(y) - \frac{c'(y)F(y)}{p - c(y)} = \delta$$

or, equivalently,

$$\frac{d}{dy}[F(y)\{p - c(y)\}] = \delta[p - c(y)]. \tag{6.9}$$

We then choose the effort E so that

$$h = EG(y^*) = F(y^*).$$

This still leaves open the question of the optimal way to reach this *optimal equilibrium population* level y^*. The answer to this question is that we should use the harvest rate $h^*(t)$ at time t which drives $y(t)$ to y^* as rapidly as possible,

$$h^*(t) = \begin{cases} h_{\max} & (y > y^*) \\ F(y^*) & (y = y^*) \\ 0 & (y < y^*). \end{cases}$$

This type of control, taking the extreme values 0 and h_{max} and switching when $y = y^*$ is called a *bang-bang control.* It may appear to be an extreme policy in practice as it involves closing down a fishery completely, possibly for a long period. Such a policy is a consequence of the assumed linear dependence on E. In the next section we shall consider a nonlinear control problem for which the optimal control is less extreme.

At equilibrium of the fishery model (6.8), $F(y) = h$. Thus the sustainable economic rent at y is equal to

$$\rho(y) = [p - c(y)]F(y).$$

With this definition of $\rho(y)$, we may write the optimizing condition (6.9) as

$$\rho'(y) = \delta[p - c(y)]. \tag{6.10}$$

A marginal decrease of Δy in y gives an immediate rent of $[p - c(y)]\Delta y$ and decrease in sustained rent

$$\Delta p \approx \rho'(y)\Delta y$$

with a present value of

$$\int_0^\infty e^{-\delta t}\rho'(y)\Delta y dt = \frac{1}{\delta}\rho'(y)\Delta y.$$

Thus for the optimal solution, the marginal immediate gain is equal to the present value of the marginal future loss.

From (6.4) and the assumption that $G(y)$ is a non-decreasing function of y, it follows that $c(y)$ is non-increasing, and $p - c(y)$ is non-decreasing. If $\rho'(y)$ is a decreasing function, the equation (6.10) has a unique solution (Figure 6.20).

$\delta = 0$ means that we should maximize the sustained economic rent, that is, we should maximize ρ by setting $\rho'(y) = 0$. The corresponding population level is y_o. If we let $\delta \to \infty$, we set zero value on future revenues and set $p - c(y) = 0$ (no sustainable rent). The corresponding population level is y_∞. Here y_0 is defined by $\rho'(y) = 0$ and y_∞ by $c(y_\infty) = p$; we have $y_\infty < y^* < y_0$.

The process described above is for a harvest $h(t) = EG\{y(t)\}$, which is not necessarily proportional to population size. In the special case $h(t) = Ey(t)$ which we have considered earlier, (6.4) implies $c(y) = \frac{c}{y}$ and this is used in the determination of y^*. Once y^* has been determined, it is easy to calculate $F(y^*)$ and thus to determine the switching point in $h^*(t)$.

Example 1: Consider a logistic model with constant effort harvest, so that

$$F(y) = ry\left(1 - \frac{y}{K}\right), \quad G(Y) = y, \quad c(y) = \frac{c}{y}.$$

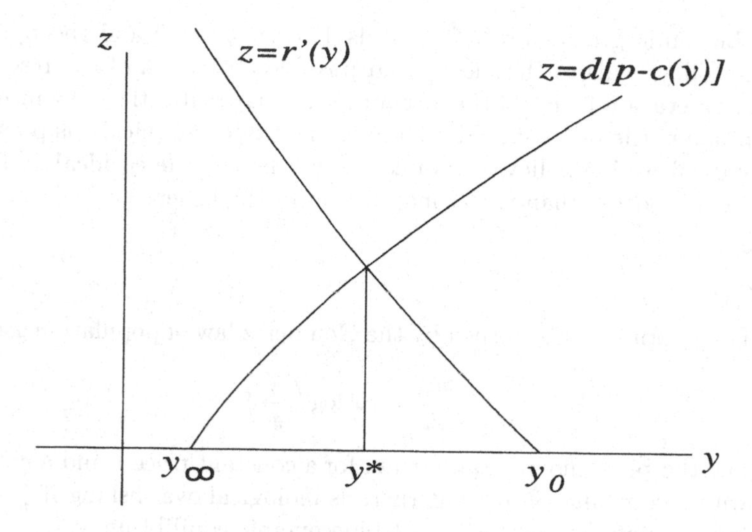

FIGURE 6.20.

The condition (6.10) becomes

$$r\frac{d}{dy}\left[y\left(p-\frac{c}{y}\right)\left(1-\frac{y}{K}\right)\right]=\delta\left(p-\frac{c}{y}\right)$$

$$r\frac{d}{dy}\left[y\left(p-\frac{c}{y}\right)-\frac{p}{K}y^2+\frac{c}{K}y\right]=\delta\left(p-\frac{c}{y}\right)$$

$$2rpy^2+y(\delta pK-rpK-rc)-\delta cK=0.$$

This is a quadratic equation having one positive root and one negative root, and we take y^* to be the positive root,

$$y^*=\frac{K}{4}\left[\left(\frac{c}{pK}+1-\frac{\delta}{r}\right)+\left\{\left(\frac{c}{pK}+1-\frac{\delta}{r}\right)^2+\frac{8c\delta}{prK}\right\}^{\frac{1}{2}}\right].$$

For zero discount, $\delta=0$, we obtain $y_0=\frac{K}{2}\left(\frac{c}{pK}+1\right)$ and since $y_\infty=\frac{c}{p}$, we have

$$y_0=\frac{1}{2}(y_\infty+K)=\frac{1}{2}y_\infty+y_{MSY}>y_{MSY}.$$

Thus we do not have biological overfishing if $\delta=0$. However, increasing δ will decrease y^* from y_0. An extreme case is $y_\infty=0$, when fishing has zero cost, and then

$$y^*=\frac{K}{2}\left(1-\frac{\delta}{r}\right)<y_{MSY}.$$

If the bionomic growth ratio $\frac{\delta}{r}$ exceeds 1, then $y^* = 0$ and the optimal harvest policy leads to the most rapid possible extinction of the resource. The interpretation is that if the discount rate δ is greater than the intrinsic growth rate r, the fisher should cash in the resources as quickly as possible. This optimal fishing policy may of course not be considered ideal by those with concerns other than the economic rent of the fishery.

Exercise

1. For a population governed by the Gompertz law of population growth

$$\frac{dx}{dt} = rx \log\left(\frac{K}{x}\right)$$

find the bioeconomic equilibrium for a constant price p and a cost cE with c constant. Show that there is biological over-fishing if $\frac{c}{p} < \frac{K}{e}$, but no biological extinction at bioeconomic equilibrium.

6.6 Justification of the Optimization Result

In the preceding section we stated a result which we used to provide the solution of an optimization problem. In this section we shall justify this result. In fact, we consider the more general

$$J[u] = \int_{t_0}^{t'} [f_0(t, y(t)) + g_0(t, y(t))u(t)]dt, \tag{6.11}$$

subject to a state equation

$$y' = f_1(t, y) + g_1(t, y)u, \tag{6.12}$$

with given values $y(t_0) = y_0$, $y(t_1) = y_1$, and with the *control* $u(t)$ chosen from an *admissible set* of functions which are piecewise continuous and satisfy

$$u_m \leq u(t) \leq u_M.$$

The problem considered in Section 6.6 was the special case

$$\begin{aligned} f_0(t, y) &= 0 & g_0(t, y) &= e^{-\delta t}[p - c(y)] \\ f_1(t, y) &= F(y) & g_1(t, y) &= -1 \\ u(t) &= h(t). \end{aligned}$$

However, in this problem we did not impose a *terminal condition* $y(t_1) = y_1$; the terminal condition is essentially replaced by the requirement that $y(t)$ approach an equilibrium as $t \to \infty$.

We begin the solution by solving the state equation (6.12) for u and substituting the result into (6.11), obtaining a form

$$\int_{t_0}^{t'} [G(t,y) + H(t,y)y']dt \tag{6.13}$$

to be maximized under constraints

$$A(y,t) \le y'(t) \le B(y,t). \tag{6.14}$$

Theorem 6.3. *Let $y^*(t)$ be the unique solution of*

$$\frac{\partial G}{\partial y} = \frac{\partial H}{\partial t} \tag{6.15}$$

and suppose

$$\frac{\partial G}{\partial y} \le \frac{\partial H}{\partial t} \quad \text{if} \quad y < y^*(t)$$
$$\frac{\partial G}{\partial y} \ge \frac{\partial H}{\partial t} \quad \text{if} \quad y > y^*(t).$$

Then the optimal solution is the "Closest possible trajectory to $y^*(t)$", defined as follows:

(i) If $y_0 > y^*(t_0)$, $y(t)$ uses the fastest possible descent $y' = A(y,t)$, $y(t_0) = y_0$ until $y^*(t)$ is reached.

(ii) If $y_0 < y^*(t_0)$, $y(t)$ uses the fastest possible ascent $y' = B(y,t)$, $y(t_0) = y_0$ until $y^*(t)$ is reached.

Similarly, if $y_1 \neq y^*(t_1)$, use the same scheme backwards from t_1, until $y^*(t)$ is reached. Between these two points where $y^*(t)$ is reached, say t_a when $y(t)$ starting at t_0 reaches $y^*(t)$ aud t_b when $y(t)$ ending at t_1 reaches $y^*(t)$, take $y(t) = y^*(t)$ [Figure 6.21].

Proof. We write the objective functional as a line integral

$$\int_{t_0}^{t_1} [G(t,y) + H(t,y)y']dt = \int_C (Gdt + Hdy),$$

where C is the curve $y = y(t)$, $t_0 \le t \le t_1$. Suppose $y(t)$ is the claimed optimal trajectory and $y_1(t)$ is an arbitrary admissible curve from (t_0, y_0) to (t_1, y_1). Because of the constraint (6.14) satisfied by $y_1(t)$, we must have $y_1(t) \le y(t)$ at least for $t_0 \le t \le t_a$. Suppose $y_1(t) < y(t)$ on $t_0 < t < t_R$, $y_1(t_R) = y(t_R)$. Then

$$\int_{t_0}^{t_R} [G(t,y) + H(t,y)y']dt - \int_{t_0}^{t_R} [G(t,y) + H(t,y)y']dt = \int_\Gamma (Gdt + Hdy), \tag{6.16}$$

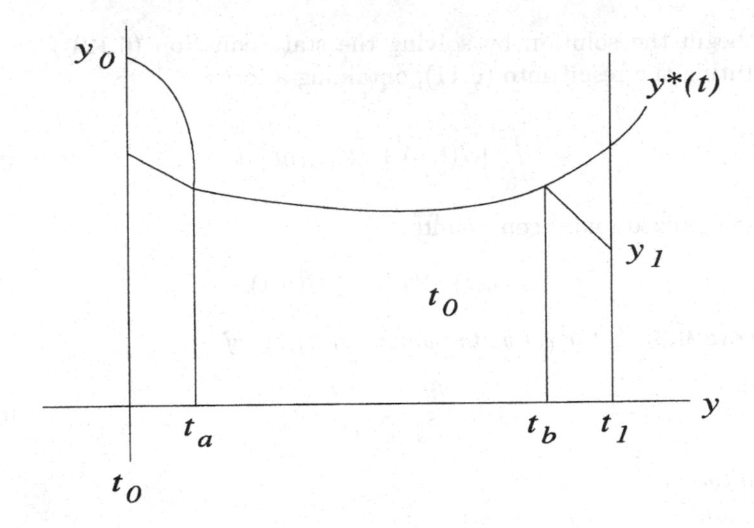

FIGURE 6.21.

where Γ is the closed curve PRS as shown in Figure 6.22.

By Green's Theorem in the plane, this is equal to

$$\int_A \int \left(\frac{\partial G}{\partial y} - \frac{\partial H}{\partial t} \right) dydt,$$

where A is the region in the interior of the curve Γ. Since this region lies below $y^*(t)$, $\frac{\partial G}{\partial y} \geq \frac{\partial H}{\partial t}$ in A, the integral (6.16) is non-negative. This shows that $y(t)$ does better than $y_1(t)$ in optimizing on the interval $t_0 \leq t \leq t_R$, and a similar argument shows the optimality from t_R to t_1. This completes the proof of the theorem. In fact, we have established the stronger result that an optimal path must be optimal on every subinterval. □

In the problem considered in Section 6.5, we used

$$u = h = F(y) - y'$$

$$J[h] = \int_0^\infty e^{-\delta t}[p - c\{y(t)\}][F\{y(t)\} - y'(t)]dt$$

so that

$$G(t, y) = e^{-\delta t}[p - c(y)]F(y)$$

$$H(t, y) = -e^{-\delta t}[p - c(y)].$$

Then the equation (6.15) becomes

$$e^{-\delta t}\frac{d}{dt}[\{p - c(y)\}F(y)] = \delta e^{-\delta t}[p - c(y)],$$

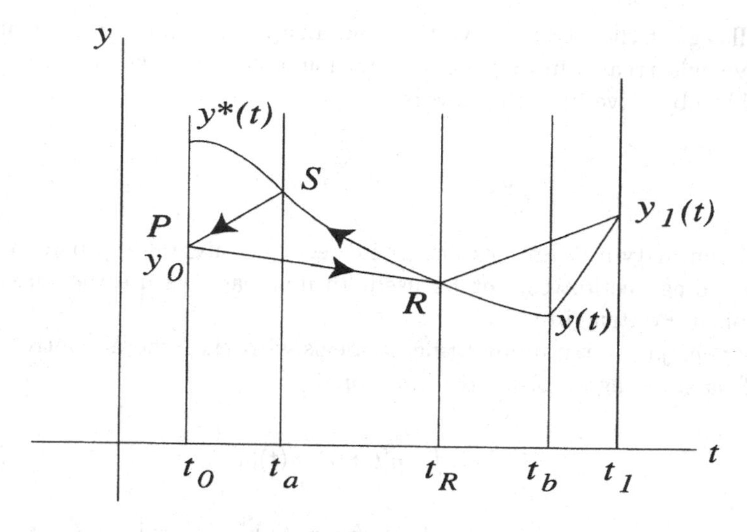

FIGURE 6.22.

which is the same as the equation (6.2) of Section 6.5 which we solved to determine the final value y^*.

If in the problem of Section 6.5 we had used a finite time horizon T instead of trying to optimize over $0 \leq t < \infty$ and had specified a terminal condition for $y(T)$, the theorem of this section would imply that it would be necessary to leave the path $y^*(t)$ before T and use a bang-bang control. Obviously, $h = 0$ would not be optimal because $h = h_{max}$ for t near T would provide a positive contribution to the present value. The "horizon effect" of rapid resource depletion for t near T is an artificial consequence of the finite time horizon. The practical effect of this effect can be eliminated by taking a sufficiently remote horizon (letting $T \to \infty$). In nonlinear models, such as will be considered in the next section, horizon effects typically tend to zero exponentially for finite t as $T \to \infty$ and are effectively eliminated by taking T large.

6.7 A Nonlinear Optimization Problem

In this section we consider a model

$$y' = F(y) - h(t), \quad y(0) = y_0, \quad y(T) = y_T \tag{6.17}$$

on a finite interval for which the revenue is not a linear funetion of h. In most real market situations the price p is not a constant but depends on the supply h; it is reasonable to assume that $p(h)$ is a non-increasing function of h. The revenue $R(h) = hp(h)$ is then not proportional to h. We will assume that the function $R(h)$ has the properties

$$R(h) > 0, \quad R'(h) \geq 0, \quad R''(h) \leq 0.$$

We will neglect the cost of harvesting, but an approach similar to the one we employ could treat a more general situation which does include harvesting cost. The objective function now is

$$J[h] = \int_0^T e^{-\delta t} R(h) dt. \tag{6.18}$$

The nonlinearity in h implies that the direct line integral approach used in the preceding section cannot be used. In its stead, we use the maximum principle of Pontrjagin (1962).

The Pontrjagin maximum principle deals with the general control problem of maximizing an objective functional

$$J[u] = \int_0^T g[t, y(t), u(t)] dt, \tag{6.19}$$

where $u(t)$ is chosen from some class of admissible controls u on $0 \leq t \leq T$, with a state equation

$$y'(t) = f[t, y(t), u(t)], \quad y(0) = y_0. \tag{6.20}$$

In addition, the control u must also be such that the response $y(t)$ satisfies the terminal condition

$$y(T) = y_t. \tag{6.21}$$

We define the *Hamiltonian*

$$\mathcal{H}[t, y(t), u(t), \lambda(t)] = g[t, y(t), u(t)] + \lambda(t) f[t, y(t), u(t)], \tag{6.22}$$

where λ is an unknown function called the *adjoint variable.* The maximum principle says that if u is the optimal control and y the corresponding response, then there is an adjoint variable λ such that

$$\frac{d\lambda}{dt} = -\frac{\partial \mathcal{H}}{\partial y} = -\frac{\partial g}{\partial y} - \lambda \frac{\partial f}{\partial y} \tag{6.23}$$

and the optimal control maximizes the value of the Hamiltonian over all admissible controls for each t, that is,

$$\mathcal{H}[t, y(t), u(t), \lambda(t)] = \max_{u \in U} \mathcal{H}[t, y(t), u(t), \lambda(t)]. \tag{6.24}$$

If the optimal control is in the interior of the control interval, that is, if the control constraints are not *binding*, then this maximization statement is

$$\frac{\partial \mathcal{H}}{\partial u} = 0. \tag{6.25}$$

In general, we have three functions $y(t)$, $u(t)$, and $\lambda(t)$ to determine, and three equations, namely the state equation (6.20), the adjoint equation (6.23), and the maximum statement (6.24) or (6.25) can be used for their determination.

For the linear problem solved in Section 6.5 of maximizing

$$\int_0^\infty e^{-\delta t}[p - c(y)]h(t)dt$$

with the state equation

$$y' = F(y) - h(t),$$

we have

$$g(t, y, h) = e^{-\delta t}[p - c(y)]h, \quad f(t, y, h) = F(y) - h(t)$$

in (6.19) and (6.20). Then (6.22) gives the Hamiltonian

$$\mathcal{H}[t, y, h, \lambda] = e^{-\delta t}[p - c(y)]h(t) + \lambda(t)[F(y) - h(t)]$$

and the adjoint equation (6.23) is

$$\frac{d\lambda}{dt} = -c'(y)h(t)e^{-\delta t} + \lambda F'(y). \tag{6.26}$$

If the constraints are not binding, that is, if we can maximize by taking $\frac{\partial \mathcal{H}}{\partial u} = 0$ without being concerned that the resulting u will not be an admissible control, the maximum condition (6.25) gives

$$e^{-\delta t}[p - c(y)] - \delta(t) = 0,$$

or

$$\lambda(t) = e^{-\delta t}[p - c(y)]. \tag{6.27}$$

Substituting the expression (6.27) into the adjoint equation (6.26), we have

$$\begin{aligned} -\delta e^{-\delta t}[p - c(y)] &= c'(y)he^{-\delta t} - \lambda F'(y) \\ &= c'(y)F(y)e^{-\delta t} - e^{-\delta t}[p - c(y)]F'(y) \end{aligned} \tag{6.28}$$

or

$$F'(y) - \frac{c'(y)F(y)}{p - c(y)} = \delta,$$

which is the solution obtained in Section 6.5. Note that we have in effect set an equilibrium value terminal condition when we replaced h by $F(y)$ in (6.28).

In order to examine the situation when the constraints are binding, we look at the linear problem of maximizing

$$J[u] = \int_{t_0}^{b_1} [f_0\{t, y(t)\} + g_0\{t, y(t)\}u(t)]dt \tag{6.29}$$

subject to a state equation

$$y' = F(y) - u(t), \tag{6.30}$$

with $y(t_0) = y_0$, $y(t_1) = y_1$, and the control constraints

$$u_m \leq u(t) \leq u_M. \tag{6.31}$$

This problem is the special case $f_1(t, y) = F(y)$, $y_1(t, y) = -1$ of the problem (6.3), (6.4) in Section 6.6, which was analyzed there by the use of line integrals and Green's Theorem. For this problem, the Hamiltonian is

$$\begin{aligned}\mathcal{H}[t, y, u, \lambda] &= f_0(t, y) + g_0(t, y)u + \lambda(t)[F(y) - u] \\ &= f_0(t, y) + \lambda(t)F(y) + u[g_0(t, y) - \lambda(t)].\end{aligned}$$

We let

$$\sigma(t) = g_0\{t, y(t)\} - \lambda(t). \tag{6.32}$$

As $\mathcal{H}$ is linear in u, it is clear that in order to maximize $\mathcal{H}$ we must take $u = u_M$ if $\sigma(t) > 0$ and $u = u_m$ if $\sigma(t) < 0$. This is a bang-bang control and σ is called the *switching function.* When the switching function vanishes, $\mathcal{H}$ is independent of u. Thus binding constraints lead to bang-bang controls.

A singular case arises if $\sigma \equiv 0$ on an interval. To determine the corresponding *singular control*, we write $\sigma'(t) = 0$, or

$$\frac{\partial g_0}{\partial t} + \frac{\partial g_0}{\partial y}\frac{dy}{dt} - \frac{d\lambda}{dt} = 0.$$

From the state equation (6.30) and the adjoint equation (6.23) in the form

$$\frac{d\lambda}{dt} = -\frac{\partial}{\partial y}(f_0 + g_0 u) - \lambda\frac{\partial}{\partial y}(F - u)$$

we have

$$\frac{\partial g_0}{\partial t} + \frac{\partial g_0}{\partial y}(F - u) + \frac{\partial f_0}{\partial y} + u\frac{\partial g_0}{\partial y} + \lambda\frac{\partial F}{\partial y} = 0.$$

From (6.32) and the switching condition $\sigma(t) = 0$, this is equivalent to

$$-\frac{\partial g_0}{\partial t} = F\frac{\partial g_0}{\partial y} + \frac{\partial f_0}{\partial y} + g_0\frac{\partial F}{\partial y} \tag{6.33}$$

In Section 6.6 we used the state equation to write the integral

$$\int_{t_0}^{t_1} [f_0\{t, y(t)\} + g_0\{t, y(t)\}u(t)]dt$$

in the form

$$\int_{t_0}^{t_1} [G(t, y) + H(t, y)y']dt$$

and here we have

$$\begin{aligned} f_0(t, y) + g_0(t, y)u &= f_0(t, y) + g_0(t, y)[F(y) - y'] \\ &= f_0(t, y) + g_0(t, y)F(y) - y'g_0(t, y). \end{aligned}$$

Thus

$$\begin{aligned} G(t, y) &= f_0(t, y) + g_0(t, y)F(y) \\ H(t, y) &= g_0(t, y) \end{aligned}$$

and we have

$$\begin{aligned} \frac{\partial G}{\partial t} &= \frac{\partial f_0}{\partial y} + g_0 \frac{\partial F}{\partial y} + F \frac{\partial g_0}{\partial y} \\ \frac{\partial H}{\partial t} &= -\frac{\partial g_0}{\partial t}. \end{aligned}$$

Now (6.33) is equivalent to

$$\frac{\partial H}{\partial t} = -\frac{\partial G}{\partial y},$$

which was the equation in Section 6.6 that was solved to obtain the function $y^*(t)$. The content of the maximum principle is that for this problem the optimal control is a combination of bang-bang and singular controls.

The optimization problem (6.17) and (6.18) formulated at the beginning of this section is not linear in h and the method used in Section 6.6 cannot be applied. Here we are forced to use the maximum principle. The Hamiltonian is

$$\mathcal{H}(t, y, u, \lambda) = e^{-\delta t}R(u) + \lambda(t)[F(y) - u].$$

If the control constraints are not binding, the maximum equation is

$$\frac{\partial \mathcal{H}}{\partial u} = e^{-\delta t}R'(u) - \lambda(t) = 0$$

or

$$\lambda(t) = e^{-\delta t}R'\{u(t)\}. \tag{6.34}$$

The adjoint equation is

$$\frac{dy}{dt} = -\frac{\partial \mathcal{H}}{\partial y} = -\lambda(t)F'(y) = -e^{-\delta t}R'\{u(t)\}F'(y).$$

From (6.34) we obtain

$$\frac{d\lambda}{dt} = e^{-\delta t}\left[R''(u)\frac{du}{dt} - \delta R'(u)\right]$$

and combination with the adjoint equation gives

$$R''(u)\frac{du}{dt} - \delta R'(u) = -R'(u)F'(y).$$

We now have a two-dimensional autonomous system for the response y and the control u, namely

$$y' = F(y) - u$$

$$u' = \frac{R'(u)}{R''(u)}[\delta - F'(y)]. \tag{6.35}$$

It is reasonable to assume that the revenue function $R(u)$ satisfies $R(u) > 0$, $R'(u) \geq 0$, and $R''(u) < 0$.

The y-isocline of the system (6.35) is the curve $u = F(y)$ and the u-isocline is the vertical line $F'(y) = \delta$. There is a unique equilibrium (y^*, u^*) given by

$$u^* = F(y^*), \quad F'(y^*) = \delta$$

and we may verify that this equilibrium is a saddle point, and that the separatrices and phase portrait are as shown in Figure 6.23

As in the linear problem of Section 6.5 there is an optimal equilibrium solution $y = y^*$, $u = u^*$. However, the optimal approach to equilibrium now is not a bang-bang control as in the linear case. The optimal trajectory must start on the line $y = y_0$ at time $t = 0$ and terminate on the line $y = y_T$ at time $t = T$. As (y^*, u^*) is an equilibrium, the velocity along a trajectory near (y^*, u^*) is small. The closer an orbit is to the separatrices, the longer the time taken to go from y_0 to y_T. Thus for small T the trajectory will be far from the saddle point and for large T the trajectory will be close to the saddle point. For each point on $y = y_0$ there is a trajectory going to the line $y = y_T$, and the optimal trajectory is the one taking the required time T (Figure 6.24).

If the time horizon $T \to \infty$, the terminal condition drops out and the optimal trajectory is simply the part of the separatrix from $y = y_0$ to (y^*, u^*). The optimal approach is more gradual than the bang-bang approach of the linear model. This reflects the dependence of prices on the harvest, which one would expect to cause more gradual changes in the optimal harvest rate.

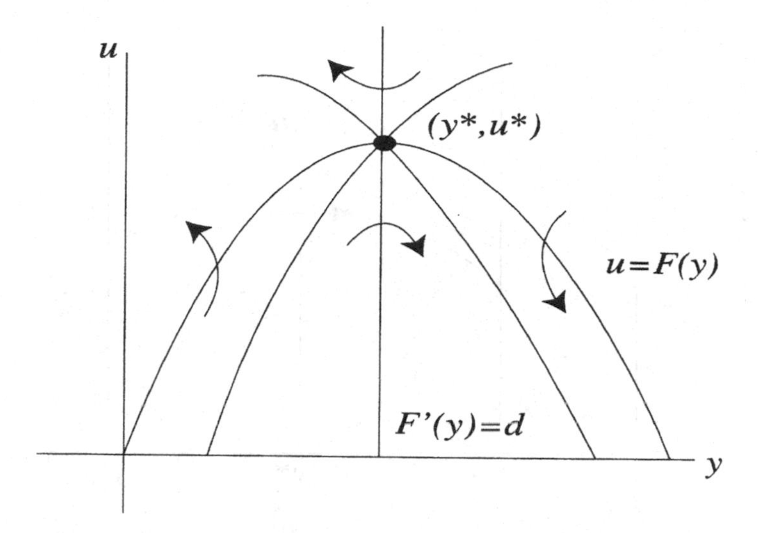

FIGURE 6.23.

6.8 Economic Interpretation of the Maximum Principle

We now give an interpretation of the maximum principle in economic terms. The computations which lead to this interpretation actually can be refined to give a proof of the maximum principle under the assumption that all functions involved are sufficiently smooth (twice differentiable). However, in practice the functions are not necessarily this smooth, as we have seen with bang-bang controls, and the general proof of the maximum principle is much more difficult. For a full proof, we refer the reader to the books of Pontrjagin *et al.* (1962) or Lee & Markus (1968).

We consider the problem of maximizing

$$J[u] = \int_{t_0}^{t_1} g[s, y(s), u(s)]ds \tag{6.36}$$

subject to a state equation

$$y' = f(t, y, u) \tag{6.37}$$

with initial and terminal conditions

$$y(t_0) = y_0 \quad \text{and} \quad y(t_1) = y_1. \tag{6.38}$$

We view the terminal point (t_1, y_1) as fixed but treat the initial point (t_0, y_0) as variable, assuming that an optimal control $u^*(t)$ exists for all initial points (t_0, y_0) under consideration. For each initial point we define the function

$$w(t_0, y_0) = J[u^*] = \max J[u]. \tag{6.39}$$

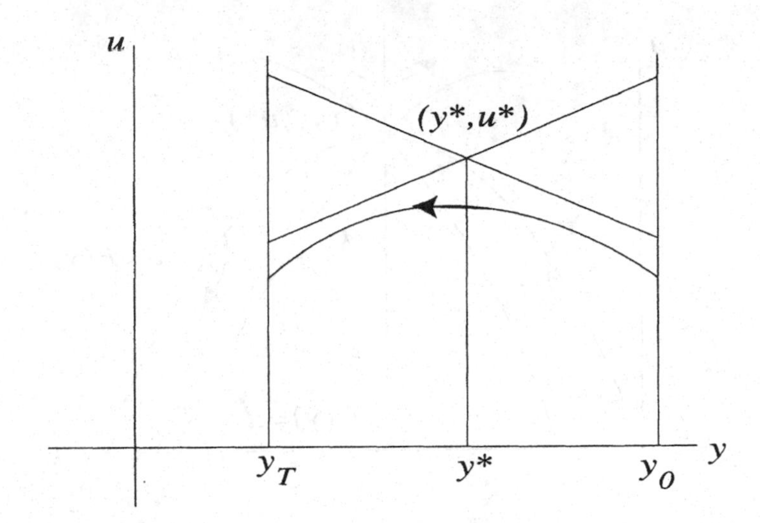

FIGURE 6.24.

This function $w(t_0, y_0)$ in economic problems represents the present value of the stock level y_0 at time t_0 under the presumption that an optimal exploitation policy will he used for $t_0 \leq t \leq t_1$. Thus $w(t_0, y_0)$ is the value of the capital asset y_0 at time t_0 with a given time horizon t_1. Because the optimal policy must he optimal at each point of the interval $t_0 \leq t \leq t_1$, we must have

$$w(t_0, y_0) = \int_{t_0}^{t} g[s, y^*(s), u^*(s)]ds + u[t, y^*(t)] \tag{6.40}$$

for every t. Differentiation of (6.39) with respect to t gives

$$\begin{aligned} \frac{d}{dt} w[t, y^*(t)] &= \frac{\partial w}{\partial t}[t, y^*(t)] + \frac{\partial w}{\partial y}[t, y^*(t)] \frac{dy}{dt} \\ &= \frac{\partial w}{\partial t}[t, y^*(t)] + \frac{\partial w}{\partial y}[t, y^*(t)] f[t, y^*(t), u^*(t)] \\ &= -g[t, y^*(t), u^*(t)]. \end{aligned} \tag{6.41}$$

If we define

$$\lambda(t) = \frac{\partial w}{\partial y}[t, y^*(t)] \tag{6.42}$$

and the Hamiltonian

$$\mathcal{H}[t, y, u, \lambda] = g(t, y, u) + \lambda f(t, y, u), \tag{6.43}$$

the maximum principle implies that

$$\mathcal{H}[t, y^*(t), u^*(t), \lambda(t)] = \max_u \mathcal{H}[t, y, u, \lambda]. \tag{6.44}$$

The function $G(t, y, u)$ defined by

$$G(t, y, u) = \frac{\partial w}{\partial t}(t, y) + \frac{\partial w}{\partial y}(t, y) f(t, y, u) + g(t, y, u) \tag{6.45}$$

has the property that

$$G(t, y^*(t), u^*(t)) = 0 \tag{6.46}$$

because of (6.41). It is possible to show, by an argument which involves considering the effect of varying the control u in (6.40), that

$$G(t, y, u) \leq 0 \tag{6.47}$$

for all (t, y, u). From (6.45) and (6.46) we see that $G(t, y, u)$ is maximized by the choice $y = y^*(t)$, $u = u^*(t)$. In particular, if we fix $y = y^*(t)$, then $G(t, y^*(t), u)$ is maximized by the choice $u = u^*(t)$, and if we fix $u = u^*(t)$ then $G(t, y, u^*(t))$ is maximized by the choice $y = y^*(t)$. As an alternative to this maximization argument, we could use the maximum principle to show that $G(t, y, u)$ is maximized by the choice $u = u^*(t)$, as

$$\begin{aligned} \mathcal{H}[t, y^*(t), u, \lambda] &= g(t, y^*(t), u) + \lambda f(t, y^*(t), u) \\ &= g(t, y^*(t), u) + \frac{\partial w}{\partial y}(t, y^*(t)) f(t, y^*(t), u) \\ &= G(t, y^*(t), u) - \frac{\partial w}{\partial t}(t, y^*(t)) \end{aligned}$$

by (6.42), (6.43), and (6.45); maximization of $G(t, y^*(t), u)$ is equivalent to maximization of $\mathcal{H}[t, y^*(t), u, \lambda]$.

Now we fix $u = u^*(t)$ and use the fact that $G(t, y^*(t), u)$ attains its maximum value when $y = y^*(t)$. If G is differentiable, we must have

$$\frac{\partial G}{\partial y}(t, y^*(t), u^*(t)) = 0. \tag{6.48}$$

Differentiation of (6.45) shows that (6.48) implies

$$\frac{\partial^2 w}{\partial y^2} f + \frac{\partial w}{\partial y}\frac{\partial f}{\partial y} + \frac{\partial^2 w}{\partial y \partial t} + \frac{\partial g}{\partial y} = 0, \tag{6.49}$$

with all functions being evaluated for $y = y^*(t)$, $u = u^*(t)$. Differentiation of (6.42) gives

$$\frac{d\lambda}{dt} = \frac{\partial^2 w}{\partial y^2}\frac{dy}{dt} f + \frac{\partial^2 w}{\partial t \partial y} = \frac{\partial^2 w}{\partial y^2} f + \frac{\partial^2 w}{\partial t \partial y}. \tag{6.50}$$

Combination of (6.49) and (6.50) gives

$$\frac{d\lambda}{dt} + \frac{\partial w}{\partial y}\frac{\partial f}{\partial y} + \frac{\partial g}{\partial y} = 0$$

or

$$\frac{d\lambda}{dt} = -\lambda\frac{\partial f}{\partial y} - \frac{\partial g}{\partial y} = -\frac{\partial \mathcal{H}}{\partial y}, \tag{6.51}$$

which is the adjoint equation.

The above calculation shows that we may interpret the adjoint variable λ in the maximum principle as being given by the expression (6.42). As $w(t, y)$ represents the value of the capital asset y at time t, the adjoint variable λ represents the *marginal value* of this capital asset at time t. Reduction of the capital by one unit would reduce the value by approximately λ; for this reason $\lambda(t)$ is called the *shadow price* of the asset.

The Hamiltonian is the sum of two terms, each of which can be interpretad as a flow of values. The term $g(t, y, u)$ is the flow of revenues obtained from resource harvesting to the objective functional (6.36). The quantity $f(t, y, u)$ is the flow of investment in capital,

$$\frac{dy}{dt} = f(t, y, u).$$

The value of this investment flow is the flow multiplied by the shadow price $\lambda(t)$. Thus, the Hamiltonian represents the rate of increase of total assets, both accumulated dividends and capital assets. The maximum principle says that an optimal control must maximize the rate of increase of total assets, but we must know the shadow price in order to determine the optimal control. In effect, the maximum principle reduces the optimal control problem to the question of determining the shadow price. Unfortunately, the analytical determination of the shadow price in practice is a formidable problem, and it is often impossible to obtain an exact expression for the shadow price.

The adjoint equation (6.48) says that the rate of depreciation of capital $-\frac{d\lambda}{dt}$ consists of two parts, the marginal flow to accumulated dividends and the marginal flow to capital assets. Along the optimal path, this rate of depreciation must be equal to the sum of these marginal flows, $\frac{\partial g}{\partial y}$ and $\lambda\frac{\partial f}{\partial y}$, respectively. The economic literature usually speaks of a *current shadow price* $\tilde{\lambda}(t) = e^{\delta t}\lambda(t)$, rather than the discounted shadow price $\lambda(t)$ and of the *current-value* Hamiltonian

$$\tilde{\mathcal{H}} = e^{\delta t}\mathcal{H}.$$

In these terms the adjoint equation $\frac{d\lambda}{dt} = -\frac{\partial \mathcal{H}}{\partial y}$ becomes $\frac{d\tilde{\lambda}}{dt} = \delta\tilde{\lambda} - \frac{\partial \tilde{\mathcal{H}}}{\partial y}$.

Part III

Structured Populations Models

7
Basic Ideas of Mathematical Epidemiology

7.1 Introduction

The idea of invisible living creatures as agents of disease goes back at least to the writings of Aristotle (384 BC-322 BC). It developed as a theory in the 16th century. The existence of microorganisms was demonstrated by Leeuwenhoek (1632-1723) with the aid of the first microscopes. The first expression of the germ theory of disease by Jacob Henle (1809-1885) came in 1840 and was developed by Robert Koch (1843-1910), Joseph Lister (1827-1912), and Louis Pasteur (1827-1875) in the latter part of the 19th century and the early part of the 20th century.

The mechanism of transmission of infections is now known for most diseases. Generally, diseases transmitted by viral agents, such as influenza, measles, rubella (German measles), and chicken pox, confer immunity against reinfection, while diseases transmitted by bacteria, such as tuberculosis, meningitis, and gonorrhea, confer no immunity against reinfection. Other diseases, such as malaria, are transmitted not directly from human to human but by vectors, agents (usually insects) who are infected by humans and who then transmit the disease to humans.

Communicable diseases such as measles, influenza, or tuberculosis, are a fact of modern life. We will be concerned both with epidemics which are sudden outbreaks of a disease, and endemic situations, in which a disease is always present. The AIDS epidemic and outbursts of diseases such as the Ebola virus are events of concern and interest to many people. The prevalence and effects of many diseases in less developed countries are probably less well-known but may be of even more importance. Every year millions of people die of measles, respiratory infections, diarrhea and other diseases that are easily treated and not considered dangerous in the Western world. Diseases such as malaria, typhus, cholera, schistosomiasis, and sleeping sickness are endemic in many parts of the world. The effects of high disease mortality on mean life span and of disease debilitation and mortality on the economy in afflicted countries are considerable.

Epidemiological models focus on the transmission dynamics of a trait or traits transmitted from individual to individual, from population to population, from community to community, from region to region, or from country to country. A "trait" can be a disease, such as (i) measles, HIV, malaria, tuberculosis, (ii) a genetic characteristic, such as gender, race, genetic diseases, (iii) a cultural "characteristic", such as language or religion, (iv) an addictive activity, such as drug use, or (v) the gain or loss of information that are communicated through gossip, rumors, and so on. The term "individual" in epidemiology is broadly understood to include various epidemiological units. The selection of an epidemiological unit is based on the question and the level of aggregation at which the investigator wishes to address such a question. In the study of disease dynamics in the immune system, types of cells provide the epidemiological unit; in the study of the spread of malaria the host (humans or other mammals) and the vectors (female mosquitoes) are usually selected as the epidemiological units; in the study of *Chagas* disease a "house" (infested houses may correspond to "infected" individuals) may be chosen as an epidemiological unit; in tuberculosis, a household or community or a group of strongly linked individuals ("cluster") may be the chosen unit.

The above are examples of *epidemiological transmission processes* which, to be understood, must be studied from various perspectives that include the study of their transmission dynamics at different spatial, temporal, or organizational scales, the level of which is dictated by the question the investigator wishes to address. Unfortunately the selection of a particular level of organization and a model may determine a priori what is and what is not relevant. Hence, two relevant questions associated with epidemiological processes are: (i) how much organizational detail such as population structure, immune response, or genetic variability must be included in epidemiological models? and (ii) how is relevant detail modeled?

In this chapter we provide an introduction to mathematical epidemiology which includes the development of mathematical models for the spread of disease as well as for their analysis.

An epidemic, which acts on a short temporal scale, may be described as a sudden outbreak of a disease that infects a substantial portion of the population in a region before it disappears. Epidemics usually leave many members untouched. Often these attacks recur with intervals of several years between outbreaks, possibly diminishing in severity as populations develop some immunity. This is an important aspect of the connection between epidemics and disease evolution.

The Book of Exodus describes the plagues that Moses brought down upon Egypt, and there are several other biblical descriptions of epidemic outbreaks. Descriptions of epidemics in ancient and medieval times frequently used the term "plague" because of a general belief that epidemics represented divine retribution for sinful living. More recently some have described AIDS as punishment for sinful activities. Such views have often hampered or delayed attempts to control this modern epidemic.

The historian W.H. McNeill argues, especially in his book *Plagues and Peoples* (1976), that the spread of communicable diseases frequently has been an important influence in history. For example, there was a sharp population increase throughout the world in the 18th century; the population of China increased from 150 million in 1760 to 313 million in 1794 and the population of Europe increased from 118 million in 1700 to 187 million in 1800. There were many factors involved in this increase, including changes in marriage age and technological improvements leading to increased food supplies, but these factors are not sufficient to explain the increase. Demographic studies indicate that a satisfactory explanation requires recognition of a decrease in the mortality caused by periodic epidemic infections. This decrease came about partly through improvements in medicine, but a more important influence was probably the fact that more people developed immunities against infection as increased travel intensified the circulation and co-circulation of diseases.

There are many biblical references to diseases as historical influences, such as the decision of Sennacherib, the king of Assyria, to abandon his attempt to capture Jerusalem about 700 BC because of the illness of his soldiers (Isaiah 37,36-38). The fall of empires has been attributed directly or indirectly to epidemic diseases. In the 2nd century AD the so-called Antonine plagues (possibly measles and smallpox) invaded the Roman Empire, causing drastic population reductions and economic hardships leading to disintegration of the empire because of disorganization, which facilitated invasions of barbarians. The Han empire in China collapsed in the 3rd century AD after a very similar sequence of events. The defeat of a population of millions of Aztecs by Cortez and his 600 followers can be explained, in part, by a smallpox epidemic that devastated the Aztecs but had almost no effect on the invading Spaniards thanks to their built-in immunities. The Aztecs were not only weakened by disease but also confounded by what they interpreted as a divine force favoring the invaders. Smallpox then spread southward to the Incas in Peru and was an important factor in the success of Pizarro's invasion a few years later. Smallpox was followed by other diseases such as measles and diphtheria imported from Europe to North America. In some regions, the indigenous populations were reduced to one tenth of their previous levels by these diseases: Between 1519 and

1530 the Indian population of Mexico was reduced from 30 million to 3 million.

The Black Death (bubonic plague) spread from Asia throughout Europe in several waves during the 14th century, beginning in 1346, and is estimated to have caused the death of as much as one-third of the population of Europe between 1346 and 1350. The disease recurred regularly in various parts of Europe for more than 300 years, notably as the Great Plague of London of 1665-1666. It then gradually withdrew from Europe. As the plague struck some regions harshly while avoiding others, it had a profound effect on political and economic developments in medieval times. In the last bubonic plague epidemic in France (1720-1722), half the population of Marseilles, 60 percent of the population in nearby Toulon, 44 per cent of the population of Arles and 30 percent of the population of Aix and Avignon died, but the epidemic did not spread beyond Provence. Expansions and interpretations of these historical remarks may be found in McNeill (1976), which was our primary source on the history of the spread and effects of diseases.

The above examples depict the *sudden* dramatic impact that disease has had on the demography of human populations via disease-induced mortality. In considering the combined role of diseases, war, and natural disasters on mortality rates, one may conclude that *historically* humans who are more likely to survive and reproduce have either a good immune system, a propensity to avoid war and disasters, or, nowadays, excellent medical care and/or health insurance.

There are many questions of interest to public health physicians confronted with a possible epidemic. For example, how severe will an epidemic be? This question may be interpreted in a variety of ways. For example, how many individuals will be affected all together and thus require treatment? What is the maximum number of people needing care at any particular time? How long will the epidemic last? How much good would quarantine of victims do in reducing the severity of the epidemic?

For diseases that are endemic in some region public health physicians need to be able to estimate the number of infectives at a given time as well as the rate at which new infections arise. The effects of quarantine or vaccine in reducing the number of victims are of importance, just as in the treatment of epidemics. In addition, the possibility of defeating the endemic nature of the disease and thus controlling or limiting the disease in a population is worthy of study. How can such questions be answered?

Scientific experiments usually are designed to obtain information and to test hypotheses. Experiments in epidemiology with controls are often difficult or impossible to design and even if it is possible to arrange an experiment there are serious ethical questions involved in withholding treatment from a control group. Sometimes data may be obtained after the fact from reports of epidemics or of endemic disease levels, but the data may be incomplete or inaccurate. In addition, data may contain enough irregularities to raise serious questions of interpretation, such as whether there is evidence of chaotic behavior [Ellner, Gallant, and Theiler (1995)]. Hence, parameter estimation and model fitting are very difficult. These issues raise the question of whether mathematical modeling in epidemiology is of value.

Mathematical modeling in epidemiology provides understanding of the underlying mechanisms that influence the spread of disease and, in the process, it suggests control strategies. In fact, models often identify behaviors that are unclear in experimental data–often because data are non-reproducible and the number of data points is limited and subject to errors in measurement. For example, one of the fundamental results in mathematical epidemiology is that *most* mathematical epidemic models, including those that include a high degree of heterogeneity, usually exhibit "threshold" behavior which in epidemiological terms can be stated as follows: *If the average number of secondary infections caused by an average infective is less than one a disease will die out, while if it exceeds one there will be an epidemic.* This broad principle, consistent with observations and quantified via epidemiological models, has been consistently used to estimate the effectiveness of vaccination policies and the likelihood that a disease may be eliminated or eradicated. Hence, even if it is not possible to verify hypotheses accurately, agreement with hypotheses of a qualitative nature is often valuable. Expressions for the basic reproductive number for HIV in various populations is being used to test the possible effectiveness of vaccines that may provide temporary protection by reducing either HIV-infectiousness or suceptibility to HIV. Models are used to estimate how widespread a vaccination plan must be to prevent or reduce the spread of HIV.

In the mathematical modeling of disease transmission, as in most other areas of mathematical modeling, there is always a trade-off between simple models, which omit most details and are designed only to highlight general qualitative behavior, and detailed models, usually designed for specific situations including short-term quantitative predictions. Detailed models are generally difficult or impossible to solve analytically and hence their usefulness for theoretical purposes is limited, although their strategic value may be high. This chapter begins with simple models in order to establish broad principles. Furthermore, these simple models have additional value as they are the building blocks of models that include detailed structure. A specific goal of this chapter is to compare the dynamics of simple and slightly more detailed models primarily to see whether slightly different assumptions can lead to significant differences in qualitative behavior.

We will often think of a disease as an invasion of a host population, consisting of separate patches (individuals), by a pathogen. An epidemic is a successful invasion if the number of occupied patches increases over time after the initial introduction of the pathogen into the host (patch) population.

Many of the early developments in the mathematical modeling of communicable diseases are due to public health physicians. The first known result in mathematical epidemiology is a defense of the practice of inoculation against smallpox in 1760 by Daniel Bernouilli, a member of a famous family of mathematicians (eight spread over three generations) who had been trained as a physician. The first contributions to modern mathematical epidemiology are due to P.D. En'ko between 1873 and 1894 [Dietz (1988)]), and the foundations of the entire approach to epidemiology based on compartmental models were laid by public health physicians such as Sir R.A. Ross, W.H. Hamer, A.G. McKendrick, and W.O. Kermack between 1900 and 1935, along with important contributions from a statistical perspective by J. Brownlee. A particularly instructive example is the work of Ross on malaria. Dr. Ross was awarded the second Nobel Prize in Medicine for his demonstration of the dynamics of the transmission of malaria between mosquitoes and humans. Although his work received immediate acceptance in the medical community, his conclusion that malaria could be controlled by controlling mosquitoes was dismissed on the grounds that it would be impossible to rid a region of mosquitoes completely and that in any case mosquitoes would soon reinvade the region. After Ross formulated a mathematical model that predicted that malaria outbreaks could be avoided if the mosquito population could be reduced below a critical threshold level, field trials supported his conclusions and led to sometimes brilliant successes in malaria control. However, the Garki project provides a dramatic counterexample. This project worked to eradicate malaria from a region temporarily. However, people who have recovered from an attack of malaria have a temporary immunity against reinfection. Thus elimination of malaria from a region leaves the inhabitants of this region without immunity when the campaign ends, and the result can be a serious outbreak of malaria.

We formulate our descriptions as *compartmental models*, with the population under study being divided into compartments and with assumptions about the nature and time rate of transfer from one compartment to another. Diseases that confer immunity have a different compartmental structure from diseases without immunity and from diseases transmitted by vectors. The rates of transfer between compartments are expressed mathematically as derivatives with respect to time of the sizes of the compartments, and as a result our models are formulated initially as *differential equations*. Later, when we study models in which the rates of transfer depend on the sizes of compartments over the past as well as at the instant of transfer, more general types of functional equation, such as differential-difference equations or integral equations, will be used.

7.2 A Simple Epidemic Model

Perhaps the first epidemic to be examined from a modeling point of view was the Great Plague in London (1665-1666). The plague was one of a sequence of attacks beginning in the year 1346 of what came to be known as the Black Death. It is now identified as the bubonic plague, which had actually invaded Europe as early as the 6th century during the reign of the Emperor Justinian of the Roman Empire and continued for more than three centuries after the Black Death. The Great Plague killed about one-sixth of the population of London. One of the few "benefits" of the plague was that it caused Cambridge University to be closed for two years. Isaac Newton, who was a student at Cambridge at the time, was sent to his home and while "in exile" he had one of the most productive scientific periods of any human in history. He discovered his law of gravitation, among other things, during this period.

The characteristic features of the Great Plague were that it appeared quite suddenly, grew in intensity, and then disappeared, leaving part of the population untouched. The same features have been observed in many other epidemics, both of fatal diseases and of diseases whose victims recovered with immunity against reinfection.

In the 19th century recurrent invasions of cholera killed millions in India. The influenza epidemic of 1918-1919 killed 20 million people overall, more than half a million in the United States. One of the questions that first attracted the attention of scientists interested in the study of the spread of communicable diseases was why diseases would suddenly develop in a community and then disappear just as suddenly without infecting the entire community. One of the early triumphs of mathematical epidemiology [Kermack and McKendrick (1927)] was the formulation of a simple model that predicted behavior very similar to the behavior observed in countless epidemics.

In order to model such an epidemic we divide the population being studied into three classes labeled S, I, and R. We let $S(t)$ denote the number of individuals who are susceptible to the disease, that is, who are not (yet) infected at time t. $I(t)$ denotes the number of infected individuals, assumed infectious and able to spread the disease by contact with susceptibles. $R(t)$ denotes the number of individuals who have been infected and then removed from the possibility of being infected again or of spreading infection. Removal is carried out either through isolation from the rest of the population or through immunization against infection or through recovery from the disease with full immunity against reinfection or through death caused by the disease. These characterizations of removed members are different from an epidemiological perspective but are often equivalent from a modeling point of view which takes into account only the state of an individual with respect to the disease.

In formulating models in terms of the derivatives of the sizes of each compartment we are assuming that the number of members in a compartment is a differentiable function of time. This may be a reasonable approximation if there are many members in a compartment, but it is certainly suspect otherwise. In formulating models as differential equations, we are assuming that the epidemic process is *deterministic*, that is, the behavior of a population is determined completely by its history and by the rules which govern the development of the model. For small compartment sizes the behavior of the compartment size may be strongly influenced by random perturbations, and other types of models, like stochastic models, may be more appropriate.

The model proposed by Kermack and McKendrick in 1927 to describe this situation is

$$\begin{aligned} S' &= -\beta SI \\ I' &= \beta SI - \alpha I \\ R' &= \alpha I. \end{aligned}$$

It is based on the following assumptions:

(i) An average infective makes contact sufficient to transmit infection with βN others per unit time, where N represents total population size.

(ii) A fraction α of infectives leave the infective class per unit time.

(iii) There is no entry into or departure from the population, except possibly through death from the disease.

According to (i), since the probability that a random contact by an infective is with a susceptible, who can then transmit infection, is S/N, the number of new infections in unit time is $(\beta N)(S/N)I = \beta SI$. We need not give an algebraic expression for N since it cancels out of the final model, but we should note that for a disease that is fatal to all who are infected $N = S + I$; while, for a disease from which all infected members recover with immunity, $N = S + I + R$. The hypothesis (iii) really says that the time scale of the disease is much faster than the time scale of births and deaths so that demographic effects on the population may be ignored. An alternative view is that we are only interested in studying the dynamics of a single epidemic outbreak. In a later section we shall consider a model that is the same as the one considered here except for the incorporation of demographic effects (births and deaths) along with the corresponding epidemiological assumptions.

The assumption (ii) requires a fuller mathematical explanation, since the assumption of a recovery rate proportional to the number of infectives has no clear epidemiological meaning. We consider the "cohort" of members who were all infected at one time and let $u(s)$ denote the number of these who are still infective s time units after having been infected. If a fraction α of these leave the infective class in unit time then

$$u' = -\alpha u,$$

and the solution of this elementary differential equation is

$$u(s) = u(0)\, e^{-\alpha s}.$$

Thus, the fraction of infectives remaining infective s time units after having become infective is $e^{-\alpha s}$, so that the length of the infective period is distributed exponentially with mean $\int_0^\infty e^{-\alpha s} ds = 1/\alpha$, and this is what (ii) really assumes (see Section 1.7).

The assumptions of a rate of contacts proportional to population size N with constant of proportionality β, and of an exponentially distributed recovery rate are unrealistically simple. Later we shall consider more general models, but our goal here is to show what may be deduced from extremely simple models. It will turn out that that many more realistic models exhibit very similar qualitative behaviors.

In our model R is determined once S and I are known, and we can drop the R' equation from our model, leaving the system of two equations

$$\begin{aligned} S' &= -\beta S I \\ I' &= \left(\beta S - \alpha\right) I. \end{aligned} \qquad (7.1)$$

We are unable to solve this system analytically but we learn a great deal about the behavior of its solutions by the following qualitative approach. We observe that $S' < 0$ for all t and $I' > 0$ if and only if $S > \alpha/\beta$. Thus I increases so long as $S > \alpha/\beta$ but since S decreases for all t, I ultimately decreases and approaches zero. If $S(0) < \alpha/\beta$, I decreases to zero (no epidemic), while if $S(0) > \alpha/\beta$, I first increases to a maximum attained when $S = \alpha/\beta$ and then decreases to zero (epidemic). We think of introducing a small number of infectives into a population of susceptibles and ask whether there will be an epidemic. The quantity $\beta S(0)/\alpha$ is a threshold quantity, called the *basic reproductive number* and denoted by R_0, which determines whether there is an epidemic or not. If $R_0 < 1$ the infection dies out, while if $R_0 > 1$ there is an epidemic.

Instead of trying to solve for S and I as functions of t, we divide the two equations of the model to give

$$\frac{I'}{S'} = \frac{dI}{dS} = \frac{\left(\beta S - \alpha\right) I}{-\beta S I} = -1 + \frac{\alpha}{\beta S},$$

and integrate to find the orbits (curves in the (S, I)-plane, or phase plane)

$$I = -S + \frac{\alpha}{\beta} \log S + c, \qquad (7.2)$$

with c an arbitrary constant of integration. Another way to describe the orbits is to define the function

$$V(S,I) = S + I - \frac{\alpha}{\beta}\log S$$

and note that each orbit is a curve given implicitly by the equation $V(S,I) = c$ for some choice of the constant c. The constant c is determined by the initial values S_0, I_0 of S and I, respectively, because $c = V(S_0, I_0) = S_0 + I_0 - \alpha/\beta \log S_0$.

Let us think of a population of size K into which a small number of infectives is introduced, so that $S_0 \approx K$, $I_0 \approx 0$, and $R_0 = \beta K/\alpha$. If we use the fact that $\lim_{t\to\infty} I(t) = 0$, and let $S_\infty = \lim_{t\to\infty} S(t)$, then the relation $V(S_0, I_0) = V(S_\infty, 0)$ gives

$$K - \frac{\alpha}{\beta}\log S_0 = S_\infty - \frac{\alpha}{\beta}\log S_\infty,$$

from which we obtain an expression for β/α in terms of the measurable quantities S_0 and S_∞, namely

$$\frac{\beta}{\alpha} = \frac{\log\frac{S_0}{S_\infty}}{K - S_\infty}. \tag{7.3}$$

We note from (7.3) that $0 < S_\infty < K$; that is, part of the population escapes infection. It is generally difficult to estimate the contact rate β which depends on the particular disease being studied but may also depend on social and behavioral factors. The quantities S_0 and S_∞ may be estimated by serological studies (measurements of immune responses in blood samples) before and after an epidemic, and from these data the basic reproductive number R_0 may be estimated by using (7.3).

The maximum number of infectives at any time is the number of infectives when the derivative of I is zero, that is, when $S = \alpha/\beta$. This maximum is given by

$$I_{max} = S_0 + I_0 - \frac{\alpha}{\beta}\log S_0 - \frac{\alpha}{\beta} + \frac{\alpha}{\beta}\log\frac{\alpha}{\beta}, \tag{7.4}$$

obtained by substituting $S = \alpha/\beta$, $I = I_{max}$ into (7.2).

Example 1. A study of Yale University freshmen [Evans (1982), reported by Hethcote(1989)] described an influenza epidemic with $S_0 = 0.911$, $S_\infty = 0.513$. Here we are measuring the number of susceptibles as a fraction of the total population size, or using the population size K as the unit of size. Substitution into (7.3) gives the estimate $\beta/\alpha = 1.44$ and $R_0 = 1.44$. Since we know that τ is approximately 3 days for influenza, we see that β is approximately 0.48 contacts per day per member of the population.

Example 2. (The Great Plague in Eyam) The village of Eyam near Sheffield, England suffered an outbreak of bubonic plague in 1665-1666 the source of which is generally believed to be the Great Plague of London. The Eyam plague was survived by only 83 of an initial population of 350 persons. As detailed records were preserved and as the community was persuaded to quarantine itself to try to prevent the spread of disease to other communities, the disease in Eyam has been used as a case study for modeling [Raggett (1982)]. Detailed examination of the data indicates that there were actually two outbreaks of which the first was relatively mild. Thus we shall try to fit the model (7.1) over the period from mid-May to mid-October 1666, measuring time in months with an initial population of 7 infectives and 254 susceptibles, and a final population of 83. Raggett (1982) gives values of susceptibles and infectives in Eyam on various dates, beginning with $S(0) = 254, I(0) = 7$, shown in Table 7.1.

Date (1666)	**Susceptibles**	**Infectives**
July 3/4	235	14.5
July 19	201	22
August 3/4	153.5	29
August 19	121	21
September 3/4	108	8
September 19	97	8
October 4/5	Unknown	Unknown
October 20	83	0

TABLE 7.1. Eyam Plague data

The relation (7.3) with $S_0 = 254$, $I_0 = 7$, $S_\infty = 83$ gives $\beta/\alpha = 6.54 \times 10^{-3}$, $\alpha/\beta = 153$. The infective period was 11 days, or 0.3667 month, so that $\alpha = 2.73$. Then $\beta = 0.0178$. The relation (7.4) gives an estimate of 30.4 for the maximum number of infectives. We use the values obtained here for the parameters β and τ in the model 7.1 for simulations of both the phase plane, the (S, I)-plane, and for graphs of S and I as functions of t (Figures 7.1, 7.2, 7.3). Figure 7.4 plots these data points together with the phase portrait given in Figure 7.1 for the model (7.1).

The actual data for the Eyam epidemic are remarkably close to the predictions of this very simple model. However, the model is really too good to be true. Our model assumes that infection is transmitted directly between people. While this is possible, bubonic plague is transmitted mainly by rat fleas. When an infected rat is bitten by a flea, the flea becomes extremely hungry and bites the host rat repeatedly, spreading the infection in the rat. When the host rat dies its fleas move on to other rats, spreading the disease further. As the number of available rats decreases the fleas move to human hosts, and this is how plague starts in a human population (although the second phase of the epidemic may have been the pneumonic form of bubonic plague, which can be spread from person to person.) One

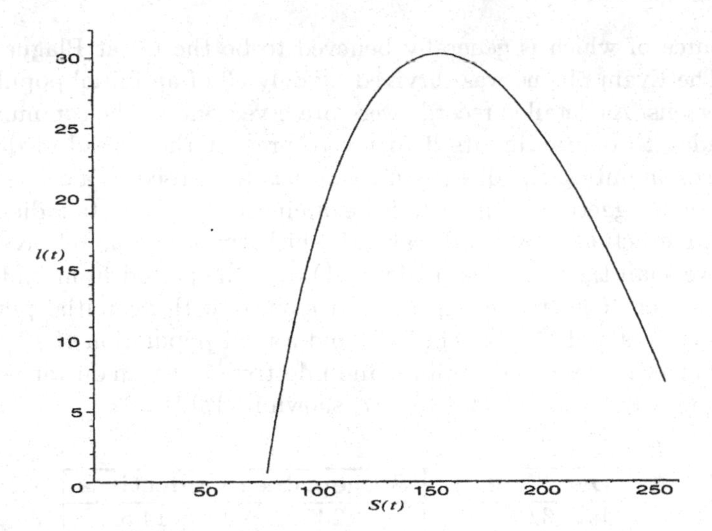

FIGURE 7.1. The S-I plane.

of the main reasons for the spread of plague from Asia into Europe was the passage of many trading ships; in medieval times ships were invariably infested with rats. An accurate model of plague transmission would have to include flea and rat populations, as well as movement in space. Such a model would be extremely complicated and its predictions might well not be any closer to observations than our simple unrealistic model. Raggett (1982) also used a stochastic model to fit the data, but the fit was rather poorer than the fit for the simple deterministic model (7.1).

In the village of Eyam the rector persuaded the entire community to quarantine itself to prevent the spread of disease to other communities. One effect of this policy was to increase the infection rate in the village by keeping fleas, rats, and people in close contact with one another, and the mortality rate from bubonic plague was much higher in Eyam than in London. Further, the quarantine could do nothing to prevent the travel of rats and thus did little to prevent the spread of disease to other communities. One message this suggests to mathematical modelers is that control strategies based on false models may be harmful, and it is essential to distinguish between assumptions that simplify but do not alter the predicted effects substantially, and wrong assumptions which make an important difference.

In order to prevent the occurrence of an epidemic if infectives are introduced into a population it is necessary to reduce the basic reproductive number R_0 below one. This may sometimes be achieved by immunization, which has the effect of transferring members of the population from the susceptible class to the removed class and thus of reducing $S(0)$. If a fraction p of the population is successfully immunized the effect is to replace $S(0)$ by $S(0)(1-p)$, and thus to reduce the basic reproductive number to $\beta S(0)(1-p)/\alpha$. The requirement $\beta S(0)(1-p)/\alpha < 1$ gives $1-p < \alpha/\beta S(0)$,

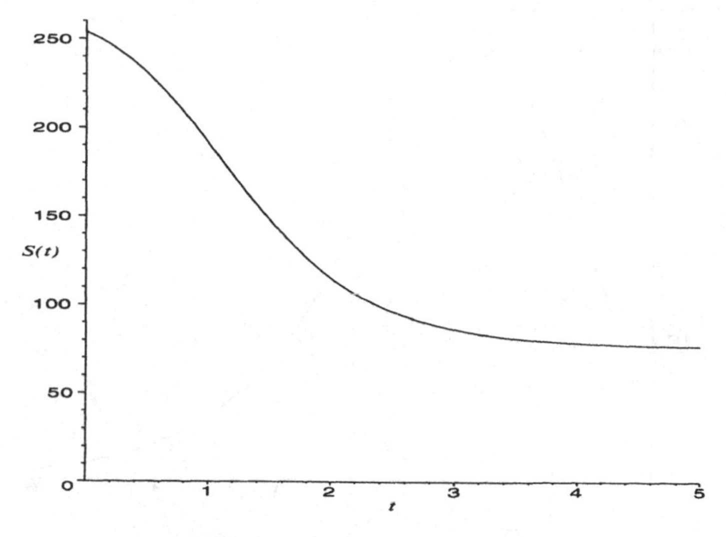

FIGURE 7.2. S as a function of t.

or

$$p > 1 - \frac{\alpha}{\beta S(0)} = 1 - \frac{1}{R_0}.$$

A large basic reproductive number means that the fraction that must be immunized to prevent the spread of infection is large. This relation is connected to the idea of herd immunity, which we shall introduce in Section 7.4.

Exercises

1. The same survey of Yale students described in Example 1 reported that 91.1 percent were susceptible to influenza at the beginning of the year and 51.4 percent were susceptible at the end of the year. Estimate the basic reproductive number β/α and decide whether there was an epidemic.

2. What fraction of Yale students in Exercise 1 would have had to be immunized to prevent an epidemic?

3. What was the maximum number of Yale students in Exercises 1 and 2 suffering from influenza at any time?

4. An influenza epidemic was reported at an English boarding school in 1970 which spread to 512 of the 763 students. Estimate the basic reproductive number β/α.

5. What fraction of the boarding school students in Exercise 4 would have had to be immunized to prevent an epidemic?

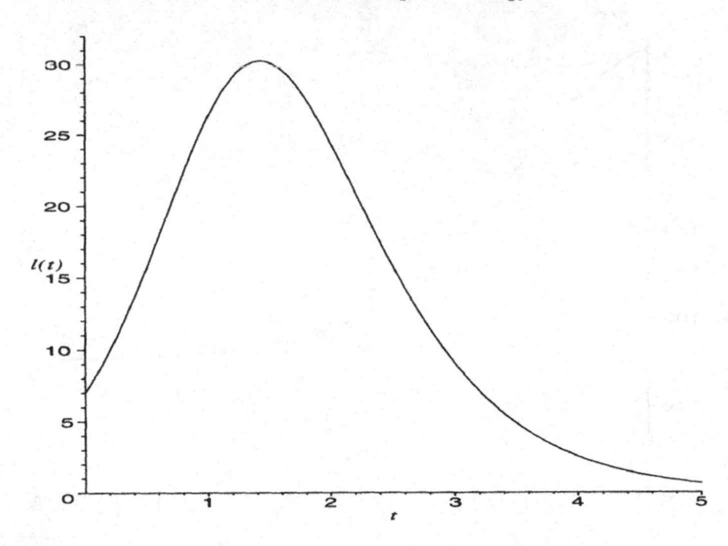

FIGURE 7.3. I as a function of t.

6. What was the maximum number of boarding school students in Exercises 4 and 5 suffering from influenza at any time?

7.3 A Model for Diseases with No Immunity

The model in the preceding section is an example of a model in which the population is divided into different compartments and the transitions between compartments are modeled. It is described as an SIR model because the transitions are from susceptible to infective to removed, with the removal coming through recovery with full immunity (as in measles) or through death from the disease (as in plague, rabies and many other animal diseases). Another type of model is an SIS model in which infectives return to the susceptible class on recovery because the disease confers no immunity against reinfection. Such models are appropriate for most diseases transmitted by bacterial or helminth agents, and most sexually transmitted diseases (including gonorrhea, but not such diseases as AIDS from which there is no recovery).

The simplest SIS model, due to Kermack and McKendrick (1932), is

$$\begin{aligned} S' &= -\beta SI + \gamma I \\ I' &= \beta SI - \gamma I. \end{aligned} \tag{7.5}$$

This differs from the SIR model only in that the recovered members return to the class S at a rate γI instead of passing to the class R. The total population $S + I$ is a constant, since $(S + I)' = 0$. We call this constant K; sometimes population size is measured using K as the unit so that the

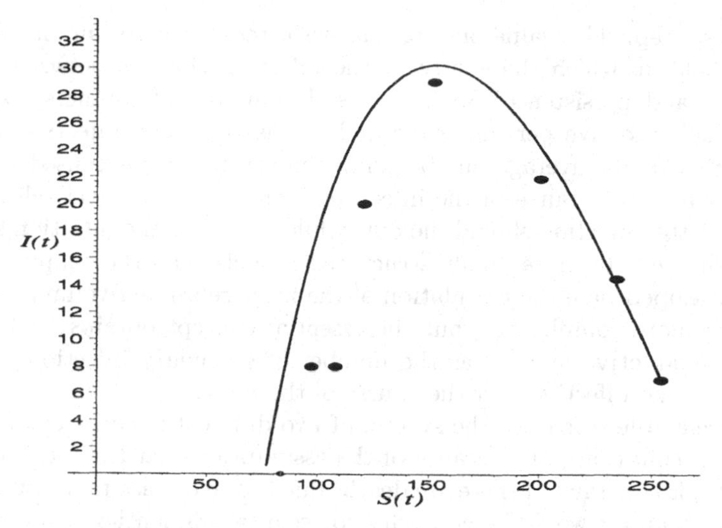

FIGURE 7.4. The S-I plane, model and data.

total population size is one. We may reduce the model to a single differential equation by replacing S by $K - I$ to give the single differential equation

$$\begin{aligned} I' &= \beta I(K - I) - \gamma I = (\beta K - \gamma)I - \beta I^2 \\ &= (\beta K - \gamma)I\left(1 - \frac{I}{K - \frac{\gamma}{\beta}}\right), \end{aligned} \tag{7.6}$$

which was studied in Chapter 1.

As (7.6) is a logistic differential equation with $r = \beta K - \gamma$ and with $L = K - \gamma/\beta$, our qualitative result tells us that if $\beta K - \gamma < 0$ or $\beta K/\gamma < 1$, then all solutions of the model (7.6) with non-negative initial values except the constant solution $I = K - \beta/\gamma$ approach the limit zero as $t \to \infty$, while if $\beta K/\gamma > 1$ then all solutions with nonnegative initial values except the constant solution $I = 0$ approach the limit $K - \gamma/\beta > 0$ as $t \to \infty$. Thus there is always a single limiting value for I but the value of the quantity $\beta K/\gamma$ determines which limiting value is approached, regardless of the initial state of the disease. In epidemiological terms this says that if the quantity $\beta K/\gamma$ is less than one the infection dies out in the sense that the number of infectives approaches zero. For this reason the equilibrium $I = 0$, which corresponds to $S = K$, is called the *disease-free equilibrium*. On the other hand, if the quantity $\beta K/\gamma$ exceeds one the infection persists. The equilibrium $I = K - \gamma/\beta$, which corresponds to $S = \gamma/\beta$, is called an *endemic* equilibrium.

As previously pointed out in Section 7.2, the dimensionless quantity $\beta K/\gamma$ is called the basic reproductive number or contact number for the disease, and it is usually denoted by R_0. In studying an infectious disease, the determination of the basic reproductive number is invariably a vital first step. The value one for the basic reproductive number defines a threshold at which the course of the infection changes between disappearance and persistence. Since βK is the number of contacts made by an average infective per unit time and $1/\gamma$ is the mean infective period, R_0 represents the average number of secondary infections caused by each infective over the course of the infection. Thus, it is intuitively clear that if $R_0 < 1$ the infection should die out, while if $R_0 > 1$ the infection should establish itself. In more highly structured models than the simple one we have developed here the calculation of the basic reproductive number may be much more complicated, but the essential concept obtains–that of the basic reproductive number as the number of secondary infections caused by an average infective over the course of the disease.

We were able to reduce the system of two differential equations (7.5) to the single equation (7.6) because of the assumption that the total population $S+I$ is constant. If there are deaths due to the disease this assumption is violated, and it would be necessary to use a two-dimensional system as a model. We shall consider this in a more general context in the next section.

Exercises

1. A disease is introduced by two visitors into a town with 1200 inhabitants. An average infective is in contact with 0.4 inhabitant per day. The average duration of the infective period is 6 days, and recovered infectives are immune against reinfection. How many inhabitants would have to be immunized to avoid an epidemic?

2. In Example 3, Section 1.3 a disease was described with $\beta = 1/3000$, $1/\alpha = 6$ days in a population of 1200 members. Suppose the disease conferred immunity on recovered infectives. How many members would have to be immunized to avoid an epidemic?

3. A survey of freshman students at Yale University reported in 1982 found that 25 percent were susceptible to rubella at the beginning of the year and 9.65 percent were susceptible at the end of the year. What fraction would have had to be immunized at the beginning of the year to avoid the spread of rubella (rubella confers immunity against reinfection)?

4. A disease begins to spread in a population of 800. The infective period has an average duration of 14 days and the average infective is in contact with 0.1 person per day. What is the basic reproductive number? To what level must the average rate of contact be reduced so that the disease will die out?

5.* European fox rabies is estimated to have a transmission coefficient β of 80 km^2 years/fox and an average infective period of 5 days.

There is a critical carrying capacity K_c measured in foxes per km^2, such that in regions with fox density less than K_c rabies tends to die out while in regions with fox density greater than K_c rabies tends to persist. Estimate K_c. [Remark: It has been suggested in Great Britain that hunting to reduce the density of foxes below the critical carrying capacity would be a way to control the spread of rabies.]

6.* A communicable disease from which infectives do not recover may be modeled by the pair of differential equations

$$S' = -\beta SI, \quad I' = \beta SI.$$

Show that in a population of fixed size K such a disease will eventually spread to the entire population.

7.* Consider a disease spread by carriers who transmit the disease without exhibiting symptoms themselves. Let $C(t)$ be the number of carriers and suppose that carriers are identified and isolated from contact with others at a constant per capita rate α, so that $C' = -\alpha C$. The rate at which susceptibles become infected is proportional to the number of carriers and to the number of susceptibles, so that $S' = -\beta SC$. Let C_0 and S_0 be the number of carriers and susceptibles, respectively, at time $t = 0$.

(i) Determine the number of carriers at time t from the first equation.

(ii) Substitute the solution to part (a) into the second equation and determine the number of susceptibles at time t.

(iii) Find $\lim_{t\to\infty} S(t)$, the number of members of the population who escape the disease.

8.* Consider a population of fixed size K in which a rumor is being spread by word of mouth. Let $y(t)$ be the number of people who have heard the rumor at time t and assume that everyone who has heard the rumor passes it on to r others in unit time. Thus, from time t to time $(t+h)$ the rumor is passed on $hry(t)$ times, but a fraction $y(t)/K$ of the people who hear it have already heard it, and thus there are only $hry(t)\left(\frac{K-y(t)}{K}\right)$ people who hear the rumor for the first time. Use these assumptions to obtain an expression for $y(t+h) - y(t)$, divide by h, and take the limit as $h \to 0$ to obtain a differential equation satisfied by $y(t)$.

9. At 9 AM one person in a village of 100 inhabitants starts a rumor. Suppose that everyone who hears the rumor tells one other person per hour. Using the model of Exercise 8, determine how long it will take until half the village has heard the rumor.

10.* If a fraction λ of the population susceptible to a disease that provides immunity against reinfection moves out of the region of an epidemic, the situation may be modeled by a system

$$S' = -\beta SI - \lambda S, \quad I' = \beta SI - \alpha I.$$

Show that both S and I approach zero as $t \to \infty$.

7.4 Models with Demographic Effects

Measles is a disease for which endemic equilibria have been observed in many places, frequently with sustained oscillations about the equilibrium. As recovery from measles provides immunity against reinfection, the SIS model of the preceding section is inappropriate. The epidemic model of Section 7.2 assumes that the epidemic time scale is so short relative to the demographic time scale that demographic effects may be ignored. For measles, however, the reason for the endemic nature of the disease is that there is a flow of new susceptible members into the population, and in order to try to model this we must include births and deaths in the model. The simplest way to incorporate births and deaths in an infectious disease model is to assume a constant number of births and an equal number of deaths per unit time so that the total population size remains constant. This is, of course, feasible only if there are no deaths due to the disease. In developed countries such an assumption is plausible because there are few deaths from measles. In less developed countries there is often a very high mortality rate for measles and therefore other assumptions are necessary. The first attempt to formulate an SIR model with births and deaths to describe measles was given by H. E. Soper (1929), who assumed a constant birth rate μK in the susceptible class and a constant death rate μK in the removed class. His model is

$$\begin{aligned} S' &= -\beta SI + \mu K \\ I' &= \beta SI - \gamma I \\ R' &= \gamma I - \mu K. \end{aligned}$$

Soper's model is unsatisfactory biologically because the linkage of births of susceptibles to deaths of removed members is unreasonable. It is also an improper model mathematically because if $R(0)$ and $I(0)$ are sufficiently small then $R(t)$ will become negative. For any disease model to be plausible it is essential that the problem be properly posed in the sense that the number of members in each class must remain non-negative. A model that does not satisfy this requirement cannot be a proper description of a disease model and therefore must contain some assumption that is biologically unreasonable. A full analysis of a model should include the verification of this property.

A model of Kermack and McKendrick (1932) includes births in the susceptible class proportional to total population size and a death rate in each class proportional to the number of members in the class. This model allows the total population size to grow exponentially or die out exponentially if the birth and death rates are unequal. It is applicable to questions of whether or not a disease will control the size of a population that would otherwise grow exponentially. We shall return to this topic, which is important in the study of many diseases in less developed countries with high birth rates. For the moment we shall follow the approach suggested by Hethcote (1976) in which the total population size is held constant by making birth and death rates equal. Such a model is

$$\begin{aligned} S' &= -\beta SI + \mu(K - S) \\ I' &= \beta SI - \gamma I - \mu I \\ R' &= \gamma I - \mu R. \end{aligned}$$

Because $S + I + R = K$, we can eliminate R and consider the two-dimensional system

$$\begin{aligned} S' &= -\beta SI + \mu(K - S) \\ I' &= \beta SI - \gamma I - \mu I. \end{aligned}$$

We shall examine a slightly more general SIR model with births and deaths for a disease that may be fatal to some infectives. For such a disease the class R of removed members should contain only recovered members, not members removed by death from the disease. It is not possible to assume that the total population size remain constant if there are deaths due to disease; a plausible model for a disease that may be fatal to some infectives must allow the total population to vary in time. The simplest assumption to allow this is constant birth rate. Let us analyze the model

$$\begin{aligned} S' &= \mu K - \beta SI - \mu S \\ I' &= \beta SI - \mu I - \alpha I - \gamma I \\ R' &= \gamma I - \mu R, \end{aligned} \tag{7.7}$$

with a constant number of births μK per unit time, a proportional death rate μ in each class, and with a fraction α of infectives dying from infection and a fraction γ of infectives recovering with immunity against reinfection. In this model if $\alpha > 0$ the total population size is not constant and K represents a carrying capacity or maximum possible population size, rather than population size. Just as in the case $\alpha = 0$ we may drop the third equation but the reasoning is somewhat different. We view the first two equations as determining S and I, and then consider the third equation as determining R once S and I are known. This is possible because R does not enter into the first two equations.

We shall analyze the model (7.7) qualitatively and then make some inferences or predictions, which follow from the model, about herd immunity, mean age at infection, and the inter-epidemic period. These inferences may be of value in testing the model or in estimating parameters. Our approach will be analogous to the method used in the previous section to study the SIS model. This method involved identifying equilibria (constant solutions) and then determining the asymptotic stability of each equilibrium, a method we have been using throughout this book. To find equilibria we set the right side of each of the two equations equal to zero. The second of the resulting algebraic equations factors, giving two alternatives. The first alternative is $I = 0$, which will give a disease-free equilibrium, and the second alternative is $\beta S = \mu + \gamma + \alpha$, which will give an endemic equilibrium, provided $\beta S = \mu + \gamma + \alpha < \beta K$. If $I = 0$ the other equation gives $S = K$. For the endemic equilibrium the first equation gives

$$I = \frac{\mu K}{\mu + \gamma + \alpha} - \frac{\mu}{\beta}. \tag{7.8}$$

We linearize about an equilibrium (S_∞, I_∞), as in Chapter 4, by letting $y = S - S_\infty$, $z = I - I_\infty$, writing the system in terms of the new variables y and z and retaining only the linear terms in a Taylor expansion. We obtain a system of two linear differential equations,

$$\begin{aligned} y' &= -(\beta I_\infty + \mu)y - \beta S_\infty z \\ z' &= \beta I_\infty y + (\beta S_\infty - \mu - \gamma - \alpha)z. \end{aligned}$$

This linear system has coefficient matrix here

$$\begin{pmatrix} -\beta I_\infty - \mu & -\beta S_\infty \\ \beta I_\infty & \beta S_\infty - \mu - \gamma - \alpha \end{pmatrix}$$

We then look for solutions whose components are constant multiples of $e^{\lambda t}$; this means that λ must be an eigenvalue of the coefficient matrix. The condition that all solutions of the linearization at an equilibrium tend to zero as $t \to \infty$ is that all eigenvalues of this coefficient matrix have negative real part. At the disease-free equilibrium the matrix is

$$\begin{pmatrix} -\mu & -\beta K \\ 0 & \beta K - \mu - \gamma - \alpha \end{pmatrix},$$

which has eigenvalues $-\mu$ and $\beta K - \mu - \gamma - \alpha$. Thus, the disease-free equilibrium is asymptotically stable if $\beta K < \mu + \gamma + \alpha$ and unstable if $\beta K > \mu + \gamma + \alpha$. Note that this condition for instability of the disease-free equilibrium is the same as the condition for the existence of an endemic equilibrium.

In general, the condition that the eigenvalues of a 2×2 matrix have negative real part is that the determinant be positive and the trace (the sum of the diagonal elements) be negative. Since $\beta S_\infty = \mu+\gamma+\alpha$ at an endemic equilibrium, the matrix of the linearization at an endemic equilibrium is

$$\begin{pmatrix} -\beta I_\infty - \mu & -\beta S_\infty \\ \beta I_\infty & 0 \end{pmatrix} \tag{7.9}$$

and this matrix has positive determinant and negative trace. Thus, the endemic equilibrium, if there is one, is always asymptotically stable. If the quantity

$$R_0 = \frac{\beta K}{\mu + \gamma + \alpha} = \frac{K}{S_\infty} \tag{7.10}$$

is less than one, the system has only the disease-free equilibrium and this equilibrium is asymptotically stable. In fact, it is not difficult to prove that this asymptotic stability is *global*, that is, that every solution approaches the disease-free equilibrium. If the quantity R_0 is greater than one then the disease-free equilibrium is unstable, but there is an endemic equilibrium that is asymptotically stable. Again, the quantity R_0 is the basic reproductive number. It depends on the particular disease (determining the parameters γ and α) and on the rate of contacts, which may depend on the population density in the community being studied. The disease model exhibits a *threshold behavior*: If the basic reproductive number is less than one the disease will die out, but if the basic reproductive number is greater than one the disease will be endemic. Our first model, in Section 7.2, also exhibited a threshold behavior but of a slightly different kind. For this model, which was an SIR model without births or natural deaths, the threshold distinguished between a dying out of the disease and an epidemic, or short term spread of disease.

The SIR model with births and deaths and the SIS model both support endemic equilibria. This suggests that a requirement for an endemic equilibrium is a flow of new susceptibles, either through births or through recovery without immunity against reinfection.

If we add the three equations of the model (7.7) we obtain

$$N' = \mu K - \mu N - \alpha I,$$

where $N = S + I + R$. From this we see that at the endemic equilibrium $N = K - \alpha/\mu I$, and the reduction in the population size from the carrying capacity K is

$$\frac{\alpha}{\mu} I = \frac{\alpha K}{\mu + \gamma + \alpha} - \frac{\alpha}{\beta}.$$

The parameter α in the SIR model may be considered as describing the pathogenicity of the disease. If α is large it is less likely that $R_0 > 1$. If α is small then the total population size at the endemic equilibrium is close to the carrying capacity K of the population. Thus, the maximum population decrease caused by disease will be for diseases of intermediate pathogenicity.

We now make some predictions implied by this model about some qualitative aspects of the spread of a disease.

1. Herd immunity. In order to prevent a disease from becoming endemic it is necessary to reduce the basic reproductive number R_0 below one. This may sometimes be achieved by immunization. If a fraction p of the μK newborn members per unit time of the population is successfully immunized, the effect is to replace K by $K(1-p)$, and thus to reduce the basic reproductive number to $R_0(1-p)$. The requirement $R_0(1-p) < 1$ gives $1-p < 1/R_0$, or

$$p > 1 - \frac{1}{R_0}.$$

A population is said to have *herd immunity* if a large enough fraction has been immunized to assure that the disease cannot become endemic. The only disease for which this has actually been achieved worldwide is smallpox for which R_0 is approximately 5, so that 80 percent immunization does provide herd immunity.

For measles, epidemiological data in the United States indicates that R_0 for rural populations ranges from 5.4 to 6.3, requiring vaccination of 81.5 percent to 84.1 percent of the population. In urban areas R_0 ranges from 8.3 to 13.0, requiring vaccination of 88.0 percent to 92.3 percent of the population. In Great Britain, R_0 ranges from 12.5 to 16.3, requiring vaccination of 92 percent to 94 percent of the population. The measles vaccine is not always effective, and vaccination campaigns are never able to reach everyone. As a result, herd immunity against measles has not been achieved (and probably never can be). Since smallpox is viewed as more serious and requires a lower percentage of the population be immunized, herd immunity is attainable for smallpox. In fact, smallpox has been eliminated; the last known case was in Somalia in 1977, and the virus is maintained now only in laboratories. The eradication of smallpox was actually more difficult than expected because high vaccination rates were achieved in some countries but not everywhere, and the disease persisted in some countries. The eradication of smallpox was possible only after an intensive campaign for worldwide vaccination [Hethcote (1978)].

2. Age at infection. In order to calculate the basic reproductive number R_0 for a disease, we need to know the values of the parameters β, μ, γ, and α. The parameters μ, γ, and α can usually be measured experimentally but the contact rate β is difficult to determine directly. There is an indirect means of estimating R_0 in terms of the life expectancy and the mean age at infection.

Consider the "age cohort" of members of a population born at some time t_0 and let a be the age of members of this cohort. If $y(a)$ represents the fraction of members of the cohort who survive to age (at least) a, then the assumption that a fraction μ of the population dies per unit time means that $y'(a) = -\mu y(a)$. Since $y(0) = 1$, we may solve this first order initial value problem to obtain $y(a) = e^{-\mu a}$. The fraction dying at (exactly) age a is $-y'(a) = \mu y(a)$. The mean life span is the average age at death, which is $\int_0^\infty a[-y'(a)]\,da$, and if we integrate by parts we find that this life expectancy is

$$\int_0^\infty a[-y'(a)]da = \Big[-ay(a)\Big]_0^\infty + \int_0^\infty y(a)\,da = \int_0^\infty y(a)\,da$$

Since $y(a) = e^{-\mu a}$, this reduces to $1/\mu$. The life expectancy is often denoted by L, so that we may write

$$L = \frac{1}{\mu}.$$

The rate at which surviving susceptible members of the population become infected at age a, time $t_0 + a$, is $\beta I(t_0 + a)$. Thus, if $z(a)$ is the fraction of the age cohort alive and still susceptible at age a, $z'(a) = -[\mu + \beta I(t_0)]z(a)$. Solution of this first linear order differential equation gives

$$z(a) = e^{-[\mu a + \int_0^a \beta I(t_0+b)db]} = y(a)e^{-\int_0^a \beta I(t_0+b)db}.$$

The mean length of time in the susceptible class for members who may become infected, as opposed to dying while still susceptible, is

$$\int_0^\infty e^{-\int_0^a \beta I(t_0+b)db}da,$$

and this is the mean age at which members become infected. If the system is at an equilibrium I_∞, this integral may be evaluated, and the mean age at infection, denoted by A, is given by

$$A = \int_0^\infty e^{-\beta I_\infty a}\,da = \frac{1}{\beta I_\infty}.$$

For our model we have found the endemic equilibrium

$$I_\infty = \frac{\mu K}{\mu + \gamma + \alpha} - \frac{\mu}{\beta},$$

and this implies

$$\frac{L}{A} = \frac{\beta I_\infty}{\mu} = R_0 - 1.$$

This relation is very useful in estimating basic reproductive numbers. For example, in some urban communities in England and Wales between 1956 and 1969 the average age of contracting measles was 4.8 years. If life expectancy is assumed to be 70 years, this indicates $R_0 = 15.6$.

The relation between age at infection and basic reproductive number indicates that measures such as inoculations, which reduce R_0, will increase the average age at infection. For diseases such as rubella (German measles), whose effects may be much more serious in adults than in children, this indicates a danger that must be taken into account: While inoculation of children will decrease the number of cases of illness, it will tend to increase the danger to those who are not inoculated or for whom the inoculation is not successful. Nevertheless, the number of older cases of infection will be reduced, although the fraction of cases in older people will increase.

3. The interepidemic period. Many common childhood diseases, such as measles, whooping cough, chicken pox, diphtheria, and rubella, exhibit variations from year to year in the number of cases. These fluctuations are frequently regular oscillations, suggesting that the solutions of a model might be periodic. This does not agree with the predictions of the model we have been using in this section; however, it would not be inconsistent with solutions of the characteristic equation, which are complex conjugate with small negative real part corresponding to lightly damped oscillations approaching the endemic equilibrium. Such behavior would look like recurring epidemics. If the eigenvalues of the matrix of the linearization at an endemic equilibrium are $-u \pm iv$, where $i^2 = -1$, then the solutions of the linearization are of the form $Be^{-ut}\cos(vt+c)$, with decreasing "amplitude" Be^{-ut} and "period" $\frac{2\pi}{v}$.

For the model (7.7) we recall from (7.8) that at the endemic equilibrium we have

$$\beta I_\infty + \mu = \mu R_0, \qquad \beta S_\infty = \mu + \gamma + \alpha$$

and from (7.9) the matrix of the linearization is

$$\begin{pmatrix} -\mu R_0 & -(\mu+\gamma+\alpha) \\ \mu(R_0 - 1) & 0 \end{pmatrix}$$

The eigenvalues are the roots of the quadratic equation

$$\lambda^2 + \mu R_0 \lambda + \mu(R_0 - 1)(\mu + \gamma + \alpha) = 0,$$

which are

$$\lambda = \frac{-\mu R_0 \pm \sqrt{\mu^2 {R_0}^2 - 4\mu(R_0 - 1)(\mu + \gamma + \alpha)}}{2}.$$

If the mean infective period $1/(\gamma+\alpha)$ is much shorter than the mean life span $1/\mu$, we may neglect the terms that are quadratic in μ. Thus, the eigenvalues are approximately

$$\frac{-\mu R_0 \pm \sqrt{-4\mu(R_0 - 1)(\gamma + \alpha)}}{2},$$

and these are complex with imaginary part $\sqrt{\mu(R_0 - 1)(\gamma + \alpha)}$. This indicates oscillations with period approximately

$$\frac{2\pi}{\sqrt{\mu(R_0 - 1)(\gamma + \alpha)}}.$$

We use the relation $\mu(R_0 - 1) = \mu L/A$ and the mean infective period $\tau = 1/(\gamma + \alpha)$ to see that the interepidemic period T is approximately $2\pi\sqrt{A\tau}$. Thus, for example, for recurring outbreaks of measles with an infective period of 2 weeks or 1/26 year in a population with a life expectancy of 70 years with R_0 estimated as 15, we would expect outbreaks spaced 2.76 years apart. Also, as the "amplitude" at time t is $e^{-\mu R_0 t}$, the maximum displacement from equilibrium is multiplied by a factor $e^{-(15)(2.76)/70} = 0.55$ over each cycle. In fact, many observations of measles outbreaks indicate less damping of the oscillations, suggesting that there may be additional influences that are not included in our simple model. To explain oscillations about the endemic equilibrium a more complicated model is needed. One possible generalization would be to assume seasonal variations in the contact rate. This is a reasonable supposition for a childhood disease most commonly transmitted through school contacts, especially in winter in cold climates. Note, however, that data from observations are never as smooth as model predictions and models are inevitably gross simplifications of reality which cannot account for random variations in the variables. It may be difficult to judge from experimental data whether an oscillation is damped or persistent.

4. "Epidemic" approach to endemic equilibrium. In the model (7.7) the demographic time scale described by the birth and natural death rates μK and μ and the epidemiological time scale described by the rate $(\alpha + \gamma)$ of departure from the infective class may differ substantially. Think, for example, of a natural death rate $\mu = 1/75$, corresponding to a human life expectancy of 75 years, and epidemiological parameters $\alpha = 0$ and $\gamma = 25$, describing a disease from which all infectives recover after a mean infective period of 1/25 year, or two weeks. Suppose we consider a carrying capacity $K = 1000$ and take $\beta = 0.1$, indicating that an average infective makes (0.1)(1000)= 100 contacts per year. Then $R_0 = 4.00$, and at the endemic equilibrium we have $S_\infty = 250.13$, $I_\infty = 0.40$, $R_\infty = 749.47$. This equilibrium is globally asymptotically stable and is approached from every initial state.

However, if we take $S(0) = 999$, $I(0) = 1$, $R(0) = 0$, simulating the introduction of a single infective into a susceptible population and solve the system numerically we find that the number of infectives rises sharply to a maximum of 400 and then decreases to almost zero in a period of 0.4 year, or about 5 months. In this time interval the susceptible population decreases to 22 and then begins to increase, while the removed (recovered and immune against reinfection) population increases to almost 1000 and then begins a gradual decrease. The size of this initial "epidemic" could not have been predicted from our qualitative analysis of the system (7.7). On the other hand, since μ is so small compared to the other parameters of the model, we might consider neglecting μ, replacing it by zero in the model. If we do this, the model reduces to the simple Kermack-McKendrick epidemic model (without births and deaths) of Section 7.2.

If we follow the model (7.7) over a longer time interval we find that the susceptible population grows to 450 after 46 years, then drops to 120 during a small epidemic with a maximum of 18 infectives, and exhibits widely spaced epidemics decreasing in size. It takes a very long time before the system comes close to the endemic equilibrium and remains clse to it. The large initial epidemic conforms to what has often been observed in practice when an infection is introduced into a population with no immunity, such as the smallpox inflicted on the Aztecs by the invasion of Cortez.

If we use the model (7.7) with the same values of β, K and μ, but take $\alpha = 25$, $\gamma = 0$ to describe a disease fatal to all infectives, we obtain very similar results. Now the total population is $S + I$, which decreases from an initial size of 1000 to a minimum of 22 and then gradually increases and eventually approaches its equilibrium size of 250.53. Thus, the disease reduces the total population size to one-fourth of its original value, suggesting that infectious diseases may have large effects on population size. This is true even for populations which would grow rapidly in the absence of infection, as we shall see in the next section.

5. The SIS model with births and deaths. In order to describe a model for a disease from which infectives recover with immunity against reinfection and that includes births and deaths as in the model (7.7), we may modify the model (7.7) by removing the equation for R' and moving the term γI describing the rate of recovery from infection to the equation for S'. This gives the model

$$\begin{aligned} S' &= -\beta SI + \mu(K - S) + \gamma I \\ I' &= \beta SI - \alpha I - \mu I - \gamma I \end{aligned} \tag{7.11}$$

describing a population with a constant number of births μK per unit time, a proportional death rate μ in each class, and with a fraction α of infectives dying from infection and a fraction γ of infectives recovering with no immunity against reinfection. In this model, if $\alpha > 0$ the total population size is not constant and K represents a *carrying capacity*, or maximum possible population size, rather than a constant population size.

The analysis of the model (7.11) is very similar to that of the SIR model (7.7), except that there is no equation for R' to be eliminated.

The only difference is the additional term γI in the equation for S', and this does not change any of the qualitative results. As in the SIR model we have a basic reproductive number

$$R_0 = \frac{\beta K}{\mu + \gamma + \alpha} = \frac{K}{S_\infty}$$

and if $R_0 < 1$ the disease-free equilibrium $S = K$, $I = 0$ is asymptotically stable, while if $R_0 > 1$ there is an endemic equilibrium (S_∞, I_∞) with $\beta S_\infty = \mu + \gamma + \alpha$ and I_∞ given by (7.8), which is asymptotically stable. There are, however, differences that are not disclosed by the qualitative analysis. If the epidemiological and demographic time scales are very different, for the SIR model we observed that the approach to endemic equilibrium is like a rapid and severe epidemic. The same happens in the SIS model, especially if there is a significant number of deaths due to disease. If there are few disease deaths the number of infectives at endemic equilibrium may be substantial, and there may be oscillations of large amplitude about the endemic equilibrium.

For both the SIR and SIS models we may write the differential equation for I as

$$I' = I\left[\beta S - (\mu + \alpha + \gamma)\right] = \beta I[S - S_\infty]$$

which implies that whenever S exceeds its endemic equilibrium value, I is increasing and epidemic-like behavior is possible. If $R_0 < 1$ and $S < K$ it follows that $I' < 0$, and thus I is decreasing. Thus, if $R_0 < 1$ I cannot increase and no epidemic can occur.

Exercise

1. Recurrent outbreaks of measles and other childhood diseases have previously been explained by an interaction of intrinsic epidemiological forces generating dampened oscillations and of seasonal and/or stochastic excitation. The following model shows that isolation or quarantine (i.e., sick individuals stay at home and have a reduced infective impact) can create self-sustained oscillations.

   In the model considered here the population is divided into Susceptibles (S), Infectives (I), Isolated or Quarantined individuals (Q), and Recovered individuals (R), for whom permanent immunity is assumed. Let N denote the total population, and $A = S + I + R$ denotes the active (nonisolated) individuals. The model takes the form:

$$\begin{aligned}
\frac{dS}{dt} &= \mu N - \mu S - \sigma S \frac{I}{A} \\
\frac{dI}{dt} &= -(\mu + \gamma) I + \sigma S \frac{I}{A} \\
\frac{dQ}{dt} &= -(\mu + \xi) Q + \gamma I \\
\frac{dR}{dt} &= -\mu R + \xi Q \\
A &= S + I + R.
\end{aligned} \tag{7.12}$$

All newborns are assumed to be susceptible. μ is the per capita mortality rate, σ is the per capita infection rate of an average susceptible individual provided everybody else is infected, γ is the rate at which individuals leave the infective class, and ξ is the rate at which individuals leave the isolated class; all are positive constants.

(i) Show that the total population size N is constant.

(ii) Give the meanings of $1/\mu$, $1/\gamma$, $1/\xi$ and their units.

(iii) Rescale the model by: $\tau = \sigma t, u = S/A, y = I/A, q = Q/A, z = R/A$. Rearrange your new model as follows:

$$\begin{aligned}
\dot{y} &= y(1 - \nu - \theta - y - z + \theta y - (\nu + \zeta) q) \\
\dot{q} &= (1 + q)(\theta y - (\nu + \zeta) q) \\
\dot{z} &= \zeta q - \nu z + z(\theta y - (\nu + \zeta) q).
\end{aligned} \tag{7.13}$$

Express the new parameters in terms of the old parameters. Check that all the new parameters and variables are dimensionless.

(iv) Study the stability of the equilibrium point $(0, 0, 0)$ and derive a basic reproductive number R_0.

(v) Use a computer algebra system to demonstrate that (7.13) has periodic trajectories. Use the parameter values $\nu = 0.0002$, $\theta = 0.0156$, and ξ close to $\theta^2(1 - \theta)$. You also need to choose proper initial values

7.5 Disease as Population Control

Many parts of the world experienced very rapid population growth in the 18th century. The population of Europe increased from 118 million in 1700 to 187 million in 1800. In the same time period the population of Great Britain increased from 5.8 million to 9.15 million, and the population of

China increased from 150 million to 313 million [McNeill (1992)]. The population of English colonies in North America grew much more rapidly than this, aided by substantial immigration from England, but the native population, which had been reduced to one tenth of their previous size by disease following the early encounters with Europeans and European diseases, grew even more rapidly. While some of these population increases may be explained by improvements in agriculture and food production, it appears that an even more important factor was the decrease in the death rate due to diseases. Disease death rates dropped sharply in the 18th century, partly from better understanding of the links between illness and sanitation and partly because the recurring invasions of bubonic plague subsided, perhaps due to reduced susceptibility. One plausible explanation for these population increases is that the bubonic plague invasions served to control the population size, and when this control was removed the population size increased rapidly.

In developing countries it is quite common to have high birth rates and high disease death rates. In fact, when disease death rates are reduced by improvements in health care and sanitation it is common for birth rates to decline as well, as families no longer need to have as many children to ensure enough survive to take care of the older generations. Again, it is plausible to assume that population size would grow exponentially in the absence of disease but is controlled by disease mortality.

The SIR model with births and deaths of Kermack and McKendrick(1932) includes births in the susceptible class proportional to population size and a natural death rate in each class proportional to the size of the class. Let us analyze a model of this type with birth rate r and a natural death rate $\mu < r$. For simplicity we assume the disease is fatal to all infectives with disease death rate α, so that there is no removed class and the total population size is $N = S + I$. Our model is

$$\begin{aligned} S' &= r(S+I) - \beta SI - \mu S \\ I' &= \beta SI - (\mu + \alpha)I \end{aligned} \qquad (7.14)$$

From the second equation we see that equilibria are given by either $I = 0$ or $\beta S = \mu + \alpha$. If $I = 0$ the first equilibrium equation is $rS = \mu S$, which implies $S = 0$ since $r > \mu$. It is easy to see that the equilibrium (0,0) is unstable. What actually would happen if $I = 0$ is that the susceptible population would grow exponentially with exponent $r - \mu > 0$. If $\beta S = \mu + \alpha$ the first equilibrium condition gives

$$r\frac{\mu + \alpha}{\beta} + rI - (\mu + \alpha)I - \frac{\mu(\mu + \alpha)}{\beta} = 0,$$

which leads to

$$(\alpha + \mu - r)I = \frac{(r - \mu)(\mu + \alpha)}{\beta}.$$

Thus, there is an endemic equilibrium provided $r < \alpha + \mu$, and it is possible to show by linearizing about this equilibrium that it is asymptotically stable (we shall not carry out this verification). On the other hand, if $r > \alpha + \mu$ there is no positive equilibrium value for I. In this case we may add the two differential equations of the model to give

$$N' = (r-\mu)N - \alpha I \geq (r-\mu)N - \alpha N = (r-\mu-\alpha)N$$

and from this we may deduce that N grows exponentially. For this model either we have an asymptotically stable endemic equilibrium or population size grows exponentially. In the case of exponential population growth we may have either vanishing of the infection or an exponentially growing number of infectives.

If only susceptibles contribute to the birth rate, as may be expected if the disease is sufficiently debilitating, the behavior of the model is quite different. Let us consider the model

$$\begin{aligned} S' &= rS - \beta SI - \mu S = S(r - \mu - \beta I) \\ I' &= \beta SI - (\mu + \alpha)I = I(\beta S - \mu - \alpha) \end{aligned} \tag{7.15}$$

which has the same form as the celebrated Lotka-Volterra predator-prey model of population dynamics (see Section 4.1). This system has two equilibria, obtained by setting the right sides of each of the equations equal to zero, namely (0,0) and an endemic equilibrium $((\mu+\alpha)/\beta, (r-\mu)/\beta)$. It turns out that the qualitative analysis approach we used for systems in the last two sections is not helpful as the equilibrium (0,0) is unstable and the eigenvalues of the coefficient matrix at the endemic equilibrium are pure imaginary. In this case the behavior of the linearization does not necessarily carry over to the full system. However, we can obtain information about the behavior of the system by a method that begins with the elementary approach of separation of variables for first order differential equations. We begin by taking the quotient of the two differential equations and using the relation

$$\frac{I'}{S'} = \frac{dI}{dS}$$

to obtain the separable first order differential equation

$$\frac{dI}{dS} = \frac{I(\beta S - \mu - \alpha)}{S(r - \beta I)}.$$

Separation of variables gives

$$\int \left(\frac{r}{I} - \beta\right) dI = \int \left(\beta - \frac{\mu + \alpha}{S}\right) dS.$$

Integration gives the relation

$$\beta(S + I) - r \log I - (\mu + \alpha) \log S = c$$

where c is a constant of integration. This relation shows that the quantity

$$V(S,I) = \beta(S+I) - r\log I - (\mu+\alpha)\log S$$

is constant on each orbit (path of a solution in the $(S,I-$ plane). Each of these orbits is a closed curve corresponding to a periodic solution. This treatment parallels our treatment of the Lotka-Volterra model in Section 4.1.

This model is the same as the epidemic model of Section 7.2 except for the birth and death terms, and in many examples the time scale of the disease is much faster than the time scale of the demographic process. We may view the model as describing an epidemic initially, leaving a susceptible population small enough that infection cannot establish itself. Then there is a steady population growth until the number of susceptibles is large enough for an epidemic to recur. During this growth stage the infective population is very small and random effects may wipe out the infection, but the immigration of a small number of infectives will eventually restart the process. As a result, we would expect recurrent epidemics. In fact, bubonic plague epidemics did recur in Europe for several hundred years. If we modify the demographic part of the model to assume limited population growth rather than exponential growth in the absence of disease, the effect would be to give behavior like that of the model studied in the previous section, with an endemic equilibrium that is approached slowly in an oscillatory manner if $R_0 > 1$.

Example 1. (Fox rabies) Rabies is a viral infection to which many animals, especially foxes, coyotes, wolves, and rats, are highly susceptible. While dogs are only moderately susceptible, they are the main source of rabies in humans. Although deaths of humans from rabies are few, the disease is still of concern because it is invariably fatal. However, the disease is endemic in animals in many parts of the world. A European epidemic of fox rabies thought to have begun in Poland in 1939 and spread through much of Europe has been modeled. We present here a simplified version of a model due to R.M. Anderson and coworkers [Anderson, Jackson, May, and Smith (1981)].

We begin with the demographic assumptions that foxes have a birth rate proportional to population size but that infected foxes do not produce offspring because the disease is highly debilitating, and that there is a natural death rate proportional to population size. Experimental data indicate a birth rate of approximately 1 per capita per year and a death rate of approximately 0.5 per capita per year, corresponding to a life expectancy of 2 years. The fox population is divided into susceptibles and infectives, and the epidemiological assumptions are that the rate of acquisition of infection is proportional to the number of encounters between susceptibles and infectives. We will assume a contact parameter $\beta = 80$, in rough agreement with observations of frequency of contact in regions where the fox density is approximately 1 fox/km^2, and we assume that all infected foxes die with a mean infective period of approximately 5 days or 1/73 year. These assumptions lead to the model

$$\begin{aligned} S' &= -\beta SI + rS - \mu S \\ I' &= \beta SI - (\mu + \alpha)I \end{aligned}$$

with $\beta = 80$, $r = 1.0$, $\mu = 0.5$, $\alpha = 73$. As this is of the form (7.15), we know that the orbits are closed curves in the (S, I) plane, and that both S and I are periodic functions of t. We illustrate with some simulations obtained using Maple (Figures 7.5, 7.6, 7.7). It should be noted from the graphs of I in terms of t that the period of the oscillation depends on the amplitude, and thus on the initial conditions, with larger amplitudes corresponding to longer periods.

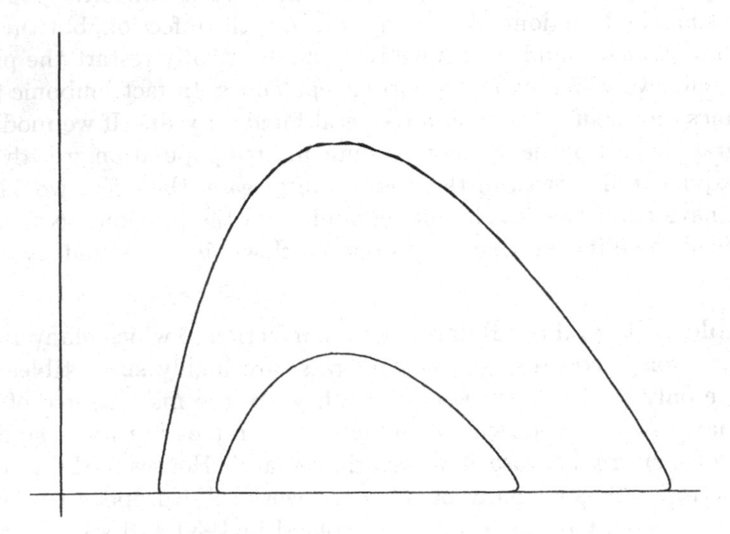

FIGURE 7.5. The (S, I) plane.

A warning is in order here. The model predicts that for long time intervals the number of infected foxes is extremely small. With such small numbers, the continuous deterministic models we have been using (which assume that population sizes are differentiable functions) are quite inappropriate. If the density of foxes is extremely small an encounter between foxes is a random event, and the number of contacts cannot be described properly by a function of population densities. To describe disease transmission properly when population sizes are very small we would need to use a stochastic model.

Now let us modify the demographic assumptions by assuming that the birth rate decreases as population size increases. We replace the birth rate of r per susceptible per year by a birth rate of re^{-aS} per susceptible per year, with a a positive constant. Then, in the absence of infection, the fox population is given by the first order differential equation

$$N' = N\left(re^{-aN} - \mu\right)$$

and equilibria of this equation are given by $N = 0$ and $re^{-aN} = \mu$, which reduces to $e^{aN} = \mu/r$, or

$$N = \frac{1}{a}\log\frac{r}{\mu}.$$

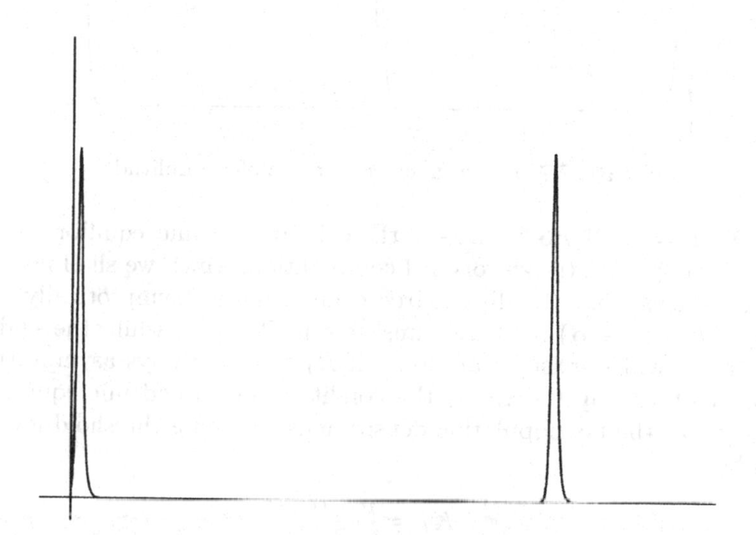

FIGURE 7.6. I as a function of t (larger amplitude).

We shall omit the verification that the equilibrium $N = 0$ is unstable while the positive equilibrium $N = (1/a)\log(r/\mu)$ is asymptotically stable. Thus, the population has a *carrying capacity* given by

$$K = \frac{1}{a}\log\frac{r}{\mu}.$$

The model now becomes

$$\begin{aligned} S' &= rSe^{-aS} - \beta SI - \mu S \\ I' &= \beta SI - (\mu + \alpha)I. \end{aligned}$$

We examine this by looking for equilibria and analyzing their stability. From the second equation, equilibria satisfy either $I = 0$ or $\beta S = \mu + \alpha$. If $I = 0$ the first equilibrium condition reduces to the same equation that determined the carrying capacity, and we have a disease-free equilibrium

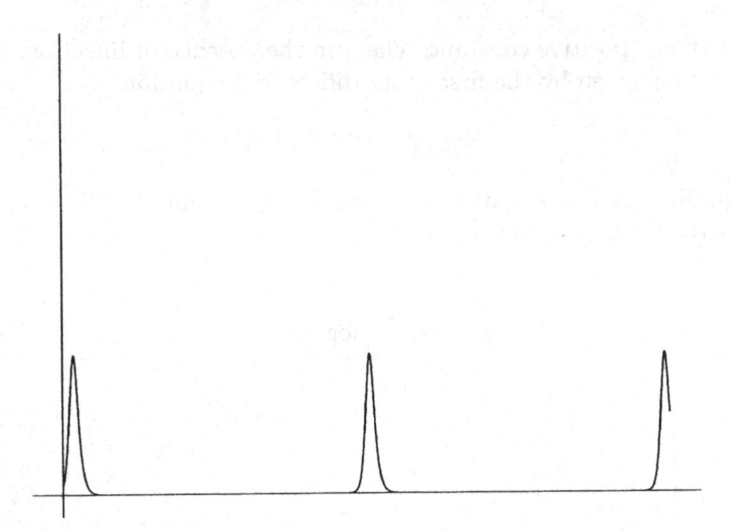

FIGURE 7.7. I as a function of t (smaller amplitude).

$S = K$, $I = 0$. If $\beta S = \mu + \alpha$ there is an endemic equilibrium with $\beta I + \mu = re^{-aS}$. A straightforward computation, which we shall not carry out here shows, that the disease-free equilibrium is asymptotically stable if $R_0 = \beta K/(\mu + \alpha) < 1$ and unstable if $R_0 > 1$, while the endemic equilibrium, which exists if and only if $R_0 > 1$, is always asymptotically stable. Another way to express the condition for an endemic equilibrium is to say that the fox population density must exceed a threshold level K_T given by

$$K_T = \frac{\mu + \alpha}{\beta}.$$

With the parameter values we have been using, this gives a threshold fox density of 0.92 fox/km^2. If the fox density is below this threshold value, the fox population will approach its carrying capacity and the disease will die out. Above the threshold density, rabies will persist and will regulate the fox population to a level below its carrying capacity. This level may be approached in an oscillatory manner for large R_0.

Exercises

1. A large English estate has a population of foxes with a density of 1.3 foxes/km^2. A large fox hunt is planned to reduce the fox population enough to prevent an outbreak of rabies. Assuming that the contact number β/α is 1 km^2/fox, find what fraction of the fox population must be caught.

2. Following a complaint from the SPCA, organizers decide to replace the fox hunt of Exercise 1 by a mass inoculation of foxes for rabies.

What fraction of the fox population must be inoculated to prevent a rabies outbreak?

3. What actually occurs on the estate of Exercises 1 and 2 is that 10 percent of the foxes are killed and 15 percent are inoculated. Is there danger of a rabies outbreak.

7.6 Infective Periods of Fixed Length

Until now we have assumed that all rates of transition out of a compartment (except for the acquisition of infection) are proportional to the number of members in the compartment. This is equivalent, as we have seen in Section 7.4 or Section 1.7, to an exponential distribution of times in the compartment. For example, the assumption that a fraction $(\mu+\gamma+\alpha)$ of members of a population leaves the infective class per unit time is equivalent to the assumption that a fraction $e^{-(\mu+\gamma+\alpha)t}$ of the population remains in the infective class until (at least) time t, so that the mean stay in the infective class is $1/(\mu+\gamma+\alpha)$. It may be considerably more realistic to assume a different distribution of infective periods. For example, an infective period of fixed length may be much closer to the actual course of many diseases.

In this section, we shall describe how to formulate models for diseases with infective periods of fixed length. These models are more complicated mathematically than the systems of differential equations we have studied until now. While the analysis of these more complicated models is of mathematical interest, their epidemiological interest depends on the possibility of differences in qualitative behavior for different infective period distributions; there are some situations in which there are such differences.

A fixed infective period will lead to a differential-difference equation model. In Section 7.8 we shall examine models for diseases with a more general infective period distribution which will lead to Volterra integral equation models. The theory governing the analysis of differential-difference equation and Volterra integral equation models is considerably more complicated than for ordinary differential equation models, as we have seen in Chapter 3. We will describe this theory in a form that exhibits the analysis of the qualitative behavior of systems of differential equations as a special case.

We begin by finding equilibria of the model, that is, constant solutions (just as in the ordinary differential equation case). We then linearize about an equilibrium by translating the equilibrium to the origin and expanding. The constant terms will always cancel at an equilibrium, and we retain all linear terms but discard all terms of higher order. In the ordinary differential equation case this gives a linear system of differential equations whose coefficient matrix is the matrix of partial derivatives of the right sides of the equations of the original system with respect to the variables, but in the more general situation it is necessary to carry out the expansion.

The characteristic equation at an equilibrium is the condition that the linearization at the equilibrium have a solution whose components are constant multiples of $e^{\lambda t}$. In the ordinary differential equation case this is just the equation that determines the eigenvalues of the coefficient matrix, a polynomial equation, but in the general case it is a transcendental equation. *The result on which our analysis depends is that an equilibrium is asymptotically stable if all roots of the characteristic equation have negative real part, or equivalently that the characteristic equation have no roots with real part greater than or equal to zero.* We will illustrate with some examples, concentrating on SIS models with an infective period of fixed length, because these have fewer compartments than other models and are thus simpler to analyze.

We begin with the simple situation of an SIS model with no births or deaths and no disease-related deaths, making the following assumptions:

(i) An average infective makes contact sufficient to transmit infection with βN others per unit time, where N represents total population size.

(ii) The population size is a constant K; there are no births, deaths, or entrants into the population.

(iii) Infectives remain infective for a fixed time τ and then return to the susceptible class.

Because the total population size is a constant K, we have $S + I = K$ and we may formulate the model in terms of the single variable I, replacing S by $K-I$. The rate of new infections at time t is $\beta S(t)I(t) = \beta I(t)[K-I(t)]$, and the rate at which infectives recover and return to the susceptible class is $\beta S(t-\tau)I(t-\tau) = \beta I(t-\tau)[K - I(t-\tau)]$. As the only changes in the infective population size are new infections and recoveries, we have

$$I'(t) = \beta I(t)[K - I(t)] - \beta I(t-\tau)[K - I(t-\tau)]. \qquad (7.16)$$

When we try to find equilibria of (7.16), we find that we have no condition to solve. The reason for this is that every constant value of I is a solution of (7.16). In other words, (7.16) by itself is not a proper description and some additional conditions must be added. We accomplish this by incorporating initial data for $-\tau \leq t \leq 0$ into the model by writing it in the integrated form,

$$I(t) = \int_{t-\tau}^{t} \beta I(x)[K - I(x)]dx \qquad (7.17)$$

describing $I(t)$ as the integral (sum) of members infected at time x between $t-\tau$ and t and therefore still infective. The equilibrium condition for (7.17) is

$$I = 0 \quad \text{or} \quad 1 = \beta\tau(K - I).$$

This gives a disease-free equilibrium $I = 0$ and an endemic equilibrium $I = K - 1/\beta\tau$. To linearize about an equilibrium I_∞ we substitute $I = I_\infty + u(t)$, giving

$$I_\infty + u(t) = \int_{t-\tau}^{t} \beta[I_\infty + u(x)][K - I_\infty - u(x)]dx$$

and neglect the quadratic term, giving the linearization

$$u(t) = \int_{t-\tau}^{t} \beta(K - 2I_\infty)u(x)dx.$$

The characteristic equation is the condition on λ that this linearization have a solution that is a constant multiple of $e^{\lambda t}$, namely

$$e^{\lambda t} = \int_{t-\tau}^{t} \beta(K - 2I_\infty)e^{\lambda x}dx = \frac{\beta(K - 2I_\infty)[e^{\lambda t} - e^{\lambda(t-\tau)}]}{\lambda}$$

or

$$\frac{\beta(K - 2I_\infty)[1 - e^{-\lambda\tau}]}{\lambda} = 1 \tag{7.18}$$

If we had linearized at an equilibrium using the differentiated form (7.16) of the model, we would have obtained as the characteristic equation the equation (7.18) multiplied through by λ. This would have $\lambda = 0$ as a root, but this root is extraneous, resulting from the use of a model that has every constant as a solution. At the disease-free equilibrium $I_\infty = 0$ and the characteristic equation (7.18) becomes

$$\frac{\beta K(1 - e^{-\lambda\tau})}{\lambda} = 1. \tag{7.19}$$

In order to establish asymptotic stability of the disease-free equilibrium we need to make use of the following fact, whose proof we delay because it is a special case of a result we will need shortly in the study of *SIS* models with births and deaths:

Lemma 7.6.1. If $\tau > 0$ and $\Re\lambda \geq 0$ then

$$\left|\frac{1 - e^{-\lambda\tau}}{\lambda}\right| \leq \tau.$$

It follows that if $\beta\tau K < 1$ then, for $\Re\lambda \geq 0$, the absolute value of the left side of the equation (7.19) is no greater than 1 and thus there can be no root (except possibly the extraneous root $\lambda = 0$) of the equation (7.19) with $\Re\lambda \geq 0$ and thus that the disease-free equilibrium is asymptotically stable if

$$R_0 = \beta\tau K < 1.$$

At the endemic equilibrium $I = K - 1/\beta\tau$, $K - 2I = K - 2K + 2/\beta\tau = 2/\beta\tau - K$, and the characteristic equation (7.18) becomes

$$\beta\Big(\frac{2}{\beta\tau} - K\Big)\frac{1 - e^{-\lambda\tau}}{\lambda} = 1.$$

If $\Re\lambda \geq 0$ the absolute value of the left side of this characteristic equation is at most $\beta\,(2/\beta\tau - K)\,\tau = 2 - R_0$. Thus, if $R_0 > 1$ there is no root with $\Re\lambda \geq 0$ and the equilibrium is asymptotically stable. Qualitatively, the situation is exactly the same as for the SIS model with exponential transition rates studied in Section 7.3 The disease-free equilibrium is asymptotically stable if and only if the basic reproductive number R_0 is less than one and the endemic equilibrium exists and is asymptotically stable if and only if the basic reproductive number exceeds one.

The obvious next step is to incorporate births and deaths into the model as in Section 7.4. Now the total population size is not necessarily constant and we need a two-dimensional model. While we could use S and I as variables, it is preferable to use I and N to facilitate an extension to models which incorporate a density-dependent contact rate. We begin with the assumption that there are no deaths due to disease; this implies that if the birth rate is μK and the death rate is μN, the differential equation for N is $N' = \mu K - \mu N$. In order to derive the differential equation for I, we use the facts that (i) the rate of new infections at time t is $\beta S(t)I(t) = \beta I(t)[N(t) - I(t)]$ and (ii), the rate of recoveries at time t is the rate of new infections $\beta S(t-\tau)I(t-\tau)$ at time $t - \tau$ multiplied by the fraction $e^{\mu\tau}$ of these new infectives who survive until time t. This gives the differential-difference equation

$$I'(t) = \beta I(t)[N(t) - I(t)] - e^{-\mu\tau}\beta I(t-\tau)[N(t-\tau) - I(t-\tau)]$$

and we now have the model

$$\begin{aligned} I'(t) &= \beta I(t)[N(t) - I(t)] - e^{-\mu\tau}\beta I(t-\tau)[N(t-\tau) - I(t-\tau)] \\ N'(t) &= \mu K - \mu N(t). \end{aligned} \tag{7.20}$$

We shall omit the details of the equilibrium analysis, which are analogous to the calculations we have made in the case $\mu = 0$ but slightly more complicated because we now have a two-dimensional system. On the other hand, since not all constants are solutions of the system we do not need to use an integrated form of the model in order to perform the equilibrium analysis. There is a disease-free equilibrium $I = 0,\ N = K$, and an endemic equilibrium $I = K - \mu/\beta(1 - e^{-\mu\tau})$, $N = K$, which exists if

$$\mathcal{R}_0 = \frac{\beta K(1 - e^{-\mu\tau})}{\mu} > 1.$$

The characteristic equation at an equilibrium $I = I_\infty$, $N = K$ is

$$\frac{\beta(K - 2I_\infty)[1 - e^{-(\lambda+\mu)\tau}]}{\lambda + \mu} = 1. \tag{7.21}$$

The analysis of this characteristic equation requires the following generalization of Lemma 7.6.1, introduced in the analysis of the SIS model without births and deaths.

Lemma 7.6.2. If $\tau > 0$ and $\Re\lambda \geq 0$ then

$$\left|\frac{1 - e^{-(\lambda+\mu)\tau}}{\lambda + \mu}\right| \leq \frac{1 - e^{-\mu\tau}}{\mu}.$$

It follows that if

$$\mathcal{R}_0 = \frac{\beta K(1 - e^{-\mu\tau})}{\mu} < 1$$

then for $\Re\lambda \geq 0$ the absolute value of the left side of the equation (7.21) with $I_\infty = 0$ is no greater than one and thus there can be no root of the equation (7.21) with $\Re\lambda \geq 0$ and thus that the disease-free equilibrium is asymptotically stable if $\mathcal{R}_0 < 1$. At the endemic equilibrium

$$\beta(K - 2I_\infty) = \frac{2\mu}{1 - e^{-\mu\tau}} - \beta K$$

and the absolute value of the left side of the equation (7.21) is no greater than $2 - \mathcal{R}_0$, which is no greater than one since $\mathcal{R}_0 > 1$. Again, there can be no root of the characteristic equation with $\Re\lambda \geq 0$ and thus the endemic equilibrium is asymptotically stable if $\mathcal{R}_0 > 1$.

If we wish to incorporate deaths due to the disease into the model (7.20) we may assume that at the end of the fixed infective period a fraction p of infectives die while the remainder return to the susceptible class. Then the death rate from disease is

$$pe^{-\mu\tau}\beta I(t-\tau)[N(t-\tau) - I(t-\tau)]$$

and this term must be subtracted in the equation for $N'(t)$, giving a model

$$\begin{aligned} I'(t) &= \beta I(t)[N(t) - I(t)] - e^{-\mu\tau}\beta I(t-\tau)[N(t-\tau) - I(t-\tau)] - \mu I(t) \\ N'(t) &= \mu K - \mu N(t) - pe^{-\mu\tau}\beta I(t-\tau)[N(t-\tau) - I(t-\tau)]. \end{aligned} \tag{7.22}$$

The analysis of this model leads to the same value for R_0 as for the model (7.20) and a characteristic equation

$$\frac{\beta[N_\infty - (2-p)I_\infty][1-e^{(\lambda+\mu)\tau}]}{\lambda+\mu} = 1 + \frac{p\beta I_\infty}{\lambda+\mu}. \tag{7.23}$$

If $R_0 < 1, I_\infty = 0$ and the right side of the equation (7.23) is one, while the left side has absolute value at most one. If $R_0 > 1$ at the endemic equilibrium for which $\beta(N-I)\frac{1-e^{-\mu\tau}}{\mu} = \mu$, we may deduce that the left side of the equation (7.23) has absolute value at most one, using $N_\infty - (2-p)I_\infty \leq N_\infty - I_\infty$. It may be shown that if $\Re\lambda \geq 0$ the real part of $\frac{p\beta I_\infty}{\lambda+\mu}$ is non-negative and it follows that the right side of the equation (7.23) has absolute value greater than one. From this we may deduce that the disease-free equilibrium is asymptotically stable if and only if $R_0 < 1$ and the endemic equilibrium, which exists if and only if $R_0 > 1$, is asymptotically stable. This is true for every p, $0 \leq p \leq 1$.

To complete the analysis of SIS models, we must establish Lemma 7.6.2. (We need not give a separate justification for Lemma 7.6.1 because it is the limiting case of Lemma 7.6.2 as $\mu \to 0$). Lemma 7.6.2 follows from the estimate, valid for $\mu > 0$ and $\Re\lambda \geq 0$,

$$\begin{aligned} \left|\int_0^\tau e^{-(\lambda+\mu)x}dx\right| &\leq \int_0^\tau |e^{-(\lambda+\mu)x}|dx = \int_0^\tau |e^{-\lambda x}|e^{-\mu x}dx \\ &\leq \int_0^\tau e^{-\mu x}dx = \frac{1-e^{-\mu\tau}}{\mu} \end{aligned}$$

and the integral expression

$$\int_0^\tau e^{-(\lambda+\mu)x}dx = \left.\frac{e^{-(\lambda+\mu)x}}{-(\lambda+\mu)}\right|_0^\tau = \frac{1-e^{-(\lambda+\mu)\tau}}{\lambda+\mu}.$$

The case $p = 1$ of (7.22), in which all infectives die from the disease at the end of the infective period, is equivalent to a universally fatal SIR model. We shall not carry out the formulation and analysis of SIR models for diseases from which some or all infectives recover with immunity, but the general ideas are similar to those for SIS models.

In this section we have examined some models in which the exponentially distributed infective periods and differential equation models studied in Section 7.1 through 7.5 are replaced by fixed infective periods and differential-difference equation models. In all cases the qualitative behavior is the same. Essentially, the only difference is in the expression for the length of the infective period corrected for natural mortality, which is $1/(\mu+\gamma+\alpha)$ in the exponentially distributed case and $(1-e^{-\mu\tau})/\mu$ in the fixed infective period case. If the behaviors were the same in all cases the results would be of little epidemiological interest. However, in the next section we shall describe a model with a temporary immunity period of fixed length for which the endemic equilibrium is not necessarily asymptotically stable. This is quite different from the analogous model with an exponentially distributed temporary immunity.

7.7 A Model with a Fixed Period of Temporary Immunity

In the models of Sections 7.6 we generalized the models introduced in Sections 7.1 through 7.5 but these generalizations did not affect the qualitative behavior of the models. There is some mathematical interest in some of the methods, but if every more general model behaves in the same way as the simple model that it is intended to replace there is little interest from an epidemiological view. In this section we shall consider an $SIRS$ model, which assumes a constant period of temporary immunity following recovery from the infection in place of an exponentially distributed period of temporary immunity. It will turn out that the endemic equilibrium for this model may be unstable, thus giving an example of a generalization which leads to new possibilities for the behavior of a model.

We shall assume, as in the models studied previously, a birth rate μK, a proportional death rate μ in each class, a rate of new infections βSI, a proportional recovery rate γ and a proportional disease death rate α per unit time. To these assumptions we add the assumption that there is a temporary immunity period of fixed length ω, after which recovered infectives revert to the susceptible class. The resulting model is described by the system

$$\begin{aligned} S'(t) &= \mu K - \mu S(t) - \beta S(t)I(t) + \gamma I(t-\omega)e^{-\mu(t-\omega)} \\ I'(t) &= \beta S(t)I(t) - (\mu+\gamma+\alpha)I(t) \\ R'(t) &= \gamma I(t) - \gamma I(t-\omega)e^{-\mu(t-\omega)} - \mu R(t). \end{aligned}$$

For simplicity we shall analyze only the special case $\mu = 0$, $\alpha = 0$ with no births, natural deaths, or disease deaths, but the general case could be analyzed in a similar way. Our special case is

$$\begin{aligned} S'(t) &= -\beta S(t)I(t) + \gamma I(t-\omega) \\ I'(t) &= \beta S(t)I(t) - (\gamma+\alpha)I(t) \\ R'(t) &= \gamma I(t) - \gamma I(t\omega). \end{aligned}$$

Since $S + I + R$ is constant, we may discard the equation for R and use a two-dimensional model

$$\begin{aligned} S'(t) &= -\beta S(t)I(t) + \gamma I(t-\omega) \\ I'(t) &= \beta S(t)I(t) - (\gamma+\alpha)I(t). \end{aligned} \tag{7.24}$$

Equilibria are given by $I = 0$ or $\beta S = \gamma$. The characteristic equation at an equilibrium (S_∞, I_∞) is obtained by the same approach used in the preceding section and is

$$\beta\gamma I_\infty \frac{1 - e^{-\omega\lambda}}{\lambda} = -\lambda - (\gamma + \beta I_\infty - \beta S_\infty). \tag{7.25}$$

At the disease-free equilibrium $S_\infty = K$, $I_\infty = 0$, this reduces to a linear equation with a single root $\lambda = \beta K - \gamma$, which is negative if and only if $R_0 = \beta K/\gamma < 1$. In order to analyze the characteristic equation (7.25) at the endemic equilibrium we need to solve for S_∞ and I_∞. In order to do this we need an additional equation because the two equations for S and I give only a single equilibrium condition. To accomplish this we must write the equation for R in the integrated form

$$R(t) = \int_{t-\omega}^{t} \gamma I(x)dx$$

to give $R_\infty = \omega\gamma I_\infty$. We also have $\beta S_\infty = \gamma$, and from $S_\infty + I_\infty + R_\infty = K$ we obtain

$$\begin{aligned} \beta K &= \beta S_\infty + \beta(1 + \omega\gamma)I_\infty = \gamma + \beta(1 + \omega\gamma)I_\infty \\ \beta I_\infty &= \frac{\beta K - \gamma}{1 + \omega\gamma}. \end{aligned}$$

Now we may rewrite the characteristic equation (4.2) at the endemic equilibrium in terms of the parameters of the model as

$$\gamma \frac{\beta K - \gamma}{1 + \omega\gamma} \frac{1 - e^{-\omega\lambda}}{\lambda} = -\lambda \frac{\beta K - \gamma}{1 + \omega\gamma}. \tag{7.26}$$

We think of ω and K as fixed and consider β and γ as parameters. If $\gamma = 0$ the equation (7.26) is linear and its only root is $-\beta K < 0$. Thus, there is a region in the (β, γ) parameter space containing the β-axis in which all roots of (7.26) have negative real part. In order to find how large this stability region is we make use of the fact that the roots of (7.26) depend continuously on β and γ. A root can move into the right half-plane only by passing through the value zero or by crossing the imaginary axis as β and γ vary. Thus, the stability region contains the β-axis and extends into the plane until there is a root $\lambda = 0$ or until there is a pair of pure imaginary roots $\lambda = \pm iy$ with $y > 0$. Since the left side of (7.26) is positive and the right side of (7.26) is negative for real $\lambda \geq 0$, there can not be a root $\lambda = 0$.

The condition that there is a root $\lambda = iy$ is

$$\gamma \frac{\beta K - \gamma}{1 + \omega\gamma} \cdot \frac{1 - e^{-i\omega\gamma}}{iy} = -iy - \frac{\beta K - \gamma}{1 + \omega\gamma}, \tag{7.27}$$

and separation into real and imaginary parts gives the pair of equations

$$\gamma\frac{\sin \omega y}{y} = -1, \qquad \gamma\frac{\beta K - \gamma}{1 + \omega\gamma} \cdot \frac{1 - \cos \omega y}{y} = y. \tag{7.28}$$

To satisfy the first condition it is necessary to have $\omega\gamma > 1$ since $|\sin \omega y| \leq |\omega y|$ for all y. This implies, in particular, that the endemic equilibrium is asymptotically stable if $\omega\gamma < 1$. In addition, it is necessary to have $\sin \omega y < 0$. There is an infinite sequence of intervals on which $\sin \omega y < 0$, the first being $\pi < \omega y < 2\pi$. For each of these intervals the equations (7.28) define a curve in the (β, γ) plane parametrically with y as parameter. The region in the plane below the first of these curves is the region of asymptotic stability, that is, the set of values of β and γ for which the endemic equilibrium is asymptotically stable. This curve is shown for $\omega = 1$, $K = 1$ in Figure 7.8. Since $R_0 = \beta/\gamma > 1$, only the portion of the (β,γ) plane below the line $\gamma = \beta$ is relevant.

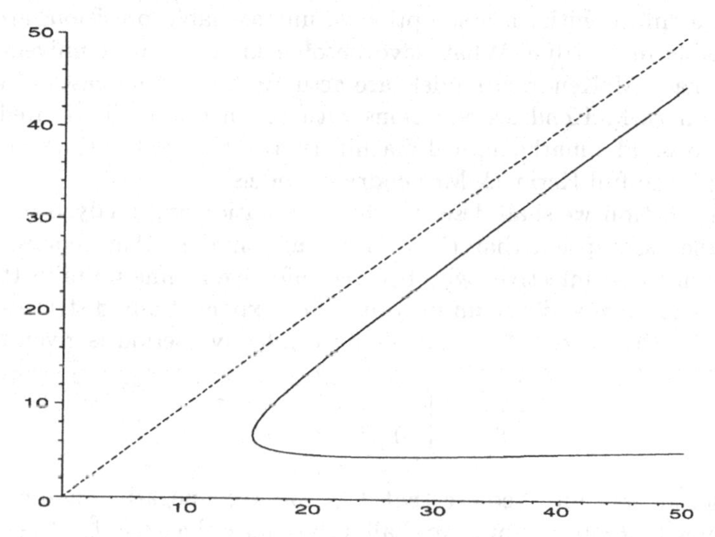

FIGURE 7.8. Region of asymptotic stability for endemic equilibria, ($\omega = 1$, $K = 1$).

The new feature of the model of this section is that the endemic equilibrium is not asymptotically stable for all parameter values. What is the behavior of the model if the parameters are such that the endemic equilibrium is unstable? A plausible suggestion is that since the loss of stability corresponds to a root $\lambda = iy$ of the characteristic equation there are solutions of the model behaving like the real part of e^{iyt}, that is, there are periodic solutions. This is exactly what does happen according to a very general result called the Hopf bifurcation theorem, which says that when roots of the characteristic equation cross the imaginary axis a stable periodic orbit arises.

From an epidemiological point of view periodic behavior is unpleasant. It implies fluctuations in the number of infectives which makes it difficult to allocate resources for treatment. It is also possible for oscillations to have a long period. This means that if data are measured over only a small time interval the actual behavior may not be displayed. Thus, the identification of situations in which an endemic equilibrium is unstable is an important problem.

7.8 Arbitrarily Distributed Infective Periods

The assumption of exponentially distributed infective periods made in Sections 7.1 through 7.5 and the assumption of an infective period of fixed length introduced in Section 7.6 are only two of many possibilities. In fact, exponential distributions, with a possibility of infinite time of stay, and fixed times, with an assumption of unreasonable precision, are quite implausible in real life. What have become known almost universally as the Kermack-McKendrick models are actually the special cases of the actual Kermack-McKendrick equations with exponentially distributed infective periods. The mathematical machinery available at the time could not cope with the full Kermack-McKendrick models.

In this section we shall describe the formulation and analysis of models under the assumption that there is a function $P(s)$ that represents the fraction of those infectives who became infective s time units in the past and are still alive and remain infective. The exponentially distributed case is given by $P(s) = e^{-(\alpha+\gamma)s}$ and the fixed infective period is given by

$$P(s) = \begin{cases} 1 & \text{for } 0 \leq s \leq \tau \\ 0 & \text{for } s > \tau \end{cases} .$$

Obviously, we must require that $P(s)$ be a nonnegative, nonincreasing function with $P(0) = 1$, and we shall also require that $\tau = \int_0^\infty P(s)\,ds$, the mean infective period, is finite. If we assume that there are no births or deaths the equation for $I(t)$ is

$$\begin{aligned} I(t) &= \int_{-\infty}^{t} \beta S(x) I(x) P(t-x) dx \\ &= \int_0^\infty \beta S(t-u) I(t-u) P(u) du \end{aligned} \tag{7.29}$$

and if we assume a proportional death rate μ in each class the equation for $I(t)$ is

$$I(t) = \int_{-\infty}^{t} \beta S(x) I(x) P(t-x) e^{-\mu(t-x)} dx = \int_0^\infty \beta S(t-u) I(t-u) P(u) e^{-\mu u} du.$$

The other equations in the model depend on the type of model. We shall consider only SIS models in this section in order to describe the methods with as little technical detail as possible. For an SIS model with no births or deaths either from natural causes or from disease the total population size is a constant K and $S+I=K$. Thus, the model is a single equation for I with S replaced by $K-I$, namely,

$$I(t) = \int_{-\infty}^{t} \beta I(x)[K - I(x)]P(t-x)dx. \tag{7.30}$$

Equilibria are given by

$$I = \beta I(K-I)\int_{-\infty}^{t} P(t-x)dx = \beta I(K-I)\int_{0}^{\infty} P(u)du = \beta I(K-I)\tau.$$

Thus, there is a disease-free equilibrium $I=0$ and an endemic equilibrium given by $\beta\tau(K-I)=1$, or $I = K - 1/\beta\tau$. The linearization at an equilibrium and the derivation of the characteristic equation follows the same pattern as for the SIS model with no births or deaths and an infective period of fixed length, except that instead of an integral of the kernel 1 over the interval $(t-\tau, t)$ we now have the integral of $P(t-x)$ from $-\infty$ to t, giving a characteristic equation

$$\beta(K-2I)\int_{-\infty}^{t} e^{-\lambda(t-x)}P(t-x)dx = \beta(K-2I)\int_{0}^{\infty} e^{-\lambda u}P(u)du = 1 \tag{7.31}$$

The expression $\int_0^\infty e^{-\lambda u}P(u)du$ is called the *Laplace transform* of P. It is a function of the (complex) variable λ and is denoted by $\widehat{P}(\lambda)$. Thus,

$$\widehat{P}(\lambda) = \int_{0}^{\infty} e^{-\lambda u}P(u)du. \tag{7.32}$$

The Laplace transform has the property that if $P(u)$ is a nonnegative function and if $\Re\lambda \geq 0$ then

$$|\widehat{P}(\lambda)| \leq \int_{0}^{\infty} |e^{-\lambda u}|P(u)du \leq \int_{0}^{\infty} P(u)du = \widehat{P}(0) = \tau.$$

This property enables us to analyze the characteristic equation (7.31) easily. If $\Re\lambda \geq 0$ the absolute value of the left side of (7.31) is no greater than $\beta(K-2I)\tau$ and thus the roots of (7.31) all satisfy $\Re\lambda < 0$ if $\beta(K-2I)\tau < 1$. For the disease-free equilibrium $I=0$, this condition is $\beta\tau K < 1$ and for the endemic equilibrium $I = K - 1/\beta\tau$ this condition reduces to $\beta\tau K > 1$. Thus, the disease-free equilibrium is asymptotically stable if and only if $R_0 = \beta\tau K < 1$ and the endemic equilibrium, which exists only if $R_0 > 1$, is asymptotically stable if $R_0 > 1$.

If we incorporate a birth rate μK and a natural proportional death rate μ in each class into an SIS model the total population size is no longer constant, and we must use two variables in the model. If we use S as one of the variables the equation for S would have to include a recovery term obtained by differentiating the integral equation for I. Since this is quite complicated we shall not carry it out here. It is more convenient to use I and N as variables, with S replaced by $N - I$, since N satisfies the differential equation $N' = \mu K - \mu N$. Thus, we have the model

$$\begin{aligned} I(t) &= \int_{-\infty}^{t} \beta I(x)[N(x) - I(x)]P(t-x)e^{-\mu(t-x)}dx \\ N' &= \mu K - \mu N. \end{aligned} \tag{7.33}$$

For the analysis of this model it is convenient to define the function

$$P_\mu(u) = P(u)e^{-\mu u}$$

representing the fraction of infectives who remain alive and infective a time u after becoming infective.

The equilibrium conditions are

$$N = K \quad \text{and either} \quad I = 0 \quad \text{or} \quad \beta I \int_0^\infty P_\mu(u)du = 1.$$

The same procedure that we have used frequently in this chapter leads to the characteristic equation

$$\det \begin{pmatrix} \beta(K-2I)\widehat{P}_\mu(\lambda) - 1 & \beta I \widehat{P}_\mu(\lambda) \\ 0 & -(\lambda+\mu) \end{pmatrix} = 0$$

and this reduces to

$$\beta(K - 2I)\widehat{P}_\mu(\lambda) = 1$$

which is exactly the same as the characteristic equation obtained for the SIS model without births and deaths with $\widehat{P}(\lambda)$ replaced by $\widehat{P}_\mu(\lambda)$. Thus, we again have asymptotic stability of the disease-free equilibrium if $R_0 < 1$ and asymptotic stability of the endemic equilibrium if $R_0 > 1$, with

$$R_0 = \beta K \int_0^\infty P_\mu(u)du.$$

The quantity $\tau_\mu = \int_0^\infty P_\mu(u)du$ represents the mean infective period corrected for natural mortality, that is the mean infective period averaged over all infectives, including those who die while still infective.

To model a universally fatal disease with no recovered members returning to the susceptible class we may use the above equation for I together with the differential equation $S' = \mu K - \mu S - \beta SI$, giving the model

$$\begin{aligned} S' &= \mu K - \mu S - \beta SI \\ I(t) &= \int_{-\infty}^{t} \beta S(x) I(x) P(t-x) e^{-\mu(t-x)} dx \end{aligned} \tag{7.34}$$

whose analysis is much the same as that of (7.33). The characteristic equation at an equilibrium (S, I) is

$$\det \begin{pmatrix} -(\lambda+\mu) - \beta I & -\beta S \\ \beta I \widehat{P}_{\mu}(\lambda) & \beta S \widehat{P}_{\mu}(\lambda) - 1 \end{pmatrix} = 0$$

which reduces to

$$\beta S \widehat{P}_{\mu}(\lambda) = 1 + \frac{\beta I}{\lambda + \mu}.$$

At the disease-free equilibrium $I = 0$ this is almost the same equation as the characteristic equation (7.31) analyzed earlier. We may infer that the disease-free equilibrium is asymptotically stable if and only if $R_0 = \beta \widehat{P}_{\mu}(0) K < 1$. For the endemic equilibrium we need only note that the absolute value of the left side is less than one if $\Re\lambda \geq 0$ since $|\beta S \widehat{P}_{\mu}(\lambda)| \leq \beta S \widehat{P}_{\mu}(0) = 1$, while the real part of the right side is greater than one, as in the analysis of (7.23) in Section 7.6. Thus, there can be no roots of the characteristic equation with $\Re\lambda \geq 0$, and the endemic equilibrium is asymptotically stable if it exists, that is, if $R_0 > 1$.

The analysis of SIR models with arbitrary infective periods is more complicated because of the increased dimension, but the methods used in this section are easily adapted to the formulation of the SIR models. There are no surprises in the results of the analysis: The disease-free equilibrium is asymptotically stable if and only if $R_0 < 1$, and the endemic equilibrium, which exists only if $R_0 > 1$, is asymptotically stable.

7.9 Directions for Generalization

We have been examining some general models for infectious diseases, looking for general classes of models rather than descriptions of specific diseases. As a result, the models we have been considering so far have been relatively simple. There are many effects of importance which we have not introduced yet. Detailed descriptions of the dynamics of many diseases may be found in such books as *Infectious Diseases of Humans* by R.M. Anderson and R.M. May (1991). Recent advances associated with the dynamics of emergent diseases can be found in Castillo-Chavez, Blower, Kirschner, van den Driessche, and Yakubu (2001a, 2001b).

Here we list some of the possible generalizations. Some of these require more advanced mathematical techniques than those used in this chapter.

1. *Models with more compartments.* One direction for generalization is the inclusion of more compartments in a compartmental model. For example, in most infectious diseases there is an exposed period after the transmission of infection from susceptible to potential infective members but before these potential infectives can retransmit infection. If the exposed period is short it is often neglected in modeling, but a longer exposed period could lead to significantly different model predictions. Fox rabies, for example, has an exposed period of about a month, considerably longer than the actual infective period. The inclusion of an exposed period would mean an additional compartment in the model; thus, we may speak of $SEIS$ or $SEIR$ models. Some diseases, such as malaria, provide a temporary immunity on recovery. This also would mean an additional compartment, giving an $SIRS$ model. For some recent applications to tuberculosis that lead to interesting dynamics (backward bifurcations) see Huang, Cooke and Castillo-Chavez (1990), Hadeler and Castillo-Chavez (1995), Feng, Castillo-Chavez, and Capurro (2000).

2. *Vertical transmission.* In many diseases, it is possible for some offspring of infected parents to be born infective. This effect, called vertical transmission, is significant in many diseases, including AIDS, Chagas' disease, hepatitis B, and rinderpest (a cattle disease, which may have been the plague of murrain inflicted on Egypt in the time of Moses). Vertical transmission may be incorporated into a model by including a fraction of the newborn members in the infective class. The book *Vertically Transmitted Diseases: Models and Dynamics*[Busenberg and Cooke (1993)] describes many aspects of infectious disease models with vertical transmission.

3. *Vector transmission.* Malaria is one of many diseases that are not transmitted directly from human to human but are transmitted through a *vector*. Malaria is acquired by humans through bites by infected mosquitoes; other mosquitoes become infective by biting infected humans. Thus, the infection goes back and forth between mosquitoes and humans, and a model for vector transmission must include both species. The two species may have radically different demographic time scales, which may allow some simplification of the model, but in general a model for vector transmission requires twice as many compartments as a model for direct transmission. One of the earliest infectious disease models to be applied practically was the model of Sir R.A. Ross for malaria [Ross (1911)]. In the introductory section we described the triumph of mathematical epidemiology represented by Ross's work. Ross's relatively simple model is still the basis for malaria models, although many other factors such as temporary immunity, superinfection, and the development by mosquitoes of immunity against mosquito controls are important in malaria control. A good account of some of the specific

features of malaria models may be found in the chapter on the population dynamics of malaria by J.L.Aron and R.M.May in the book *Population Dynamics of Infectious Diseases* [Anderson (1982)].

For sexually transmitted diseases transmitted by heterosexual contact it is necessary to use models with separate compartments for males and females. These compartments may be of very different sizes. Gonorrhea has been modeled in this way, and the predictions of the models have been used to good effect for disease control. A good reference for such models is the book *Gonorrhea Transmission Dynamics: Dynamics and Control* [Hethcote and Yorke (1984)]. The books on the transmission dynamics of HIV by Castillo-Chavez (1989) and Hethcote and Van Ark (1992) are also quite relevant. Two-sex models for sexually transmitted diseases have many properties in common with vector transmission models, since the transmission of disease goes back and forth between the two groups. Efforts to model interactions between multiple groups (contact structures) have been the source of great research activity over the last decade and a half. Some relevant references include Blythe and Castillo-Chavez (1990), Busenberg and Castillo-Chavez (1991), Blythe, Busenberg, and Castillo-Chavez (1995), and Castillo-Chavez, Busenberg, and Gerow (1991).

4. *Nonhomogeneous mixing.* We have confined ourselves to models in which the mixing of members of the population is homogeneous and the number of contacts per member per unit time is proportional to total population size. It may be more realistic to assume that the number of contacts per member per unit time grows less rapidly as population size increases. An extreme case, which may be quite realistic for sexually transmitted disease models, would assume a constant number of contacts per member per unit time. More generally, it may be appropriate to replace the proportionality constant β in our models by a nonnegative, nonincreasing function of total population size.

In sexually transmitted disease models it is important to look closely at mixing patterns, as members may have very different levels of sexual activity and the mixing may include preferences for contacts with members having a specific activity level. A model that attempts to give a precise description of a sexually transmitted disease must incorporate an understanding of the sociological patterns in the population. Models that have incorporated these ideas include the HIV models found in Castillo-Chavez, Cooke, Huang, and Levin (1989a, 1989b). Gender is an important source of heterogeneity and efforts in this direction have been carried out by Dietz (1988), Dietz and Hadeler (1988), Hadeler, Waldstätter, and Wörz-Busekros (1988), Hadeler (1989b), Waldstätter (1989, 1990), Castillo-Chavez, Busenberg and Gerow (1991), and many others.

5. *Age-structured populations.* Populations are not necessarily homogeneous, and the mixing of members may be represented better if some structure is

included in the populations by subdividing the basic compartments. One of the most obvious examples of inhomogeneity is age structure. For example, childhood diseases may be transmitted mainly between members of the population of the same age, and an age-structured population model with contact rates depending on age may be a better description than a simpler model. The classical "childhood" diseases, such as measles, chicken pox, and rubella (German measles), are spread mainly by contacts between children of similar ages. To incorporate this into a model one could consider an age-structured population with the rate of contact between members dependent on the ages of the members. Such a model would be a system of partial differential equations whose independent variables are time and age and whose dependent variables are the number of members of each class at a given time of a given age. Models that have incorporated these ideas include the influenza models found in Castillo-Chavez, Hethcote, Andreasen, Levin and Liu (1988, 1989) and the TB models found in Castillo-Chavez and Feng (1998a, 1998b). Two-sex models with age structure have been developed by Hadeler (1989a, 1989b) and others.

6. *Distributions that are spatially non-uniform.* The history of bubonic plague describes the movement of the disease from place to place carried by rats. The course of an infection usually cannot be modeled accurately without some attention to its spatial spread. To model this would require partial differential equations, possibly leading to descriptions of population waves analogous to the waves of disease that have often been observed. The article by Durrett and Levin (1994) will give the reader an overview of some of the complexities associated with the study of spatially explicit models.

7. *Variable infectivity.* One of the important features of HIV/AIDS is that the infectivity of infective members depends on the time since becoming infected, sometimes called the *infection age* (as opposed to chronological age). Infected members are highly infective for a short period after having become infected and then have relatively low infectivity for a long period (possibly on the order of ten years) before again becoming highly infective shortly before developing full-blown AIDS. This would have to be incorporated into models to describe the spread of the disease. The article by Thieme and Castillo-Chavez (1993) incorporates variable infectivity and age of infection in the study of HIV dynamics.

8. *Macroparasitic infections.* We have been describing diseases caused by microparasites, including viruses and bacteria. The main feature of the models is that members of the population are either infective or not infective, and there is no distinction by level of infection. There are also infections caused by macroparasites, such as parasitic helminths and arthropods, where the number of infection agents in the host makes a difference. Often the parasites tend to aggregate in a part of the host population.

Such infections require different types of models that take account of the distribution of macroparasites in the hosts. There has been extensive research in the study of macroparasitic infections, particularly in the study of host-helminth parasite interactions. Some important references include Anderson (1974), Dobson (1988), Hadeler (1982), Hadeler and Dietz (1984), Hudson and Dobson (1995), Kretzschmar (1989, 1993), May and Anderson (1978).

9. *Stochastic models.* Our models have been deterministic, with the progress of the disease determined entirely by the current state and possibly by the past history, but with no allowance for random effects. When population sizes are large this is a reasonable approximation, but for small population sizes individual differences and random effects are important. Stochastic models are needed to describe such situations. There is an extensive literature in this field. The books by Mollison (1995) and Castillo-Chavez, Blower, van den Driessche, Kirschner and Yakubu (2001a and 2001b) provide an excellent starting point.

The idea of a threshold may be important in other disciplines as well. There are many phenomena in which the rate at which individuals move from one compartment to another is proportional to both the number of individuals who have already made the transition and to the number of individuals who have not yet moved. This rate is the term in the model equations that produces the threshold phenomenon. It has been suggested that some social problems are contagious in the same sense [Crane (1991); Gladwell (2000)]. Sociologists use the term "tipping point" for what we are calling the threshold; their models are descriptive rather than mathematical but there is some qualitative evidence to support their idea. The basic assumption is that social problems are spread through peer influence, so that rates of development of such problems as teenage pregnancy and juvenile crime depend on the reservoir of potential victims and on the number of cases. One inference, if this idea is accepted, would be that if a level of preventive measures can be instituted in some neighborhood to reduce the basic reproductive number below one, then maintenance of this state should be relatively easy. Thus, it would make sense to target neighborhoods that are experiencing such an epidemic or are at high risk.

A variation on this idea may be applied to situations in which a researcher wishes to encourage a particular kind of behavior. In such a situation the goal would be to increase the basic reproductive number. [See, for example, Project 4.2.]

There has been extensive research on the development of approaches for the computation and interpretation of thresholds in epidemiology. The thesis by Heesterbeek (1992) and the recent book by Diekmann and Heesterbeek (2000) give a comprehensive view of the current research in this area. The article by Heesterbeek and Roberts (1995) outlines a possible extension of the notion of threshold to macroparasitic infections. The work of Huang, Cooke and Castillo-Chavez (1990), Hadeler and Castillo-Chavez (1995), and Feng, Castillo-Chavez, and Capurro (2000) illustrate the care that must be exercised in the interpretation of these thresholds.

7.10 Project 7.1: Pulse Vaccination

Consider an SIR model (7.7). For measles typical parameter choices are $\mu = 0.02$, $\beta = 1800$, $\xi = 100$, $K = 1$ (to normalize carrying capacity to 1) [Engbert and Drepper (1984)].

Question 1.
Show that for these parameter choices $R_0 \approx 18$ and to achieve herd immunity would require vaccination of about 95 percent of the susceptible population.

In practice, it is not possible to vaccinate 95 percent of a population because not all members of the population would come to be vaccinated and not all vaccinations are successful. One way to avoid recurring outbreaks of disease is "pulse vaccination" [Agur, Mazor, Anderson and Danon (1993); Shulgin, Stone and Agur (1998); Stone, Shulgin and Agur (2000)]. The basic idea behind pulse vaccination is to vaccinate a given fraction p of the susceptible population at intervals of time T with T (depending on p) chosen to ensure that the number of infectives remains small and approaches zero. In this project we will give two approaches to the calculation of a suitable function $T(p)$.

The first approach depends on the observation that I decreases so long as $S < \Gamma < (\mu + \gamma)/\beta$. We begin by vaccinating $p\Gamma$ members, beginning with $S(0) = (1 - p)\Gamma$. From (8.7),

$$S' = \mu K - \mu S - \beta SI \geq \mu K - \mu S.$$

Then $S(t)$ is greater than the solution of the initial value problem

$$S' = \mu K - \mu S, \quad S(0) = (1 - p)\Gamma.$$

Question 2.
Solve this initial value problem and show that the solution obeys $S(t) < \Gamma$ for $0 \leq t < \frac{1}{\mu} \log \frac{K - (1 - p)\Gamma}{K - \Gamma}$. Thus a suitable choice for $T(p)$ is

$$T(p) = \frac{1}{\mu} \log \frac{K - (1 - p)\Gamma}{K - \Gamma} = \frac{1}{\mu} \log \left[1 + \frac{p\Gamma}{K - \Gamma} \right].$$

Calculate $T(p)$ for $p = m/10$ $(m = 1, 2, \ldots, 10)$.

The second approach is more sophisticated. Start with $I = 0$, $S' = \mu K - \mu S$. We let $t_n = nT$ $(n = 0, 1, 2, \ldots)$, run the system for $0 \leq t \leq t_1 = T$ and let $S_1 = (1-p)S(t_1)$. We then repeat, i.e., for $t_1 \leq t \leq t_2$, $S(t)$ is the solution of $S' = \mu K - \mu S$, $S(t_1) = S_1$, and $S_2 = (1-p)S_1$. We obtain a sequence S_n in this way.

Question 3.
Show that

$$S_{n+1} = (1-p)K(1-e^{-\mu T}) + (1-p)S_n e^{-\mu T}$$

and, for $t_n \leq t \leq t_{n+1}$

$$S(t) = K\left[1 - e^{-\mu(t-t_n)}\right] + S_n e^{-\mu(t-t_n)}$$

Question 4
Show that the solution is periodic if

$$S_{n+1} = S_n = S^* \quad (n = 0, 1, 2, \ldots)$$

with

$$S^* = K\left[1 - \frac{pe^{\mu T}}{e^{\mu T} - (1-p)}\right]$$

and that the periodic solution is

$$S(t) = \begin{cases} K\left[1 - \frac{pe^{\mu T}}{e^{\mu T} - (1-p)} e^{-\mu(t-t_n)}\right] \; for & t_n \leq t \leq t_{n+1} \\ S^* \; for & t = t_{n_1} \end{cases}$$

$$I(t) = 0$$

It is possible to show by linearizing about this periodic solution that the periodic solution is asymptotically stable if

$$\frac{1}{T}\int_0^T S(t)\,dt < \frac{\mu + \xi}{\beta}.$$

If this condition is satisfied the infective population will remain close to zero.

Question 5
Show that this stability condition reduces to

$$\frac{K(\mu T - p)(e^{\mu T} - 1) + pK\mu T}{\mu T\,[e^{\mu} - (1-p)]} < \frac{\mu + \xi}{\beta}.$$

Question 6
Use a computer algebra system to graph $T(p)$, where T is defined implicitly by

$$\frac{K(\mu T - p)(e^{\mu T} - 1) + pK\mu T}{\mu T\left[e^{\mu} - (1-p)\right]} = \frac{\mu + \xi}{\beta}.$$

Compare this expression for T with the one obtained earlier in Question 2 in this project. A larger estimate for a safe value of T would save money by allowing less frequent vaccination pulses.

7.11 Project 7.2: A Model with Competing Disease Strains

We model a general discrete time SIS model with two competing strains in a population with discrete and nonoverlapping generations. This model arises from a particular discretization in time of the corresponding SIS continuous time stochastic model for two competing strains.

State variables

S_n	population of susceptible individuals in generation n
I_n^1	population of infected individuals with strain 1 in generation n
I_n^2	population of infected individuals with strain 2 in generation n
T_n	total population in generation n
f	recruitment function

Parameters

μ	per capita natural death rate
γ_i	per capita recovery rate for strain i
α_i	per capita infection rate for strain i

Construction of the model equations
The model assumes that: (i) the disease is not fatal; (ii) all recruits are susceptible and the recruitment function depends only on T_n; (iii) there are no coinfections; (iv) death, infections, and recoveries are modeled as a Poisson processes with rates μ, α_i, γ_i $(i = 1, 2)$; (v) the time step is measured in generations; (vi) the populations change only because of "births" (given by the recruitment function), deaths, recovery, and infection of a susceptible individual for each strain; (vii) individuals recover but do not

develop permanent or temporary immunity, that is, they immediately become susceptible again.
By assumption we have that the probability of k successful encounters is a Poisson distribution, which in general has the form $p(k) = e^{-\beta}\beta^k/k!$, where β is the parameter of the Poisson distribution. In our context only one success is necessary. Therefore, when there are no successful encounters, the expression $p(0) = e^{-\beta}$ represents the probability that a given event does not occur. For example, the probability that a susceptible individual does not become infective is Prob(not being infected by strain i) $= e^{-\alpha_i I_n^i}$; and, Prob(not recovering from strain i) $= e^{-\gamma_i I_n^i}$. Hence, Prob(not being infected)= Prob(not being infected by strain 1)Prob(not being infected by strain 2) $= e^{-\alpha_1 I_n^1} e^{-\alpha_2 I_n^2}$.

Now the probability that a susceptible does become infected is given by $1-e^{-\alpha_i I_n^i}$. Then, Prob(infected by strain i) = Prob(infected). Prob(infected by strain i | infected) $=(1 - e^{-(\alpha_1 I_n^1+\alpha_2 I_n^2)})\frac{\alpha_i I_n^i}{\alpha_1 I_n^1+\alpha_2 I_n^2}$.

(a) Using the above discussion show that the dynamics are governed by the system

$$\begin{aligned} S_{n+1} &= f(T_n) + S_n e^{-\mu} e^{-(\alpha_1 I_n^1+\alpha_2 I_n^2)} \\ &\quad + I_n^1 e^{-\mu}(1-e^{-\gamma_1}) + I_n^2 e^{-\mu}\left(1-e^{-\gamma_2}\right), \qquad (7.35) \\ I_{n+1}^1 &= \frac{\alpha_1 S_n I_n^1}{\alpha_1 I_n^1 + \alpha_2 I_n^2} e^{-\mu}(1-e^{-(\alpha_1 I_n^1+\alpha_2 I_n^2)}) + I_n^1 e^{-\mu} e^{-\gamma_1}, \\ I_{n+1}^2 &= \frac{\alpha_2 S_n I_n^2}{\alpha_1 I_n^1 + \alpha_2 I_n^2} e^{-\mu}(1-e^{-(\alpha_1 I_n^1+\alpha_2 I_n^2)}) + I_n^2 e^{-\mu} e^{-\gamma_2}. \end{aligned}$$

(b) Show that

$$T_{n+1} = f(T_n) + T_n e^{-\mu},$$

where

$$T_n = S_n + I_n^1 + I_n^2. \qquad (7.36)$$

This equation is called the *demographic equation.* This equation describes the total population dynamics.

(c) If we set $I_{n+1}^1 = I_{n+1}^2 = 0$ then model (7.35) reduces to the demographic model

$$T_{n+1} = f(T_n) + T_n e^{-\mu}.$$

Check that this is the case.

(d) Study the disease dynamics at a demographic equilibrium, that is, at a point where $T_\infty = T_\infty e^{-\mu} + f(T_\infty)$. Substitute $S_n = T_\infty - {I_n}^1 - I_n^2$

where T_∞ is a stable demographic equilibrium, that is, assume $T_0 = T_\infty$ to get the following equations:

$$\begin{aligned} I^1_{n+1} &= \frac{\alpha_1 I^1_n}{\alpha_1 I^1_n + \alpha_2 I^2_n}(T_\infty - I^1_n - I^2_n)e^{-\mu}\left(1 - e^{-(\alpha_1 I^1_n + \alpha_2 I^2_n)}\right) \\ &\quad + I^1_n e^{-\mu} e^{-\gamma_1} \qquad (7.37) \\ I^2_{n+1} &= \frac{\alpha_2 I^2_n}{\alpha_1 I^1_n + \alpha_2 I^2_n}(T_\infty - I^1_n - I^2_n)e^{-\mu}\left(1 - e^{-(\alpha_1 I^1_n + \alpha_2 I^2_n)}\right) \\ &\quad + I^2_n e^{-\mu} e^{-\gamma_2} \end{aligned}$$

System (7.37) describes the dynamics of a population infected with the two strains at a demographic equilibrium.

Show that in system (7.37) if $R_1 = \frac{e^{-\mu}T_\infty \alpha_1}{1-e^{-(\mu+\gamma_1)}} < 1$ and $R_2 = \frac{e^{-\mu}T_\infty \alpha_2}{1-e^{-(\mu+\gamma_2)}} < 1$, then the equilibrium point $(0,0)$ is asymptotically stable.

(e) Interpret biologically the numbers R_i, $i = 1, 2$.

(f) Consider $f(T_n) = \Lambda$, where Λ is a constant. Show that

$$T_{n+1} = \Lambda + T_n e^{-\mu}$$

and that

$$T_\infty = \frac{\Lambda}{1 - e^{-\mu}}.$$

(g) Consider $f(T_n) = rT_n(1 - T_n)/k)$ and show that in this case the total population dynamic is given by,

$$T_{n+1} = rT_n(1 - \frac{T_n}{k}) + T_n e^{-\mu}$$

and that the fixed points are

$$T_n{}^* = 0,$$

$$T_n{}^{**} = \frac{k(r + e^{-\mu} - 1)}{r}$$

whenever $r + e^{-\mu} > 1$.

(h) Assume that one of the strains is missing, that is, let $I^i_n = 0$ for either $i = 1$ or 2. Equation (7.37) reduces to

$$I_{n+1} = (T_\infty - I_n)e^{-\mu}(1 - e^{\alpha_1 I_n}) + I_n e^{-(\mu+\gamma)}$$

Establish necesary and sufficient conditions for the stability and/or instability of boundary equilibria for the system (7.37). Compare your results with simulations of the system (7.37) and of the full system (7.35).

(i) Does the system (7.37) have endemic ($I_1^* > 0, I_2^* > 0$) equilibria?

(j) Simulate the full system (7.35) when the demographic equation is in the period doubling regime. What are your conclusions?

References: Perez-Velazquez (2000) and Castillo-Chavez, Huang, and Li (1996a, 1997). (2000).

7.12 Project 7.3: An Epidemic Model in Two Patches

Consider the following SIS model with dispersion between 2 patches, Patch 1 and Patch 2, where in Patch $i \in \{1,2\}$ at generation t, $S_i(t)$ denotes the population of susceptible individuals; $I_i(t)$ denotes the population of infecteds assumed infectious; $T_i(t) \equiv S_i(t) + I_i(t)$ denotes the total population size. The constant dispersion coefficients D_S and D_I measure the probability of dispersion by the susceptible and infective individuals, respectively. Observe that we are using a different notation from what we have used elsewhere, writing variables as a function of t rather than using a subscript for the independent variable in order to avoid needing double subscripts.

$$\begin{aligned}
S_1(t+1) &= (1-D_S)\widetilde{S}_1(t) + D_S\widetilde{S}_2(t)\\
I_1(t+1) &= (1-D_I)\widetilde{I}_1(t) + D_I\widetilde{I}_2(t)\\
S_2(t+1) &= D_S\widetilde{S}_1(t) + (1-D_S)\widetilde{S}_2(t)\\
I_2(t+1) &= D_I\widetilde{I}_1(t) + (1-D_I)\widetilde{I}_2(t)
\end{aligned}$$

where

$$\widetilde{S}_i(t) = f_i(T_i(t)) + \gamma_i S_i(t)\exp(\frac{-\alpha_i I_i(t)}{T_i(t)}) + \gamma_i I_i(t)(1-\sigma_i),$$

$$\widetilde{I}_i(t) = \gamma_i(1-\exp(\frac{-\alpha_i I_i(t)}{T_i(t)}))S_i(t) + \gamma_i\sigma_i I_i(t)$$

and

$$0 \le \gamma_i, \sigma_i, \alpha_i, D_S, D_I \le 1.$$

Let

$$f_i(T_i(t)) = T_i(t)\exp(r_i - T_i(t)),$$

where r_i is a positive constant.

(a)Using computer explorations determine if it is possible to have a globally stable disease-free equilibrium on a patch (without dispersal) where the full system with dispersal has a stable endemic equilibrium. Do you have a conjecture?

(b) Using computer explorations determine if it is possible to have a globally stable endemic equilibrium on a patch (without dispersal) where the full system with dispersal has a stable disease-free equilibrium. Do you have a conjecture?

References: Gonzalez, Sanchez and Saenz (2000), Arreola, Crossa, and Velasco (2000), Castillo-Chavez and Yakubu (2000a, 2000g).

7.13 Project 7.4: Population Growth and Epidemics

When one tries to fit epidemiological data over a long time interval to a model, it is necessary to include births and deaths in the population. Throughout the book we have considered population models with birth and death rates which are constant in time. However, population growth often may be fit better by assuming a linear population model with a time-dependent growth rate, even though this does not have a model-based interpretation. There could be many reasons for variations in birth and death rates; we could not quantify the variations even if we knew all of the reasons. Let $r(t) = \frac{dN}{dt}/N$ denote the time dependent per-capita growth rate. To estimate $r(t)$ from linear interpolation of census data proceed as follows:

1. Let N_i and N_{i+1} be the consecutive census measurements of population size taken at times t_i and t_{i+1} respectively. Let $\Delta N = N_{i+1} - N_i$, $\Delta t = t_{i+1} - t_i$, and $\delta N = N(t + \delta t) - N(t)$.
2. If $t_i \leq t \leq t_{i+1}$, $\frac{\Delta N}{\Delta t} = \frac{\delta N}{\delta t}$. Then we make the estimate $r(t) \simeq \frac{\Delta N}{\Delta t N(t)}$
3. A better approximation is obtained by replacing $N(t)$ by $N(t+\delta t/2)$. Why? Show that in this case $r(t) \simeq (\frac{\delta t}{2} + \frac{N(t)\Delta t}{\Delta N})^{-1}$.

Question 1

Use the data of Table 7.2 to estimate the growth rate $r(t)$ for the population of the USA.

Fig. 7.9 shows the time evolution of the USA mortality rate. This mortality rate is fit well by

$$\mu = \mu_0 + \frac{\mu_0 - \mu_f}{1 + e^{(t - t'_{1/2})/\Delta'}} \tag{7.38}$$

with $\mu_0 = 0.01948$, $\mu_f = 0.008771$, $t'_{1/2} = 1912$, and $\Delta' = 16.61$. Then the "effective birth rate", $b(t)$, is defined as the real birth rate plus the immigration rate.

1700	250888	1800	5308483	1900	75994575
1710	331711	1810	7239881	1910	91972266
1720	466185	1820	9638453	1920	105710620
1730	629445	1830	12866020	1930	122775046
1740	905563	1840	17069453	1940	131669275
1750	1170760	1850	23192876	1950	151325798
1760	1593625	1860	31443321	1960	179323175
1770	2148076	1870	39818449	1970	203302031
1780	2780369	1880	50155783	1980	226542199
1790	3929214	1890	62947714	1990	248718301
–	–	–	–	2000	274634000

TABLE 7.2. Population data growth for the USA

Question 2
Estimate $b(t)$ using $r(t) = b(t) - \mu(t)$, with $r(t)$ found in Question 1.

Consider an $SEIR$ disease transmission model. We assume that

(a) an average infective individual produces β new infections per unit of time when all contacts are with susceptibles but that otherwise this rate is reduced by the ratio S/N.

(b) individuals in the exposed class E progress to the infective class at the per capita rate k.

(c) there is no disease induced mortality, permanent immunity, and a mean infective period of $1/\gamma$.

We define $\gamma = r + \mu$. The model becomes:

$$\begin{aligned}
\frac{dS}{dt} &= bN - \mu S - \beta S\frac{I}{N} \\
\frac{dE}{dt} &= \beta S\frac{I}{N} - (k+\mu)E \\
\frac{dI}{dt} &= kE - (r+\mu)I \\
\frac{dR}{dt} &= rI - \mu R.
\end{aligned} \tag{7.39}$$

Question 3

(a) Show that the mean number of secondary infections (belonging to the exposed class) produced by one infective individual in a population of susceptibles is $Q_0 = \beta/\gamma$.

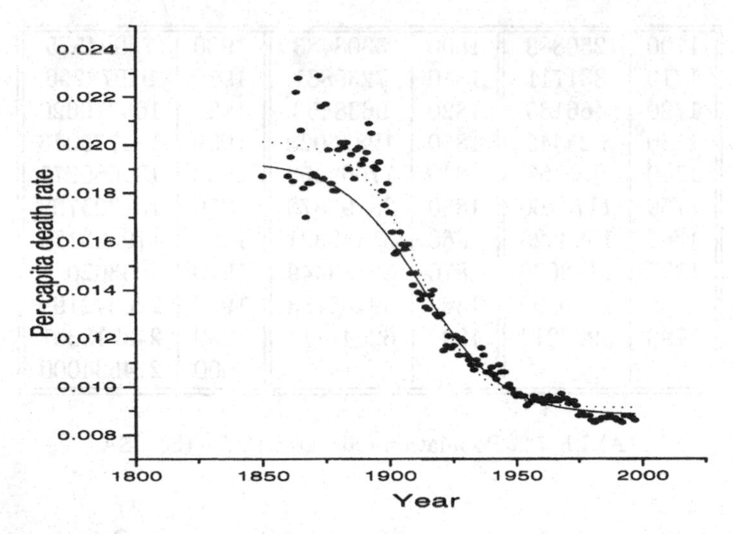

FIGURE 7.9. Observed death rate (•) and the best fit obtained with the function (7.38).

(b) Assuming that k and μ are time-independent, show that R_0 is given by $Q_0 f$ where $f = k/(k+\mu)$. What is the epidemiological interpretation of $Q_0 f$?

The usual measure of the severity of the epidemic is the incidence of infective cases. The incidence of infective cases is defined as the number of new infective individuals per year. If we take one year as the unit of time, the incidence of infective cases is given approximately by kE. The incidence rate of infective cases per 100000 population is given approximately by $10^5 kE/N$.

Tuberculosis (TB) is an example of a disease with an exposed (non-infective) stage. Infective individuals are called active TB cases. Estimated incidence of active TB in the USA was in a growing phase until around 1900 and then experienced a subsequent decline. The incidence rate of active TB exhibited a declining trend from 1850 (See Table 7.3 and Figure 7.10). The proportion of exposed individuals who survive the latency period and become infective is $f = \frac{k}{k+\mu}$. f will be used as a measure of the risk of developing active TB by exposed individuals.

Question 4

Assume that mortality varies according to the expression (7.38), and that the value of b found in Question 2 is used. Set $\gamma = 1\ yr^{-1}$ and $\beta = 10\ yr^{-1}$, both constant through time. Simulate TB epidemics starting in 1700 assuming constant values for f. Can you reproduce the observed trends (Table 7.3)?

Year	Incidence rate	Incidence	Year	Incidence rate	Incidence
1953	53	84304	1976	15	32105
1954	49.3	79775	1977	13.9	30145
1955	46.9	77368	1978	13.1	28521
1956	41.6	69895	1979	12.6	27769
1957	39.2	67149	1980	12.3	27749
1958	36.5	63534	1981	11.9	27337
1959	32.5	57535	1982	11	25520
1960	30.8	55494	1983	10.2	23846
1961	29.4	53726	1984	9.4	22255
1962	28.7	53315	1985	9.3	22201
1963	28.7	54042	1986	9.4	22768
1964	26.6	50874	1987	9.3	22517
1965	25.3	49016	1988	9.1	22436
1966	24.4	47767	1989	9.5	23495
1967	23.1	45647	1990	10.3	25701
1969	19.4	39120	1992	10.5	26673
1970	18.3	37137	1993	9.8	25287
1971	17.1	35217	1994	9.4	24361
1972	15.8	32882	1995	8.7	22860
1973	14.8	30998	1996	8	21337
1974	14.2	30122	1997	7.4	19885
1975	15.9	33989	1998	6.8	18361

TABLE 7.3. Reported incidence and incidence rate (per 100,000 population) of active-TB.

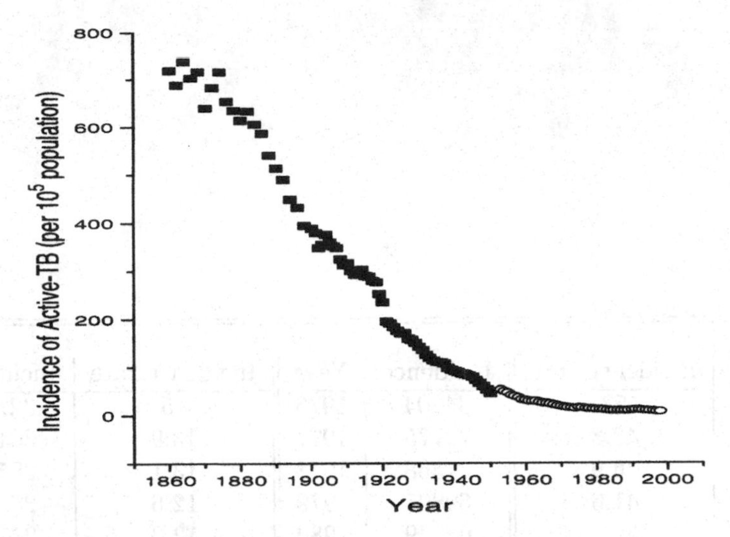

FIGURE 7.10. Incidence of Active TB

It is not possible to obtain a good fit of the data of Table 7.3 to the model (7.39). It is necessary to use a refinement of the model that includes time dependence in the parameters, and the next step is to describe such a model. The risk of progression to active TB depends strongly on the standard of living. An indirect measure of the standard of living can be obtained from the life expectancy at birth. The observed live expectancy for the USA is approximated well by the sigmoid shape function

$$\tau = \tau_f + \frac{(\tau_0 - \tau_f)}{1 + exp[(t - t_{1/2})/\Delta]}. \tag{7.40}$$

shown in Figure 7.11. Here τ_0 and τ_f are asymptotic values for life expectancy; $t_{1/2} = 1921.3$ is the time by which life expectancy reaches the value $(\tau_0 + \tau_f)/2$; and $\Delta = 18.445$ determines the width of the sigmoid.

Assume that the risk f varies exactly like life expectancy, that is, assume that f is given by

$$f(t) = f_f + \frac{(f_i - f_f)}{1 + exp[(t - t_{1/2})/\Delta]}, \tag{7.41}$$

We refine the model (7.39) by replacing the parameter k by the variable expression $\mu f(t)/(1 - f(t)$ and $k + \mu$ by $\mu/(1 - f(t)$, obtained from the relation $f = k/(k + \mu)$. Since the time scale of the disease is much faster than the demographic time scale, the recovery rate r is approximately equal to γ This gives the model

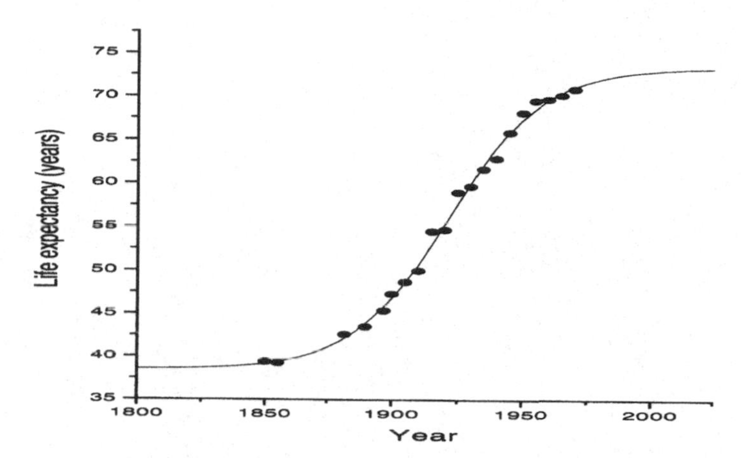

FIGURE 7.11. Observed average life expectancy at birth (•) and its best fit (continuous line) using Expression (7.40).

$$\begin{aligned}
\frac{dS}{dt} &= b(t)N - \mu(t)S - \beta S\frac{I}{N} \\
\frac{dE}{dt} &= \beta S\frac{I}{N} - \frac{\mu(t)}{1-f(t)}E \\
\frac{dI}{dt} &= \frac{\mu(t)f(t)}{1-f(t)}E - \gamma I \\
\frac{dR}{dt} &= \gamma I - \mu(t)R.
\end{aligned} \tag{7.42}$$

Question 5
Simulate TB epidemics starting in 1700 using the model (7.42) with $\gamma = 1\ yr^{-1}$ and $\beta = 10\ yr^{-1}$, both constant, and with $\mu(t)$ given by (7.38) and $f(t)$ given by (7.41). Find values of f_0 and f_f for which an accurate reproduction of the observed TB trends (Table 7.3) is achieved.

References: Aparicio, Capurro, and Castillo-Chavez (2000a, 2000b, 2001a, 2001d); U.S. Bureau of the Census (1975, 1980, 1991, 1999); Castillo-Chavez and Feng (1997, 1998a, 1998b); Feng, Castillo-Chavez, and Capurro (2000); Feng, Huang, and Castillo-Chavez (2001).

8

Models for Populations with Age Structure

8.1 Linear Discrete Models

In the preceding chapters we studied mainly models in which all members were alike, so that birth and death rates depended on total population size. However, we gave a few examples of populations with two classes of members and a birth rate that depended on the size of only one of the two classes, for discrete models in Section 2.6 and for continuous models in Section 3.3. These are examples of *structured* populations. In this chapter we shall study models for populations structured by age. In practice, animal populations are often measured by size with age structure used as an approximation to size structure. The study of age-structured models is considerably simpler than the study of general size-structured models, primarily because age increases linearly with the passage of time while the linkage of size with time may be less predictable. Age-structured models may be either discrete or continuous. We shall begin with linear models, for which total population size generally either increases or decreases over time.

We will consider a population that is divided into a finite number of age classes labeled from zero to m, and will describe this population by giving the number of members in each class at a sequence of times. Thus, we let $\rho_{j,n}$ be the number of members in the jth class at the nth time ($j = 0, 1, \ldots, m$; $n = 0, 1, 2, \ldots$). We will assume that the length of time spent in each age class is the same and that this length of time is equal to the interval between measurements of population. Thus, $\rho_{j,n+1}$, the number of members in the jth age class at the $(n+1)$st time, is equal to $\rho_{j-1,n}$, the number of members in the $(j-1)$st age class at the nth time minus the number of members of this age cohort who die before entering the next age class. We will assume that the probability of survival from one age class to the next depends only on age, that is, that there are constants $p_0, p_1, \ldots, p_{m-1}$ such that

$$\rho_{1,n+1} = p_0\rho_{0,n}, \quad \rho_{2,n+1} = p_1\rho_{1,n}, \quad \ldots, \quad \rho_{m,n+1} = p_{m-1}\rho_{m-1,n},$$

for $n = 0, 1, 2, \ldots$. Thus, p_j is the probability that a member of the jth age class survives to the $(j+1)$st age class $(j = 0, 1, 2, \ldots, m-1)$; it is assumed that no member survives beyond the mth age class.

All recruitment of new members into the population is assumed to come from a birth process, with fecundity depending only on age. Thus, we assume that there are constants $\beta_0, \beta_1, \ldots, \beta_m$ so that

$$\rho_{0,n+1} = \beta_0\rho_{0,n} + \beta_1\rho_{1,n} + \ldots + \beta_m\rho_{m,n}.$$

The fecundities of the various age classes need not all be different from zero.

We now have a population model described by a system of $(m+1)$ linear difference equations

$$\begin{aligned}
\rho_{0,n+1} &= \beta_0\rho_{0,n} + \beta_1\rho_{1,n} + \ldots + \beta_m\rho_{m,n} \\
\rho_{1,n+1} &= p_0\rho_{0,n} \\
\rho_{2,n+1} &= p_1\rho_{1,n} \\
&\vdots \\
\rho_{m,n+1} &= p_{m-1}\rho_{m-1,n},
\end{aligned}$$

to which we must add an initial condition and specify $\rho_{0,0}, \rho_{1,0}, \ldots, \rho_{m,0}$. We assume that the survival probabilities $p_0, p_1, \ldots, p_m$ are positive and that the fecundity coefficients $\beta_0, \beta_1, \ldots, \beta_m$ are non-negative and not all zero. This population model is usually known as the *Leslie matrix model*, formulated in 1945, although similar forms appeared in earlier work of H. Bernardelli (1941)and E.G. Lewis (1942). The Leslie model may be written in vector-matrix form. We define the $(m+1)$-dimensional column vector

$$\vec{\rho}_n = \begin{pmatrix} \rho_{0,n} \\ \rho_{1,n} \\ \vdots \\ \rho_{m,n} \end{pmatrix}$$

and the Leslie matrix

$$A = \begin{pmatrix} \beta_0 & \beta_1 & \beta_2 & \dots & \beta_{m-1} & \beta_m \\ p_0 & 0 & 0 & \dots & 0 & 0 \\ 0 & p_1 & 0 & \dots & 0 & 0 \\ \vdots & \vdots & \vdots & \ddots & \vdots & \vdots \\ 0 & 0 & 0 & \dots & p_{m-1} & 0 \end{pmatrix}$$

Then the model consists of the vector difference equation

$$\vec{\rho}_{n+1} = A\vec{\rho}_n$$

together with the specification of the initial vector $\vec{\rho}_0$. It is easy to solve this difference equation by induction and obtain the solution

$$\vec{\rho}_n = A^n \vec{\rho}_0.$$

This formal solution is of little use without an understanding of the nature of the matrix A^n for a given Leslie matrix A; such an understanding requires knowledge of the eigenvalues and eigenvectors of the matrix A.

In the simplest case, in which the $(m+1) \times (m+1)$ matrix A has $(m+1)$ linearly independent eigenvectors $\vec{v}_0, \vec{v}_1, \dots, \vec{v}_m$ corresponding to the eigenvalues $\lambda_0, \lambda_1, \dots, \lambda_m$, respectively we may expand

$$\vec{\rho}_n = \sum_{j=0}^{m} c_{j,n}\vec{v}_j,$$

because $\vec{v}_0, \vec{v}_1, \dots, \vec{v}_m$ span the space $\mathbb{R}^{m+1}$ of $(m+1)$-dimensional vectors with real components. Then

$$\vec{\rho}_{n+1} = \sum_{j=0}^{m} c_{j,n+1}\vec{v}_j = A\vec{\rho}_n = \sum_{j=0}^{m} c_{j,n}A\vec{v}_j = \sum_{j=0}^{m} c_{j,n}\lambda_j\vec{v}_j,$$

and therefore

$$\sum_{j=0}^{m}(c_{j,n+1} - c_{j,n}\lambda_j)\vec{v}_j = \vec{0}.$$

Because the vectors $\vec{v}_0, \vec{v}_1, \dots, \vec{v}_m$ are linearly independent, $c_{j,n+1} = \lambda_j c_{j,n}$ for $j = 0, 1, \dots, m$; $n = 0, 1, 2, \dots$, and this implies $c_{j,n} = \lambda_j^n c_{j,0}$ for $j = 0, 1, \dots, m$; $n = 0, 1, 2, \dots$. Now we may write

$$\vec{\rho}_n = \sum_{j=0}^{m} c_{j,0}\lambda_j^n\vec{v}_j, \qquad n = 0, 1, 2, \dots$$

We now suppose that the eigenvalues can be arranged in decreasing order, or at least that λ_0 is a real simple eigenvalue which is *dominant* in the sense that $|\lambda_j| < \lambda_0$ for $j = 1, 2, \ldots, m$. Then we may write

$$\vec{\rho}_n = c_{0,0}\lambda_0^n \vec{v}_0 + \vec{u}_n,$$

where $\lambda_0^{-n}\vec{u}_n \to \vec{0}$ as $n \to \infty$. In other words, $\vec{\rho}_n$ is λ_0^n multiplied by an eigenvector of A corresponding to λ_0 plus terms that are negligible compared to this term as $n \to \infty$. We define

$$P_n = \sum_{j=0}^{m} \rho_{j,n} \qquad \text{(total population at time } n\text{)}$$

$$B_n = \sum_{j=0}^{m} \beta_j \rho_{j,n} = \rho_{0,n+1} \qquad \text{(births into population at time } n\text{)}$$

Then we have

$$P_n = c_{0,0}\lambda_0^n \sum_{j=0}^{m} v_{0,j} + \sum_{j=0}^{m} u_{n,j}$$

and

$$\frac{\vec{\rho}_n}{P_n} = \frac{c_{0,0}\vec{v}_0 + \lambda_0^{-n}\vec{u}_n}{c_{0,0}\sum_{j=0}^{m} v_{0,j} + \lambda_0^{-n}\sum_{j=0}^{m} u_{n,j}}.$$

Because $\lambda_0^{-n}\vec{u}_n \to 0$ as $n \to \infty$, we have

$$\frac{\vec{\rho}_n}{P_n} \sim \frac{\vec{v}_0}{\sum_{j=0}^{m} v_{0,j}}.$$

Thus, the fraction of the population in each age class approaches a limit, and these limits are proportional to the components of the eigenvector $\vec{v}_0$. Thus, there is a *stable age distribution.*

If the matrix A does not have $(m+1)$ linearly independent eigenvectors the above reasoning must be modified, but we can still show that if there is a real simple dominant eigenvalue λ_0 with corresponding eigenvector $\vec{v}_0$ then

$$\vec{\rho}_n = c\lambda_0^n \vec{v}_0 + \vec{u}_n$$

with $\lim_{n\to\infty} \lambda_0^{-n}\vec{u}_n = \vec{0}$. This suffices to show that there is a stable age distribution. Of course, the result makes no sense biologically unless every component of the eigenvector $\vec{v}_0$ is positive.

By examining the particular form of the Leslie matrix we may show that the characteristic equation is

$$\lambda_{m+1} - \beta_0\pi_0\lambda^m - \beta_1\pi_1\lambda^{m-1} - \ldots - \beta_m\pi_m = 0,$$

where $\pi_0 = 1, \pi_1 = p_0, \pi_2 = p_0p_1, \ldots, \pi_m = p_0p_1 \ldots p_{m-1}$. Thus, $\pi_j > 0$ is the probability of survival from birth to entry into the jth age class. Division by λ^{m+1} gives the equivalent characteristic equation

$$\sum_{j=0}^{m} \beta_j \pi_j \lambda^{-(j+1)} = 1.$$

Because $\sum_{j=0}^{m} \beta_j \pi_j \lambda^{-(j+1)}$ is unbounded as $\lambda \to 0^+$, tends to zero as $\lambda \to \infty$, and is monotone decreasing, there is a unique positive real root λ_0. If $\vec{v}_0$ is the corresponding eigenvector, with components $(v_0, v_1, \ldots, v_m)$ we may again use the form of the Leslie matrix to write the equations

$$\begin{aligned}
\beta_0 v_0 + \beta_1 v_1 + \ldots + \beta_m v_m &= \lambda_0 v_0 \\
p_0 v_0 &= \lambda_0 v_1 \\
p_1 v_1 &= \lambda_0 v_2 \\
p_{m-1} v_{m-1} &= \lambda_0 v_m.
\end{aligned}$$

If we take $v_0 = 1$ then from these equations we obtain

$$\begin{aligned}
v_1 &= \frac{p_0}{\lambda_0} = \frac{\pi_1}{\lambda_0}, \\
v_2 &= \frac{p_1 v_1}{\lambda_0} = \frac{p_0 p_1}{\lambda_0^2} = \frac{\pi_2}{\lambda_0^2}, \\
&\vdots \\
v_m &= \frac{p_{m-1} v_{m-1}}{\lambda_0} = \frac{p_0 \cdots p_{m-1}}{\lambda_0^m} = \frac{\pi_m}{\lambda_0^m}.
\end{aligned}$$

Thus, the components of $\vec{v}_0$ are proportional to π_j / λ_0^j $(j = 0, 1, \ldots, m)$ and these are the proportions in the stable age distribution, obviously positive. If $\lambda_0 > 1$, or equivalently if $\sum_{j=0}^{m} \beta_j \pi_j > 1$ (because then the value of λ which makes $\sum_{j=0}^{m} \beta_j \lambda_j^{-(j+1)} = 1$ must be greater than one), the total population size grows like λ_0^n. If $\sum_{j=0}^{m} \beta_j \pi_j < 1$, so that $\lambda_0 < 1$, the total population size decreases like λ_0^n. If $\sum_{j=0}^{m} \beta_j \pi_j = 1$, so that $\lambda_0 = 1$, the total population size remains constant. In each case the proportion of the population in each age class tends to a limit.

While we have shown that for the Leslie matrix model there is a unique real simple eigenvalue λ_0 with a corresponding eigenvector whose components are all positive, we have *not* shown that this eigenvalue is dominant. In fact, additional conditions are needed to assure the dominance of the eigenvalue λ_0. The following algebraic result, known as the *Perron-Frobenius* theorem, is of some assistance.

Theorem 8.1. *Let A be a matrix all of whose elements are non-negative, and such that for some positive integer k every element of the matrix A^k is* *positive. Then A has a simple positive eigenvalue λ_0 with a corresponding eigenvector having all components positive, and $|\lambda_j| < \lambda_0$ for every other eigenvalue λ_j.*

The conclusions of the classical Perron-Frobenius theorem are exactly what we need, but the hypothesis is unsuitable for our purposes. Fortunately it is possible to show that for a Leslie matrix model with two consecutive β_j different from zero some power of the matrix has all elements positive, and thus the conclusion of the Perron-Frobenius theorem holds. Under these conditions we have as $n \to \infty$,

$$\vec{\rho}_n \sim c\lambda_0^n \vec{v}_0$$

$$\rho_n \sim c\sum_{j=0}^{m} \lambda_0^n \frac{\pi_j}{\lambda_0^j} = c\lambda_0^n \sum_{l=0}^{m} \frac{\pi_j}{\lambda_0^j},$$

so that $P_n \sim c\lambda_0^n$ if $\lambda_0 > 1$, but $P_n \sim c\pi_m \lambda_0^{n-m}$ if $\lambda_0 < 1$ and

$$B_n = c\sum_{j=0}^{m} \lambda_0^n \frac{\pi_j}{\lambda_0^j} \sim c\sum_{j=0}^{m} \beta_j \lambda_0^n \frac{\pi_j}{\lambda_0^j} = c\lambda_0^{n+1} \sum_{j=0}^{m} \beta_j \frac{\pi_j}{\lambda_0^{j+1}} = c_0 \lambda_0^{n+1},$$

because $\sum_{j=0}^{m} \beta_j \frac{\pi_j}{\lambda_0^{j+1}} = 1$.

There are age-structured populations for which the Perron-Frobenius result does not hold. The simplest example of a two-stage model with immature (non-reproducing) members and adults, which we have already encountered in our discussion of delayed-recruitment models, corresponds to a Leslie matrix of the form

$$A = \begin{pmatrix} 0 & \beta \\ \rho & 0 \end{pmatrix}$$

having two eigenvalues of equal absolute value and opposite sign.

If $\lambda = re^{i\theta}$ is a complex eigenvalue then $\lambda^n = r^n e^{in\theta} = r^n \cos n\theta + ir^n \sin n\theta$. Corresponding to the pair of complex conjugate eigenvalues $re^{\pm i\theta}$ there is a pair of real solutions of the difference equation containing terms $r^n \cos n\theta$ and $r^n \sin n\theta$ representing oscillations. If the absolute value r of this pair of eigenvalues is as large as λ_0 then these terms cannot be neglected in the asymptotic expression for the solution vector $\vec{\rho}_n$. An extreme case would arise if the fertility is concentrated in a single age class, so that the birth rate might be periodic in time. A periodic age structure is known as a *Bernardelli population wave.* Such population waves have been observed in human populations. A simple example is given by the 3×3 Leslie matrix

$$A = \begin{pmatrix} 0 & 0 & \beta_2 \\ p_0 & 0 & 0 \\ 0 & p_1 & 0 \end{pmatrix}$$

with characteristic equation $\lambda^3 - \beta_2 p_0 p_1 = 0$. This equation has three roots, one real and positive and two non-real, all having the same absolute value. The three cube roots of 1 are $-\frac{1}{2} + i\frac{\sqrt{3}}{2}$, $-\frac{1}{2} - i\frac{\sqrt{3}}{2}$, and 1. Letting $w = -\frac{1}{2} - i\frac{\sqrt{3}}{2}$, we have $w^2 = -\frac{1}{2} + i\frac{\sqrt{3}}{2}$, and $w^3 = 1$. Then the eigenvalues are $\sqrt[3]{\beta_2 p_0 p_1}w$, $\sqrt[3]{\beta_2 p_0 p_1}w^2$, and $\sqrt[3]{\beta_2 p_0 p_1}$.

Exercises

In Exercises 1 through 4, for the given Leslie matrix find the dominant eigenvalue (if there is one), the corresponding eigenvector and the stable age distribution.

1. $A = \begin{pmatrix} 0 & 1 & 1 \\ \frac{2}{3} & 0 & 0 \\ 0 & \frac{1}{3} & 0 \end{pmatrix}$

2. $A = \begin{pmatrix} 0 & 1 \\ \frac{1}{2} & 0 \end{pmatrix}$

3. $A = \begin{pmatrix} 1 & 2 & 1 \\ \frac{1}{2} & 0 & 0 \\ 0 & \frac{1}{2} & 0 \end{pmatrix}$

4. $A = \begin{pmatrix} 1 & 0 & 1 \\ 1 & 0 & 0 \\ 0 & \frac{1}{2} & 0 \end{pmatrix}$

5. A population starts with 100 members of age zero. Assume that each member of age zero produces one offspring and that 2/3 survive to age one. All members of age one produce 3 offspring and then die. Find the corresponding Leslie matrix, its dominant eigenvalue and a corresponding eigenvector. Describe the stable age distribution and the asymptotic behavior of the population.

6. Consider the Leslie matrix

$$\begin{pmatrix} 0 & 0 & 0 \\ \frac{1}{2} & 0 & 0 \\ 0 & \frac{1}{3} & 0 \end{pmatrix}$$

[Bernardelli (1941)] describing an organism that matures in 2 years, reproduces, and then dies. Begin with an initial population with 100 members in each class and find the population densities for 5 time intervals.

8.2 Linear Continuous Models

It was a physician, Lt. Col. A.G. McKendrick who first introduced age structure into the dynamics of a one-sex population (1926). The McKendrick model assumes that the female population can be described by a function of two variables: age and time. Let $\rho(a,t)$ denote the *density* of individuals of age a at time t; that is, the number of individuals with ages between a and $a+\Delta a$ at time t is approximately $\rho(a,t)\Delta a$. Then the total population at time t is approximately $\sum_a \rho(a,t)\Delta a$, whose "limit as $\Delta a \to 0$" is $\int_0^\infty \rho(a,t)da$, and we define the total population

$$P(t) = \int_0^\infty \rho(a,t)da.$$

In practice, it is reasonable to expect that $\rho(a,t) = 0$ for all sufficiently large a, so that this integral is not necessarily an infinite integral.

We will assume that members leave the population only through death, and that there is an age-dependent death rate $\mu(a)$. This means that over the time interval from t to $t+\Delta t$ a fraction $\mu(a)\Delta t$ of the members with ages between a and $a+\Delta a$ at time t die. At time t there are $\rho(a,t)\Delta a$ individuals with ages between a and $a+\Delta a$. Between the times t and $t+\Delta t$ the number of deaths from this age cohort is $\rho(a,t)\Delta a\mu(a)\Delta t$, and the remainder survive, having ages between $a+\Delta t$ and $a+\Delta t+\Delta a$ at time $t+\Delta t$. Thus,

$$\rho(a+\Delta t, t+\Delta t)\Delta a \approx \rho(a,t)\Delta a - \rho(a,t)\mu(a)\Delta a\Delta t.$$

Division by $\Delta a\Delta t$ gives

$$\frac{\rho(a+\Delta t, t+\Delta t) - \rho(a,t)}{\Delta t} + \mu(a)\rho(a,t) \approx 0.$$

We then let $\Delta t \to 0$. If $\rho(a,t)$ is a differentiable function of a and t we have

$$\begin{aligned}
\lim_{\Delta t\to 0} \frac{\rho(a+\Delta t, t+\Delta t) - \rho(a,t)}{\Delta t} &= \lim_{\Delta t\to 0} \frac{\rho(a+\Delta t, t+\Delta t) - \rho(a,t+\Delta t)}{\Delta t} \\
&\quad + \lim_{\Delta t\to 0} \frac{\rho(a, t+\Delta t) - \rho(a,t)}{\Delta t} \\
&= \lim_{\Delta t\to 0} \left(\rho_a(a,t+\Delta t) + \rho_t(a,t)\right) \\
&= \rho_a(a,t) + \rho_t(a,t).
\end{aligned}$$

Thus we obtain the *von Foerster equation*, which is more properly called the *McKendrick equation* (1926)

$$\rho_a(a,t) + \rho_t(a,t) + \mu(a)\rho(a,t) = 0. \tag{8.1}$$

The function $\mu(a) \geq 0$ is called the *mortality function* or *death modulus.* If $y(\alpha)$ is the number of individuals starting at age a who survive to age α then

$$y(\alpha + \Delta\alpha) - y(\alpha) \approx -\mu(\alpha)y(\alpha)\Delta\alpha.$$

If we divide by $\Delta\alpha$ and let $\Delta\alpha \to 0$ we obtain $y'(\alpha) = -\mu(\alpha)y(\alpha)$, and this implies $y(a_2) = y(a_1)e^{-\int_{a_1}^{a_2}\mu(\alpha)d\alpha}$. Thus the probability that an individual of age a_1 will survive to age a_2 is $e^{-\int_{a_1}^{a_2}\mu(\alpha)d\alpha}$. In particular,

$$\pi(a) = e^{-\int_0^a \mu(\alpha)d\alpha}$$

is the probability of survival from birth to age a.

Next we assume that the birth process is governed by a function $\beta(a)$ called the *birth modulus*; that is, that $\beta(a)\Delta t$ is the number of offspring produced by members with ages between a and $a + \Delta a$ in the time interval from t to $t + \Delta t$. Thus, the total number of births between time t and time $t + \Delta t$ is $\Delta t \sum \beta(a)\rho(a,t)\Delta a$, which "tends as $\Delta a \to 0$" to $\Delta t \int_0^\infty \beta(a)\rho(a,t)da$. As this quantity must also be $\rho(0,t)\Delta t$, we obtain the *renewal condition*

$$\rho(0,t) = \int_0^\infty \beta(a)\rho(a,t)da.$$

In order to complete the model we must specify an initial age distribution (at time zero). Thus, the full model is

$$\begin{aligned} \rho_a(a,t) + \rho_t(a,t) + \mu(a)\rho(a,t) &= 0 \\ \rho(0,t) &= \int_0^\infty \beta(a)\rho(a,t)da \\ \rho(a,0) &= \phi(a). \end{aligned} \tag{8.2}$$

This is analogous to the discrete Leslie model: The function $\beta(a)$ corresponds to the sequence $\beta_0, \beta_1, \ldots, \beta_m$; the function $\mu(a)$ corresponds to the values $p_0, p_1, \ldots, p_{m-1}$ (more precisely, $e^{-\int_0^a \mu(\alpha)d\alpha}$ corresponds to the sequence of survival probabilities $\pi_1, \ldots, \pi_m$).

In order to transform this problem, which consists of a partial differential equation with auxiliary conditions, one of which is an integral condition, into a more manageable form, we will integrate along characteristics of the partial differential equation. The characteristics are the lines $t = a + c$. Their importance lies in the fact that the value of the function ρ at a point (a,t) is determined by the values of ρ on the characteristic through (a,t) because a member of the population of age a at time t must have been of age $a - \alpha$ at time $t - \alpha$ for every $\alpha \geq 0$ such that $\alpha \leq a$ and $\alpha \leq t$. In other words, the points on a given characteristic $t = a + c$ all correspond to the same age cohort.

If $t \geq a$, $\rho(a,t)$ is just the number of survivors to age a of the $\rho(0, t-a)$ members born at time $(t-a)$. Since the fraction surviving to age a is $\pi(a)$, we have $\rho(a,t) = \rho(0, t-a)\pi(a)$ if $t \geq a$. If $t < a$, $\rho(a,t)$ is just the number of survivors to age a of the $\phi(a-t)$ members who were of age $(a-t)$ at time zero. Since the fraction surviving from age $(a-t)$ to age a is $\pi(a)/\pi(a-t)$, we have $\rho(a,t) = \phi(a-t)\pi(a)/\pi(a-t)$ if $t < a$. Thus

$$\rho(a,t) = \begin{cases} \rho(0, t-a)\pi(a) & \text{for } t \geq a \\ \phi(a-t)\pi(a)/\pi(a-t) & \text{for } t < a. \end{cases} \tag{8.3}$$

In terms of μ, we may write this as

$$\begin{aligned} \rho(a,t) &= \rho(0, t-a)e^{-\int_0^a \mu(\alpha)d\alpha} && \text{for } t \geq a \\ \rho(a,t) &= \phi(a-t)e^{-\int_{a-t}^a \mu(\alpha)d\alpha} && \text{for } t < a. \end{aligned}$$

It is convenient to define the function $B(t)$ representing the number of births in unit time t by

$$B(t) = \rho(0,t).$$

Then the original problem is

$$\begin{aligned} \rho_a(a,t) + \rho_t(a,t) + \mu(a)\rho(a,t) &= 0 \\ B(t) &= \int_0^\infty \beta(a)\rho(a,t)da \\ \rho(a,0) &= \phi(a). \end{aligned}$$

and we have obtained the representation

$$\begin{aligned} \rho(a,t) &= B(t-a)e^{-\int_0^a \mu(\alpha)d\alpha} && \text{for } t \geq a \\ \rho(a,t) &= \phi(a-t)e^{-\int_{a-t}^a \mu(\alpha)d\alpha} && \text{for } t < a. \end{aligned}$$

We let $\psi(t)$ be the rate of births from members who were present in the population at time zero. From (8.1) and (8.3) we have

$$\begin{aligned} \rho(0,t) &= \int_0^\infty \beta(a)\rho(a,t)\,da \\ &= \int_0^t \beta(a)\rho(a,t)\,da + \int_t^\infty \beta(a)\rho(a,t)\,da \\ &= \int_0^t \beta(a)\rho(0,t-a)e^{-\int_0^a \mu(\alpha)\,d\alpha}\,da + \int_t^\infty \beta(a)\phi(a-t)e^{-\int_{a-t}^a \mu(\alpha)\,d\alpha}\,da \\ &= \int_0^t \beta(a)e^{-\int_0^a \mu(\alpha)\,d\alpha}B(t-a)\,da + \int_t^\infty \beta(a)\phi(a-t)e^{-\int_{a-t}^a \mu(\alpha)\,d\alpha}\,da. \end{aligned}$$

Thus, we may evaluate ψ in terms of the birth and death moduli and the initial age distribution

$$\psi(t) = \int_t^\infty \beta(a)\phi(a-t)e^{-\int_{a-t}^a \mu(\alpha)\,d\alpha}\,da$$
$$= \int_0^\infty \beta(t+s)\phi(s)e^{-\int_s^{s+t} \mu(\alpha)\,d\alpha}\,ds. \tag{8.4}$$

We also write $\pi(a) = e^{-\int_0^a \mu(\alpha)d\alpha}$, and we see that $B(t)$ is a solution of the *renewal equation*

$$B(t) = \psi(t) + \int_0^t \beta(a)\pi(a)B(t-a)da,$$

a linear Volterra integral equation of convolution type with kernel $\beta(a)\pi(a)$. Conversely, if $B(t)$ is a solution of the renewal equation, then we have a solution of the original problem given by

$$\rho(a,t) = \begin{cases} B(t-a)e^{-\int_0^a \mu(\alpha)d\alpha} = B(t-a)\pi(a) & \text{for } t \geq a \\ \phi(a-t)e^{-\int_{a-t}^a \mu(\alpha)d\alpha} & \text{for } t < a \end{cases}$$

The problem which now faces us is to describe the behavior of solutions of the renewal equation under the assumptions that $\phi(t) \to 0$ as $t \to \infty$, and, in fact, $\int_0^\infty \phi(t)dt < \infty$ and $R = \int_0^\infty \beta(a)\pi(a)da < \infty$. The number R is the expected number of offspring for each individual over a lifetime, being the sum over all ages a of probability of survival to age a multiplied by number of offspring at age a.

One approach to this problem [Feller(1941)] involves taking Laplace transforms and using the convolution property to obtain

$$\hat{B}(p) = \hat{\Phi}(p) + \hat{F}(p)\hat{B}(p),$$

or

$$\hat{B}(p) = \frac{\hat{\Phi}(p)}{1-\hat{F}(p)},$$

where $\hat{B}$ is the Laplace transform of B, $\hat{\Phi}$ is the Laplace transform of Φ, and $\hat{F}$ is the Laplace transform of $\beta\pi$, that is,

$$\hat{F}(p) = \int_0^\infty \beta(a)\pi(a)e^{-pa}da.$$

Then

$$\hat{F}(0) = \int_0^\infty \beta(a)\pi(a)da = R, \qquad \lim_{p\to\infty} \hat{F}(p) = 0,$$

and $\hat{F}(p)$ is a monotone decreasing function of p. Thus there is a unique real solution p_0 of $\hat{F}(p) = 1$, which is positive if $R > 1$, zero if $R = 1$, and negative if $R < 1$. For every complex $p = \alpha + i\gamma$ we have

$$\begin{aligned}
\hat{F}(p) &= \hat{F}(\alpha + i\gamma) = \int_0^\infty \beta(a)\pi(a)e^{-(\alpha+i\gamma)a}da \\
&= \int_0^\infty \beta(a)\pi(a)e^{-\alpha a}(\cos\gamma a - i\sin\gamma a)da \\
\Re\hat{F}(\alpha + i\gamma) &= \int_0^\infty \beta(a)\pi(a)e^{-\alpha a}\cos\gamma a da.
\end{aligned}$$

If $\alpha \geq p_0$ and $\beta \neq 0$ we obtain

$$\begin{aligned}
|\Re\hat{F}(\alpha + i\gamma)| &< \int_0^\infty \beta(a)\pi(a)e^{-\alpha a}da \\
&\leq \int_0^\infty \beta(a)\pi(a)e^{-p_0 a}da = 1,
\end{aligned}$$

and thus p_0 is a *dominant root* of the characteristic equation $\hat{F}(p) = 1$. From this it is possible to conclude that $\hat{B}(p)$ is equal to the sum of two terms, one of which has the form $B/(p - p_0)$ for some constant B and the other of which is analytic for $\Re p \geq p_0$. From this it is possible to deduce that

$$B(t) = Be^{p_0 t} + E(t),$$

where $|E(t)| \leq ce^{(p_0-\epsilon)t}$ as $t \to \infty$. From the representation of $\rho(a,t)$ in terms of $B(t)$ we obtain

$$\rho(a,t) = Be^{p_0(t-a)}\pi(a) + E(t),$$

and from $P(t) = \int_0^\infty \rho(a,t)da$ we obtain

$$P(t) = Pe^{p_0 t} + E(t)$$

for some constant P. In each of these expressions, $E(t)$ represents a function that grows no faster than $e^{(p_0-\epsilon)t}$ as $t \to \infty$.

A *stable age distribution* (or *persistent age distribution*) is defined to be a solution $\rho(a,t)$ of the form $\rho(a,t) = A(a)T(t)$. We may shift a constant factor between $A(a)$ and $T(t)$ and thus assume $\int_0^\infty A(a)da = 1$; then we have

$$P(t) = \int_0^\infty \rho(a,t)da = T(t)\int_0^\infty A(a)da = T(t).$$

Thus $\rho(a,t) = P(t)A(a)$, and in a stable age distribution the proportion $\rho(a,t)/P(t)$ of age a is $A(a)$, independent of time.

Substitution of the form $\rho(a,t) = A(a)P(t)$ into the McKendrick equation (8.1) gives

$$A'(a)P(t) + A(a)P'(t) + \mu(a)A(a)P(t) = 0,$$

or

$$-\mu(a) - \frac{A'(a)}{A(a)} = \frac{P'(t)}{P(t)}.$$

Here primes denote derivatives with respect to the appropriate variable. Because the left side of this equation is a function of a only, while the right side is a function of t only, each side must be equal to a constant p_0 (as yet undetermined). Thus we have two separate ordinary differential equations:

$$P'(t) - p_0 P(t) = 0,$$

with solution

$$P(t) = P(0)e^{p_0 t},$$

and

$$A'(a) + (\mu(a) + p_0)A(a) = 0,$$

with solution

$$A(a) = A(0)e^{-p_0 a}e^{-\int_0^a \mu(\alpha)d\alpha} = A(0)e^{-p_0 a}\pi(a).$$

To satisfy the renewal condition $\rho(0,t) = \int_0^\infty \beta(a)\rho(a,t)da$ we must have

$$\begin{aligned} P(t)A(0) &= \int_0^\infty \beta(a)A(a)P(t)da \\ &= P(t)\int_0^\infty \beta(a)A(0)e^{-p_0 a}\pi(a)da \end{aligned}$$

or

$$\int_0^\infty \beta(a)\pi(a)e^{-p_0 a}da = 1,$$

which is known as the *Lotka-Sharpe equation* (1911) for p_0. As we have already remarked, this has a unique real root, which is positive if $R = \int_0^\infty \beta(a)\pi(a)da > 1$, zero if $R = 1$, and negative if $R < 1$. In a stable age distribution,

$$\rho(a,t) = ce^{p_0 t}e^{-p_0 a}\pi(a) = ce^{p_0(t-a)}\pi(a),$$

and the content of our analysis of the renewal equation earlier is that asymptotically (as $t \to \infty$) every age distribution tends to a stable age distribution. The Lotka-Sharpe equation is the analogue of the equation $\sum_{j=0}^{m} \beta_j \pi_j \lambda_0^{-(j+1)} = 1$ obtained in the discrete case, and the result that every age distribution tends to a stable age distribution is the analogue of the result that in the discrete case $\vec{p}_n \sim c\lambda_0^n \vec{v}_0$.

If $R = 1$ then the total population size $P(t)$ is constant,

$$\rho(a,t) = B\pi(a),$$

and the birth rate is also a constant,

$$B(t) = \int_0^\infty \beta(a)P(0)A(0)\pi(a)da = P(0)A(0)\int_0^\infty \beta(a)\pi(a)da = P(0)A(0).$$

In this case we have an *equilibrium age distribution.* It is easy to see that, conversely, if the total population size is constant then the birth rate is also constant and $\rho(a,t)$ is independent of t.

Example 1. In the "genesis" model we assume $\phi(a) = \delta(a)$, the Dirac delta function with $\delta(a) = 0$ for $a \neq 0$, $\int_0^\infty \delta(a)da = 1$. Thus the initial population is all at age zero. Let us assume also that the birth modulus $\beta(a)$ and the death modulus $\mu(a)$ are both constants, β and μ, respectively. Then the renewal equation takes the form

$$B(t) = \phi(t) + \int_0^t \beta e^{-\int_0^a \mu d\alpha} B(t-a)da,$$

with

$$\begin{aligned}
\psi(t) &= \int_0^\infty \beta\phi(s)e^{-\int_s^{s+t}\mu d\alpha}ds \\
&= \int_0^\infty \delta(s)e^{-\mu t}ds = \beta e^{-\mu t}.
\end{aligned}$$

Also

$$\beta e^{-\int_0^a \mu d\alpha} = \beta e^{-\mu a}.$$

Thus, $B(t)$ satisfies

$$\begin{aligned}
B(t) &= \beta e^{-\mu t} + \beta\int_0^t e^{-\mu a}B(t-a)da \\
&= \beta e^{-\mu t} + \beta\int_0^t e^{-\mu(t-s)}B(s)ds \\
&= \beta e^{-\mu t} + \beta e^{-\mu t}\int_0^t e^{\mu s}B(s)ds.
\end{aligned}$$

Differentiation gives

$$\begin{aligned} B'(t) &= -\mu\beta e^{-\mu t} + \beta e^{-\mu t} e^{\mu t} B(t) - \beta\mu e^{-\mu t} \int_0^t e^{\mu s} B(s) ds \\ &= -\mu\left(\beta e^{-\mu t} + \beta e^{-\mu t} \int_0^t e^{\mu s} B(s) ds\right) + \beta B(t) \\ &= -\mu B(t) + \beta B(t) = (\beta - \mu) B(t). \end{aligned}$$

From the renewal equation with $t = 0$ we see that $B(0) = \beta$. Now $B'(t) = (\beta - \mu)B(t)$, $B(0) = \beta$ implies $B(t) = \beta e^{(\beta-\mu)t}$, and this gives the age distribution function

$$\rho(a,t) = \begin{cases} \beta e^{(\beta-\mu)(t-a)-\mu a} & \text{for } t \geq a \\ \delta(a-t)e^{-\mu t} & \text{for } t < a. \end{cases}$$

The total population size is

$$\begin{aligned} P(t) &= \int_0^\infty \rho(a,t)da = \int_0^t \rho(a,t)da + \int_t^\infty \rho(a,t)da \\ &= \int_t^0 \beta e^{(\beta-\mu)t} e^{-\beta a} da + \int_t^\infty \delta(a-t)e^{-\mu t} da \\ &= \beta e^{(\beta-\mu)t} \int_0^t e^{-\beta a} da + e^{-\mu t} \\ &= e^{(\beta-\mu)t}(1 - e^{-\beta t}) + e^{-\mu t} = e^{(\beta-\mu)t}. \end{aligned}$$

Exercises

1. Consider a model with β and μ constant and the initial age distribution $\phi(a)$ arbitrary.

   (i) Show that $\psi(t) = \beta e^{-\mu t} \int_0^\infty \phi(s)\, ds$.

   (ii) Obtain an integral equation for $B(t)$.

   (iii) Show by differentiation of the integral equation obtained in part (b) that

   $$B'(t) = (\beta - \mu)B(t), \quad B(0) = \beta \int_0^\infty \phi(s)\, ds.$$

   (iv) Solve for $B(t)$.

2.* Consider a model with $\phi(a)$ arbitrary, μ constant, and $\beta(a) = \beta e^{-\mu a}$.

   (i) Show that $\psi(t) = \beta \int_0^\infty e^{-\mu s}\phi(s)\, ds$.

(ii) Show that $B(t)$ satisfies the integral equation

$$B'(t) = (\beta - 2\mu)B(t) + 2\beta\mu \int_0^\infty e^{-\mu s}\phi(s)\, ds.$$

(iii) Solve the initial value problem obtained in part (c) to find that

$$B(t) = \frac{\beta \int_0^\infty e^{-\mu s}\phi(s)\, ds}{\beta - 2\mu}\left[\beta e^{-(\beta-2\mu)t} - 2\mu\right].$$

3.* Consider a model with μ constant and an arbitrary initial age distribution $\phi(a)$ for which $\beta(a) = 0$ if $a \leq T$ and $\beta(a)$ is a constant β if $a > T$. Show that the integral equation satisfied by $B(t)$ for $t > T$ is

$$B(t) = \beta e^{-\mu t}\int_0^\infty \phi(s)\, ds + \beta e^{-\mu t}\int_0^{t-T} e^{\mu s} B(s)\, ds$$

and deduce that $B(t)$ satisfies the differential-difference equation

$$B'(t) = \beta e^{-\mu T} B(t - T) - \mu B(t).$$

8.3 Nonlinear Continuous Models

Before we introduced age structure in population models we consistently permitted birth and death rates to depend on population size, a possibility ruled out in the age-structured models discussed in the two preceding sections. We now consider the possibility of birth and death moduli of the form $\beta(a, P(t))$ and $\mu(a, P(t))$, depending on total population size. A variant, which can be developed by analogous methods, would be to allow the birth modulus (and possibly also the death modulus) to depend on $\rho(a,t)$, the number of members in the same age cohort. We will consider only the continuous case because the methods of linear algebra used to treat the linear discrete case have no direct adaptation to the nonlinear discrete model.

If the birth and death moduli are allowed to depend on total population size the description of the model must include the definition of $P(t)$. Thus, the full model is now

$$\begin{aligned} \rho_a(a,t) + \rho_t(a,t) + \mu(a, P(t))\rho(a,t) &= 0 \\ B(t) = \rho(0,t) &= \int_0^\infty \beta(a, P(t))\rho(a,t)da \\ P(t) &= \int_0^\infty \rho(a,t)da \\ \rho(a,0) &= \phi(a) \end{aligned} \tag{8.5}$$

We can transform the problem just as in the linear case by integrating along characteristics. If we define

$$\mu^*(\alpha) = \mu(\alpha, P(t-a+\alpha))$$

the same calculations give

$$\rho(a,t) = \begin{cases} B(t-a)e^{-\int_0^a \mu^*(\alpha)d\alpha} & \text{for } t \geq a \\ \phi(a-t)e^{-\int_{a-t}^a \mu^*(\alpha)d\alpha} & \text{for } t < a. \end{cases} \tag{8.6}$$

When we substitute these expressions into (8.5) and (8.6) we obtain a pair of functional equations for $B(t)$ and $P(t)$, whose solution gives an explicit solution for $\rho(a,t)$, namely

$$\begin{aligned} B(t) &= b(t) + \int_0^t \beta(a, P(t))e^{-\int_0^a \mu^*(\alpha)d\alpha} B(t-a)da \\ P(t) &= p(t) + \int_0^t e^{-\int_0^a \mu^*(\alpha)d\alpha} B(t-a)da \end{aligned}$$

where

$$\begin{aligned} \mu^*(\alpha) &= \mu(\alpha, P(t-a+\alpha)) \\ b(t) &= \int_t^\infty \beta(a, P(t))\phi(a-t)e^{-\int_{a-t}^a \mu^*(\alpha)d\alpha}da, \\ p(t) &= \int_t^\infty \phi(a-t)e^{-\int_{a-t}^a \mu^*(\alpha)d\alpha}da. \end{aligned}$$

It is reasonable to assume that $\int_0^\infty \phi(a)da < \infty$ and that the functions $\beta(a,P), \mu(a,P)$ are continuous and nonnegative; under these hypotheses it is easy to verify that $b(t)$ and $p(t)$ are continuous and nonnegative and tend to zero as $t \to \infty$, and that $b(0) > 0$, $p(0) > 0$. Without additional assumptions it is not necessarily true that the pair of functional equations has a solution for $0 \leq t < \infty$ but is possible to prove that if $\sup_{a \geq 0, P \geq 0} \beta(a,P) < \infty$, then there is a unique continuous nonnegative solution on $0 \leq t < \infty$. This model is due to M.L. Gurtin and R.C. MacCamy (1974).

A solution $\rho(a,t)$ that is independent of t is called an *equilibrium* age distribution. If $\rho(a,t) = \rho(a)$ is an equilibrium age distribution then both $P(t) = \int_0^\infty \rho(a)da$ and $B(t) = \int_0^\infty \beta(a, P(t))\rho(a)da$ are constant. Conversely, if $P(t)$ and $B(t)$ are constant then $\rho(a,t)$ is independent of t and thus is an equilibrium age distribution.

If $\rho(a)$ is an equilibrium age distribution the McKendrick equation becomes an ordinary differential equation $\rho'(a) + \mu(a,P)\rho(a) = 0$, with initial condition $\rho(0) = B$, whose solution is

$$\rho(a) = Be^{-\int_0^a \mu(\alpha,P)d\alpha}.$$

If we define

$$\pi(a, P) = e^{-\int_0^a \mu(\alpha, P) d\alpha} \tag{8.7}$$

the probability of survival from birth to age a when the population size is the constant P, then

$$\rho(a) = B\pi(a, P).$$

From $P = \int_0^\infty \rho(a)da$ we obtain

$$P = B \int_0^\infty \pi(a)da,$$

and from $B = \int_0^\infty \beta(a, P)\rho(a)da$, we obtain

$$P = B \int_0^\infty \beta(a, P)\pi(a, P)da.$$

Thus, for an equilibrium age distribution with birth rate B and population size P, P must satisfy the equation

$$R(P) = \int_0^\infty \beta(a, P)\pi(a, P)da = 1,$$

and then B is given by

$$B = \frac{P}{\int_0^\infty \pi(a, P)da}.$$

Then $1/\int_0^\infty \pi(a, P)da$ is the average life expectancy and the equilibrium age distribution is $\rho(a) = B\pi(a, P)$. $R(P)$, called the *reproductive number*, is the expected number of offspring that an individual has over its lifetime, when the total population size is P.

Example 1. Suppose that β is independent of age and is a function of P only. Then

$$\begin{aligned} B(t) &= \int_0^\infty \beta(P(t))\rho(a, t)da = \beta(P(t)) \int_0^\infty \rho(a, t)da \\ &= P(t)\beta(P(t)), \end{aligned} \tag{8.8}$$

and the problem is reduced to a single functional equation for $P(t)$ together with this explicit formula for $B(t)$. If we define $g(P) = P\beta(P)$ the equation for $P(t)$ is

$$P(t) = p(t) + \int_0^t e^{-\int_0^a \mu^*(\alpha)d\alpha} g(P(t-a))da.$$

Example 2. Suppose that β is independent of age and in addition that μ is independent of population size and is a function of age only. Then, instead of $\mu^*(\alpha) = \mu(\alpha, P(t-a+\alpha))$, we have $\mu(\alpha)$, and the equation for $P(t)$ is a Volterra integral equation called the *nonlinear renewal equation*

$$\begin{aligned} P(t) &= p(t) + \int_0^t e^{-\int_0^a \mu(\alpha)d\alpha} g(P(t-a))da \\ &= p(t) + \int_0^t \pi(\alpha) g(P(t-a))da. \end{aligned} \tag{8.9}$$

It is possible to prove [Londen(1974)] that unless $g(P)/P$ has the constant value $1/\int_0^\infty \pi(a)da$ on some P interval, then every nonnegative solution of this equation tends to a limit P_∞ as $t \to \infty$, with

$$P_\infty = g(P_\infty) \int_0^\infty \pi(a)da,$$

or

$$\beta(P_\infty) \int_0^\infty \pi(a)da = 1.$$

If we differentiate the integral equation after writing

$$\int_0^t \pi(a)g(P(t-a))da = \int_0^t \pi(t-\tau)g(P(\tau))d\tau$$

and using the relation

$$\begin{aligned} \frac{d}{dt}\int_0^t \pi(t-\tau)g(P(\tau))d\tau &= \pi(0)g(P(t)) + \int_0^t \pi'(t-\tau)g(P(\tau))d\tau \\ &= \pi(0)g(P(t)) + \int_0^t \pi'(a)g(P(t-a))da, \end{aligned}$$

we obtain the integro-differential equation

$$P'(t) = p'(t) + \pi(0)g(P(t)) + \int_0^t g(P(t-a))\pi'(a)da$$

whose linearization about the equilibrium P_∞ is

$$u'(t) = \pi(0)g'(P_\infty)u(t) + g'(P_\infty)\int_0^t u(t-a)\pi'(a)da.$$

We recall that for the linear integro-differential equation $u'(t) = \alpha u(t) + \beta\int_0^t u(t-a)k(a)da$ with $\int_0^\infty k(a)da = 1$ if $\alpha + \beta \geq 0$ then solutions do *not* tend to zero (Section 3.4). Here $\alpha = \pi(0)g'(P_\infty)$,

$$k(a) = \frac{\pi'(a)}{\int_0^\infty \pi'(a)da} = \frac{\pi'(a)}{\pi(0)},$$

so that $\beta = g'(P_\infty)\int_0^\infty \pi'(a)da = -g(P_\infty)\pi(0)$ and $\alpha + \beta = 0$. Thus, the "equilibrium" P_∞ of the nonlinear renewal equation cannot be asymptotically stable. However, it can be shown [Brauer(1976,1987b)] that if $g'(P_\infty)\int_0^\infty \pi(a)da < 1$, the equilibrium P_∞ is stable in a weaker sense, namely that small disturbances do not alter the solution very much, while if $g'(P_\infty)\int_0^\infty \pi(a)da > 1$ solutions tend away from P_∞. Thus, the condition $g'(P_\infty)\int_0^\infty \pi(a)da < 1$ is necessary for the limit P_∞ to be meaningful biologically. Since

$$g'(P_\infty) = \beta(P_\infty) + P_\infty\beta'(P_\infty)$$

and $\beta(P_\infty)\int_0^\infty \pi(a)da = 1$, this condition is equivalent to $\beta'(P_\infty) < 0$.

For age-structured populations with μ a function of age only and β a function of population size only, we have shown that every solution tends to an equilibrium age distribution, and we have

$$\lim_{t\to\infty} P(t) = P_\infty, \qquad \lim_{t\to\infty} B(t) = P_\infty\beta(P_\infty),$$

while

$$\rho(a,t) = B(t-a)\pi(a) \sim P_\infty\beta(P_\infty)\pi(a).$$

Such models are more realistic than models described by ordinary differential equations in which age dependence is ignored completely. However, the assumption that mortality is independent of population size is quite unrealistic, especially in populations whose members compete for resources.

The nonlinear renewal equation (8.9) as a model for a population with a birth rate depending only on total population size and a death rate depending only on age can be constructed directly, without starting from an age-structured model. If the number of births in unit time when total population size is P is $B = g(P) = P\beta(P)$, and the fraction of the population surviving to age a is $\pi(a)$, then the number of members born at time $(t-a)$ and having age between a and $a + \Delta a$ at time t is approximately $\pi(a)g(P(ta))\Delta a$. Thus, the total population size $P(t)$ at time t is the sum of two terms. The first term $p(t)$ represents the number of members born before the time $t = 0$ and still surviving at time t, and the second term $\int_0^t \pi(a)g(P(t-a))da$ represents the number of members born between time zero and time t and surviving to time t. This explains the model (8.9).

We may proceed in the other direction to derive the age distribution from the nonlinear renewal equation (8.9) under the assumptions that the birth rate depends on total population size and the death rate depends on age. From the relations (8.1), (8.2), and (8.3) we have, for $t \geq a$,

$$\begin{aligned} \rho(a,t) &= B(t-a)\pi(a) \\ &= P(t-a)\beta(P(t-a))\pi(a) \\ &= g(P(t-a))\pi(a). \end{aligned}$$

Example 3. Suppose that μ is independent of age and is a function of P only. Then the McKendrick equation is

$$\rho_a(a,t) + \rho_t(a,t) + \mu(P(t))\rho(a,t) = 0.$$

We integrate with respect to a from zero to ∞ and assume $\lim_{a\to\infty} \rho(a,t) = 0$. Then

$$\begin{aligned}\int_0^\infty \rho_a(a,t)da &= \rho(a,t)\Big|_{a=0}^{a=\infty} = -\rho(0,t) = -B(t)\\ \int_0^\infty \rho_t(a,t)da &= \frac{d}{dt}\int_0^\infty \rho(a,t)da = P'(t),\end{aligned}$$

and

$$\int_0^\infty \mu(P(t))\rho(a,t)da = \mu(P(t))\int_0^\infty \rho(a,t)da = P(t)\mu(P(t)).$$

Thus, the problem can be reformulated as a functional equation for $P(t)$ together with an ordinary differential equation $P' + \mu(P)P = B$. We now assume in addition that β is independent of population size and is a function of age only. Then the system takes the form

$$\begin{aligned}P'(t) + P(t)\mu(P(t)) &= B(t)\\ B(t) &= b(t) + \int_0^t \beta(a)e^{-\int_0^a \mu^*(\alpha)d\alpha}B(t-a)da,\end{aligned}$$

with

$$\mu^*(\alpha) = \mu(P(t-a+\alpha)).$$

It is convenient to make a change of variables and write

$$\int_0^a \mu^*(\alpha)d\alpha = \int_0^a \mu(P(t-a+\alpha))d\alpha = \int_{t-a}^t \mu(P(u))du,$$

so that

$$B(t) = b(t) + \int_0^t \beta(a)e^{-\int_{t-a}^t \mu(P(u))du}B(t-a)da.$$

Now we let

$$B^*(t) = B(t)e^{\int_0^t \mu(P(u))du}, \qquad P^*(t) = P(t)e^{\int_0^t \mu(P(u))du}$$

and substitute, obtaining the system

$$\begin{aligned}B^*(t) &= b^*(t) + \int_0^t \beta(a)B^*(t-a)da\\ P^{*'}(t) &= B^*(t),\end{aligned}$$

with

$$b^*(t) = \int_t^\infty \beta(a)\phi(a-t)da.$$

If we search for stable age distributions by substituting $\rho(a,t) = A(a)P(t)$ into the McKendrick equation we obtain

$$-\frac{A'(a)}{A(a)} = \frac{P'(t)}{P(t)} + \mu(P(t)),$$

and see as before that each side of this equation must be a constant p_0. Thus, $A'(a) + p_0 A(a) = 0$, which gives $A(a) = A(0)e^{-p_0 a}$, and $P'(t) + P(t)\mu(P(t)) = p_0 P(t)$. The renewal condition $\rho(0,t) = \int_0^\infty \beta(a)\rho(a,t)da$ gives $P(t)A(0) = \int_0^\infty \beta(a)P(t)A(0)e^{-p_0 a}da$, which implies

$$\int_0^\infty \beta(a)e^{-p_0 a}da = 1.$$

In the linear theory we showed that every solution approaches a stable age distribution. A similar but weaker result is true in the nonlinear case. By the use of the Laplace transform and the equation for $B^*(t)$ we can show that $B^*(t) \sim ce^{p_0 t}$ as $t \to \infty$ and because $P^{*'}(t) = B^*(t)$ we have $P^*(t) \sim ce^{p_0 t}/p_0$ (unless $p_0 = 0$). We now obtain $\rho(a,t) \sim A(a)Q(t), B(t) \sim p_0 Q(t)$ as $t \to \infty$, where

$$Q(t) = \frac{c}{p_0} e^{p_0 t - \int_0^t \mu(P(u))du}.$$

This differs from the result in the linear case because $Q(t)$ does not in general satisfy the differential equation $y' + y\mu(y) = p_0 y$ and thus $A(a)Q(t)$ is not a solution of the model. It is, however, true that $Q(t)/P(t) \to 1$ as $t \to \infty$.

Exercises

1.* Assume that μ is independent of age and is a function of P only. Assume also that $\lim_{a\to\infty} \rho(a,t) = 0$. Integrate (8.5) with respect to a from zero to ∞ and deduce that

$$P'(t) = B(t) - P(t)\mu(P(t)).$$

2.* Assume that μ is a function of P only and that $\beta(a,P) = \beta_0 e^{-\alpha a}$, with β_0 and α positive constants. Multiply (8.5) by $e^{-\alpha a}$ and integrate with respect to a from zero to ∞ and deduce that

$$B'(t) = \left[\beta_0 - \alpha - \mu(P(t))\right]B(t).$$

3.* Exercises 1 and 2 show that if μ is a function of P only and $\beta(a, P) = \beta_0 e^{-\alpha a}$ the model (8.5) is equivalent to the system of ordinary differential equations

$$P' = B - P\mu(P)$$
$$B' = \left[\beta_0 - \alpha - \mu(P)\right] B.$$

(i) Show that $P = 0$, $B = 0$ is an equilibrium that is asymptotically stable if and only if $\beta_0 - \alpha < \mu(0)$.

(ii) Show that there is an equilibrium (P_∞, B_∞) with $\mu(P_\infty) = \beta_0 - \alpha$ and $B_\infty = P_\infty \mu(P_\infty) = (\beta_0 - \alpha) P_\infty$.

(iii) Show that the equilibrium (P_∞, B_∞) is asymptotically stable if and only if $\alpha - \beta_0 < P_\infty \mu'(P_\infty)$, $\mu'(P_\infty) > 0$.

8.4 Numerical Methods for the McKendrick-Von Foerster Model

The introduction of high-powered computers has made it feasible to analyze, via computer simulations, large and often complex models of biological systems. Since the field of numerical analysis is immense it is impossible to discuss it in any reasonable depth, even in the context of the type of models in this book. Nevertheless, we have decided to include this section on numerical methods for the McKendrick-Von Foerster age-structured model because of the role that numerical methods now play in the study of structured populations.

Numerical analysis plays a role in a variety of settings. The following examples illustrate its use.

Example 1. We often measure continuous variables of time such as temperature but the measurements are done in discrete time. For example, if $Q(t)$ is a continuous quantity of time and discrete measurements of Q are taken at n times $(t_1, t_2, t_3, \cdots, t_n)$. Hence, we know or have estimates of only the following values of $Q(t)$:

$$Q(t_1), Q(t_2), \cdots, Q(t_n).$$

If we can recognize the pattern of Q_1, Q_2,$\cdots$,Q_n, for example, if it looks like an exponential then we can fit a continuous, most often differentiable function. It is common to select a family of functions parametrized by few parameters and then use the data, that is, the pattern, Q_1, Q_2,$\cdots$,Q_n, to estimate which function $Q(t)$ from this family best fits the data. A simpler approach to get a $Q(t)$ that does not bind us to a particular parametric family of functions is to connect Q_1, Q_2,$\cdots$,Q_n by straight lines. This process is called *interpolation*. From the points Q_1, Q_2,$\cdots$,Q_n we obtain

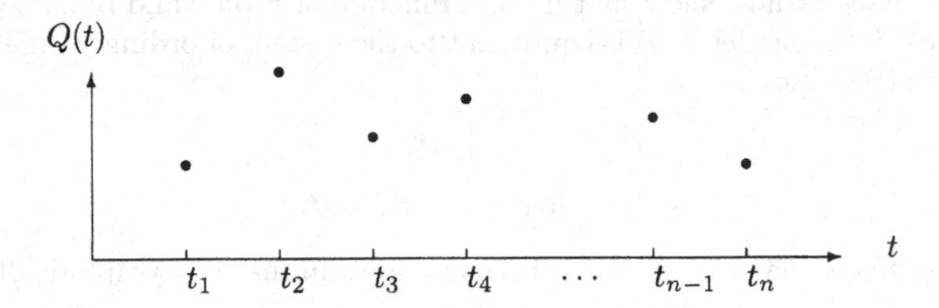

FIGURE 8.1.

a piecewise linear continuous function $\bar{Q}(t)$ and numerical analysis (and statistics) answers the question, "How far is $\bar{Q}(t)$from the real function $Q(t)$?"

Example 2. The computation of integrals also falls within the domain of numerical analysis. We know that computing $\int_a^b f(x)dx$ even for simple $f(x)$ is not always possible or easy. (For example, consider $\int_1^2 e^{x^2} dx$.) Fortunately, numerical analysis has developed methods for computing this integral (as well as others) approximately. How is it done? The interval $[a, b]$ is subdivided into subintervals and the area under the curve is approximated by a variety of methods which have given rise to approximation formulas. The area under the curve can in fact, be approximated via Riemann sums or, more efficiently, via the trapezoidal rule or Simpson's rule (which may be found in almost any calculus text). For example, the *Riemann Sum Left point rule* uses the following approximation:

$$\int_a^b f(x)dx \approx \sum_{i=1}^{n-1} f(x_i)\Delta x \text{ , where } \Delta x = x_{i+1} - x_i.$$

Of course, using an approximate formula prompts the question, "If we use the sum instead of the integral, what are we neglecting?" Answering this type of question is also part of the field of numerical analysis. Let's see how it works in this example:

$$\int_a^b f(x)dx = \sum_{i=0}^{n-1} \int_{x_i}^{x_{i+1}} f(x)dx,$$

By Taylor's formula,

$$f(x) = f(x_i) + f'(x_i)(x - x_i) + \textit{higher order terms}$$

Here, "higher order terms" means terms that involve higher powers $(x-x_i)^n$ with$(n \geq 1)$ in (Δx). We apply Taylor's formula on each interval (x_i, x_{i+1}),

$$\begin{aligned}\int_{x_i}^{x_{i+1}} f(x)dx &= \int_{x_i}^{x_{i+1}} f(x_i)dx + f'(x_i)\int_{x_i}^{x_{i+1}} (x-x_i)dx + \textit{higher order terms} \\ &= f(x_i)(x_{i+1}-x_i) + f'(x_i)\frac{(x-x_i)^2}{2}\Big|_{x_i}^{x_{i+1}} + \textit{higher order terms} \\ &= f(x_i) + \Delta x f'(x_i)\frac{(\Delta x_i)^2}{2} + \textit{higher order terms} \text{ (in } \Delta x)\end{aligned}$$

Thus on the interval (a, b) summation over all subintervals gives

$$\begin{aligned}\int_a^b f(x)dx &= \sum_{i=0}^{n-1} f(x_i)\Delta x + \sum_{i=0}^{n-1} f'(x_i)\frac{(\Delta x_i)^2}{2} \\ &= \sum_{i=0}^{n-1} f(x_i)\Delta x + O((\Delta x)^2),\end{aligned}$$

where $O(\Delta x)^2)$ signifies an error which is no greater than a constant multiple of $\Delta x)^2$

Example 3. A critical example to this section deals with the way that numerical analysis approximates derivatives. The goal is to compute approximately $f'(x_0)$. How is this done? For x closer to x_0, expand $f(x)$ around x_0 (Taylor's series)

$$f(x) = f(x_0) + f'(x_0)(x-x_0) + f''(x_0)\frac{(x-x_0)^2}{2} + \textit{higher order terms}$$

Solve for $f'(x_0)$ in the equation

$$\frac{f(x)-f(x_0)}{x-x_0} = f'(x_0) + f''(x_0)\frac{(x-x_0)}{2}$$

Letting $x - x_0 = h$, $x = x_0 + h$, and *ignoring* the quadratic and higher order terms in h, we get

$$f'(x_0) \approx \frac{f(x_0+h)-f(x_0)}{h}$$

as our approximation for $f'(x_0)$.

Now we are ready to provide an outline for the numerical solution of the McKendrick-Von Foerster model.

8.4.1 A Numerical Scheme for the McKendrick-Von Foerster Model

The McKendrick-VonFoerster model is critical to the study of demographic processes in demography and population biology. This linear, simple model cannot be explicitly solved in all cases. Fortunately, it can be solved numerically.

We consider the linear, age-structured population model

$$
\begin{aligned}
u_x + u_t &= -\mu(x,t)u \\
u(0,t) &= \int_0^A \beta(x,t)u(x,t)dx \\
u(x,0) &= u_0(x).
\end{aligned}
\tag{8.10}
$$

We want the solution of this system in the rectangle

$$
\begin{aligned}
0 \le x \le A \\
0 \le t \le T.
\end{aligned}
$$

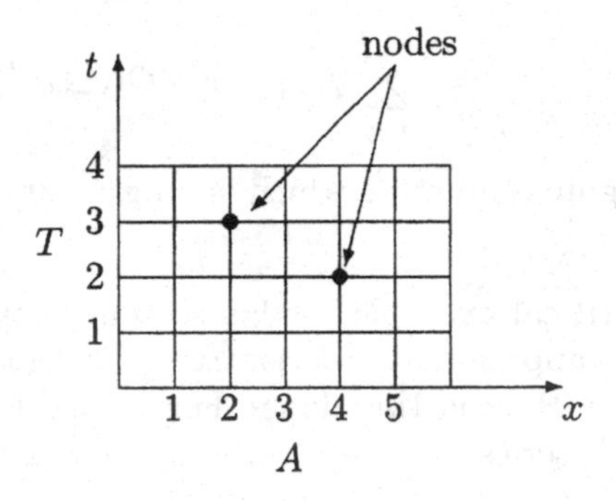

FIGURE 8.2.

We divide the x-axis into N equal intervals with length h, that is,

$$
A = Nh \text{ or } N = \frac{A}{h},
$$

we divide the t-axis into subintervals of the same length, that is,

$$
T = Mh \text{ or } M = \frac{T}{h},
$$

to obtain a quadratic (actually rectangular) mesh. It is reasonable to pick intervals of equal length in both axes because the characteristic lines of (8.10) go through the nodes (time and age are measured in the same units). Each node (as a point in the plane) has two coordinates (x_i, y_j), and, as is customary in the field of numerical analysis, we will identify the point with coordinates (x_i, y_j) with (i, j). In Figure 8.2 the two darker nodes have coordinates (2,3) and (4,2). Our first goal is to compute the solution $u(x_i, y_j)$ of (8.10) at the nodes, that is, we want to find approximations to the following set of values:

$$
u(x_i, y_j) \text{ where } i = 1, \cdots, N; \quad j = 1, \cdots, M
$$

Again, as is customary in numerical analysis, we replace the variables by their indices or, more explicitly we use the notation: $u(x_i, y_j) = U_{i,j}$; $\mu(x_i, y_j) = \mu_{i,j}$; and $\beta(x_i, y_j) = \beta_{i,j}$. Furthermore, we set $x = x_i$, $y = y_j$, in the first equation of (8.10) or explictly, we write

$$u_x(x_i, y_j) + u_t(x_i, y_j) = -\mu_{i,j} u(x_i, y_j). \tag{8.11}$$

To set up a numerical scheme that would solve our partial differential equation or better our *initial boundary value problem* (8.10), we look at the solution via a function u along a characteristic line close to (x_i, y_j), namely,

$$\begin{aligned} \bar{u}(\tau) &:= u(x_i + \tau, t_j + \tau), \\ \bar{u}(0) &= u_{i,j}, \\ \bar{u}(-h) &= u(x_i - h, t_j - h) = u(x_{i-1}, t_{j-1}) = u_{i-1,j-1}. \end{aligned}$$

Expanding $\bar{u}(-h)$ around zero we get an expression for $\bar{u}'(0)$,

$$\bar{u}(-h) = \bar{u}(0) + \bar{u}'(0)(-h) + \bar{u}''(0)\frac{h^2}{2} + \textit{higher order terms}$$

Hence, the expression for $\bar{u}'(0)$ is

$$\begin{aligned} \bar{u}'(0)h &= \bar{u}(0) - \bar{u}(-h) + O(h^2) \\ \bar{u}'(0) &= \frac{\bar{u}(0) - \bar{u}(-h)}{h} + O(h). \end{aligned}$$

Rewriting this for $u(x, t)$ we have expressions for $\bar{u}'(0)$ and $\bar{u}'(\tau)$

$$\begin{aligned} \bar{u}'(\tau) &= u_x(x_i + \tau, t_j + \tau) + u_t(x_i + \tau, t_j + \tau), \\ \bar{u}'(0) &= u_x(x_i, t_j) + u_t(x_i, t_j). \end{aligned}$$

Hence, this is exactly what we have in the left hand side of (8.10)

$$u_x(x_i, t_j) + u_t(x_i, t_j) = \frac{U_{ij} - U_{i-1,j-1}}{h} + O(h)$$

To approximate equation (8.11) we use

$$\frac{u_{ij} - u_{i-1,j-1}}{h} = -\mu_{ij} u_{i,j},$$

which guarantees that we are making an error $O(h)$. We solve the above approximation instead of equation (8.11). Here u_{ij} is the approximate solution and, therefore, we expect that

$$u_{ij} \approx u(x_i, t_j) = Uij$$

(but this has to be proved!). From the second equation of our initial boundary value problem (8.10) and setting $t = t_j$ we get that

$$u(0, t_j) = \int_0^A \beta(x, t_j) u(x, t_j) dx.$$

Rewriting the integral as a Riemann sum, we have

$$U_{0,j} = \sum_{i=0}^{N-1} \beta_{ij} u_{ij} h + O(h) + higherorderterms$$

which suggests that we can approximate the second equation in (8.10) by the sum

$$u_{0,j} = \sum_{i=0}^{N-1} \beta_{ij} u_{ij} h.$$

From the third equation in (8.10) we get

$$U_{i,0} = u_i^0.$$

Hence, we can arrange that the approximate solution satisfies

$$u_{i,0} = u_i^0$$

without making any error here. So from (8.10) we obtained the following system to compute the approximate solution u_{ij} of the true solution $U_{ij} = u(x_i, t_j)$:

$$\begin{array}{l} \frac{u_{ij}-u_{i-1,j-1}}{h} = -\mu_{ij} u_{ij} \quad i = 1, \cdots, N \;\; j = 1, \cdots, N \\ u_{0,j} = \sum_{i=0}^{N-1} \beta_{ij} u_{ij} h \quad j = 1, \cdots, M \\ u_{j,0} = u_i^0 \quad i = 0, \cdots, N. \end{array}$$

The set of discrete equations that approximate the solution of the continuous equations is called a "numerical scheme." This numerical scheme is a linear system of equations which can be solved numerically (in this case even by hand). If, $\mu_{i,j}, \beta_{i,j}, u_i^0$ are given, we solve the first equation for u_{ij}:

$$\begin{aligned} u_{ij} - u_{i-1,j-1} &= -\mu_{ij} u_{ij} h \\ (1 + \mu_{ij} h) u_{ij} &= u_{i-1,j-1} \\ \Longrightarrow u_{i,j} &= \frac{1}{1 + \mu_{ij} h} u_{i-1,j-1}. \end{aligned}$$

We solve for $u_{0,j}$ in the second equation

$$\begin{aligned} u_{0,j} &= \beta_{0j} h u_{0j} + \sum_{i=1}^{N-1} \beta_{ij} u_{ij} h \\ \Longrightarrow (1 - \beta_{0j}) u_{0j} &= \sum_{i=1}^{N-1} \beta_{ij} u_{ij} h. \end{aligned}$$

With the above two expressions we arrive at the following explicit scheme

$$u_{ij} = \frac{1}{1+\mu_{ij}h} u_{i-1,j-1}, \quad i = 1, \cdots, N; \quad j = 1, \cdots, N \qquad (8.12)$$

$$u_{0j} = \frac{1}{1-\beta_{0j}h} \sum_{i=1}^{N-1} \beta_{ij} u_{ij} h \quad j = 1, \cdots, N \qquad (8.13)$$

$$u_{i0} = u_i^0 \quad i = 1, \cdots, N \qquad (8.14)$$

Now the question is how do we compute an explicit numerical solution using (8.12), (8.13) and (8.14):

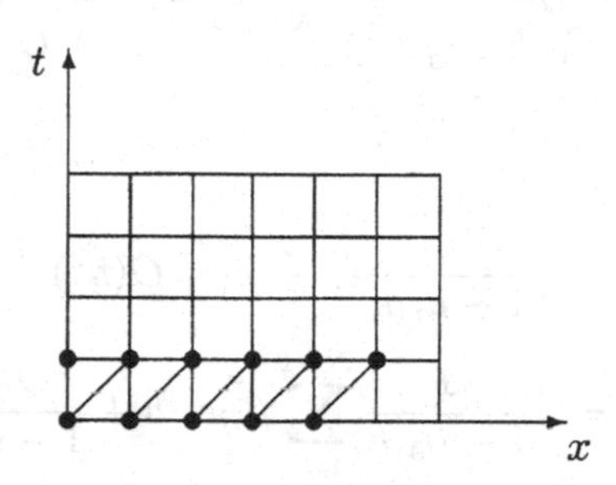

FIGURE 8.3.

General Idea. The approach is recursive. The information at each time level is computed from the information at the previous time level. Here is the algorithm in steps:

Step 1

(A) Compute $u_{i,0}$, $i = 0, \cdots, N$ from (8.14).

(B) Compute $u_{i,1}$, $i = 1, \cdots, N$ from (8.12).

(C) Compute $u_{0,1}$ from (8.13).

Step 2

(A) Assume we have computed $u_{i,j-1}$, i=0,$\cdots$,N. How do we compute $u_{i,j}$ i=0,$\cdots$,N?

(B) Compute $u_{i,j}$, $i = 1, \cdots, N$ from (8.12).

(C) Compute $u_{0,j}$ from (8.13).

Apply Step 2 until $j = M$ (cycle).

Applying this procedure we compute the solution of (8.12), (8.13), (8.14). However, it is not clear if $u_{ij} \approx U_{ij}$ and if it is in what sense. It is reasonable to expect that as $h \to 0$ u_{ij} gets closer to U_{ij}, in some sense. In order to address this issue we study how the differences or errors ε_{ij} change. Set

$$\varepsilon_{ij} = u_{i,j} - U_{ij}.$$

Subtraction of the equations in the two systems shows that the errors satisfy

$$\begin{aligned}\frac{\varepsilon_{i,j} - \varepsilon_{i-1,j-1}}{h} &= -\mu_{ij}\varepsilon_{ij} + O(h) \\ \varepsilon_{0,j} &= \sum_{i=0}^{N-1} \beta_{ij} h \varepsilon_{ij} + O(h) \\ \varepsilon_{i,0} &= 0\end{aligned}$$

Solution of the above system gives recursive formulas for the errors, namely

$$\begin{array}{l}(1+\mu_{ij}h)\varepsilon_{ij} = \varepsilon_{i-1,j-1} + O(h^2) \\ (1-\beta_{0j}h)\varepsilon_{0,j} = \sum_{i=1}^{N-1} \beta_{ij} h \varepsilon_{ij} + O(h) \\ \varepsilon_{i,0} = 0,\end{array}$$

or

$$\begin{aligned}\varepsilon_{ij} &= \frac{1}{1+\mu_{ij}h}\varepsilon_{i-1,j-1} + O(h^2) \\ \varepsilon_{0,j} &= \frac{1}{1-\beta_{0j}h}\sum_{i=1}^{N-1} \beta_{ij}\varepsilon_{ij} h + \frac{O(h)}{1-\beta_{0j}h} \\ \varepsilon_{i,0} &= 0.\end{aligned} \tag{8.15}$$

We would like to show that the errors go to zero as h approaches zero. Let

$$\max_{0 \le x \le A, 0 \le t \le T} \beta(x,t) \le \bar{\beta}$$

$$\min_{0 \le x \le A, 0 \le t \le T} \mu(x,t) \le \underline{\mu}$$

From (8.15) we have

$$|\varepsilon_{0,j}| \le \frac{1}{1-\bar{\beta}h}\bar{\beta}h \sum_{i=0}^{N-1} |\varepsilon_{ij}| + \frac{O(h)}{1-\beta_{0j}h}.$$

If we assume that $h < h^* < \frac{1}{\bar{\beta}}$, then

$$\frac{1}{1-\bar{\beta}h} < \frac{1}{1-\bar{\beta}h^*} = \text{constant} = C.$$

$$|\varepsilon_{0,j}| \le C\bar{\beta}h \sum_{i=0}^{N-1} |\varepsilon_{ij}| + O(h).$$

From (8.15) we have

$$|\varepsilon_{ij}| \le |\varepsilon_{i-1,j-1}| + O(h^2).$$

Multiplying by h and adding over the index i we get the following inequalities

$$\begin{aligned}\sum_{i=1}^{N}|\varepsilon_{ij}|h &\le \sum_{i=0}^{N}|\varepsilon_{i-1,j1}|h+O(h^2)\\ \sum_{i=1}^{N}|\varepsilon_{ij}|h &\le \sum_{i=0}^{N}|\varepsilon_{i,j-1}|h+O(h^2) \qquad (8.16)\\ |\varepsilon_{0,j}|h &\le C\bar{\beta}h\sum_{i=0}^{N}|\varepsilon_{ij}|h+O(h^2). \qquad (8.17)\end{aligned}$$

Adding inequalities (8.16) and (8.17) we obtain the following series of estimates:

$$\sum_{i=0}^{N}|\varepsilon_{ij}|h \le C\bar{\beta}h\sum_{i=0}^{N-1}|\varepsilon_{ij}|h+O(h^2)+\sum_{i=0}^{N}|\varepsilon_{i,j-1}|h+O(h^2)$$

$$(1-C\bar{\beta}h)\sum_{i=0}^{N}|\varepsilon_{ij}|h \le \sum_{i=0}^{N}|\varepsilon_{i,j-1}|h+O(h^2)$$

$$\begin{aligned}\sum_{i=0}^{N}|\varepsilon_{ij}|h &\le \frac{1}{1-C\bar{\beta}h^*}\left(\sum_{i=0}^{N}|\varepsilon_{i,j-1}|h+O(h^2)\right)\\ \sum_{i=0}^{N}|\varepsilon_{ij}|h &\le K\sum_{i=0}^{N}|\varepsilon_{i,j-1}|h+O(h^2)\\ &\le K\left(\sum_{i=0}^{N}|\varepsilon_{i,j-2}|h+O(h^2)\right)+O(h^2)\\ &= K\sum_{i=0}^{N}|\varepsilon_{i,j-2}|h+KO(h^2)+O(h^2)\\ &= \cdots\\ &= K\sum_{i=0}^{N}|\varepsilon_{i,0}|h+(K^{j-1}+\cdots+1)\big(O(h^2)\big)\\ &\le 0\end{aligned}$$

The first term in the sum is zero because of (8.17) and the sum of powers of K can be estimated as:

$$O(h^2)\frac{K^n-1}{K-1}=O(h^2)\frac{\left(\frac{1}{1-C\bar{\beta}h}\right)^M-1}{\frac{1}{1-C\bar{\beta}h}-1}=\frac{\left(\frac{1}{1-C\bar{\beta}h}\right)^{-M}-1}{\frac{1-1+C\bar{\beta}h}{1-C\bar{\beta}h}}O(h^2)$$

$$\leq \frac{(e^{C\bar{\beta}T}-1)(1-C\bar{\beta}h)}{C\bar{\beta}h}O(h^2) \leq \frac{e^{C\bar{\beta}T}-1}{C\bar{\beta}h}O(h^2) = O(h^2),$$

provided $h < 1/c\bar{\beta}$.

Therefore,

$$\sum_{i-0}^{N} |\varepsilon_{ij}| h \leq O(h),$$

and hence

$$\max_{ij} |\varepsilon_{ij}| \sum_{i=1}^{N} h \leq O(h),$$

from which we obtain

$$\max_{ij} |\varepsilon_{ij}| \leq O(h).$$

Exercises

Carry out the analysis of this section for the following two systems, that is, find the numerical schemes that would approximate their solutions:

1. **System 1**

$$\begin{aligned} &q_a + q_t + \mu(a)q(a,t) + \beta q(A,t) = 0 \\ &q(0,t) = 0 \\ &q(a,t) = \phi(a) \end{aligned}$$

2. **System 2**

$$\begin{aligned} &l_a + l_t + \mu(a)l = u(a,t) \ (u(a,t) \text{ is given}) \\ &l(0,t) = \textstyle\int_0^t \beta(a)l(a,t)da \\ &l(a,0) = \phi(a) \ (\phi(a) \text{ is given}) \end{aligned}$$

Epilogue

On Mathematical and Theoretical Biology

This book attempts to bridge the gap between mathematics and population biology. It is intended to show students of biology how to apply mathematics to the study of some questions of importance to population biology and to introduce modeling in the natural sciences to students of mathematics. It may also be used as a reference on mathematical methods for working biological scientists.

For the most part, we have given little description of the background of the subject, and we urge the reader to explore the history of population ecology in such sources as Kingsland (1985). We also suggest exploring Real and Brown (1991), a collection of 40 classic papers in ecology over the period 1887 to 1974.

Naturally, there are many topics in the mathematics of population biology that have been omitted in this book. The modeling of spatial population distributions is but one important omissions. The beginnings of this active area of research may be found in Skellam (1951) and Kierstad and Slobodkin (1955). Typically, spatial models are given by partial differential equations, particularly reaction-diffusion equations. This is a topic of considerable current research. The articles or books by Durrett and Levin (1994), Edelstein-Keshet (1988), Gurney and Nisbet (1998), Murray (1989), Nisbet and Gurney (1982), and Renshaw (1991) will give a reasonable perspective of current activity at different levels.

Another omission involves the use of stochastic models. In any real life situation there are random effects. If population sizes are large these effects are often small enough to be ignored, but when population sizes are small then their use is essential. Some references for such models are Durrett and Levin (1994), Nisbet and Gurney (1982), Pollard (1973), and Renshaw (1991).

There is also no consideration of structured models that take into account nonlinear birth, death, and infection processes. Some references in this area are Castillo-Chavez (1987), Diekmann and Metz (1986), Gurney and Nisbet (1998), Gurtin and McCamy (1974)], and Hoppensteadt (1975).

There has been a great deal of work in the mathematical theory of epidemics, see for example Castillo-Chavez (1989), Castillo-Chavez, Blower, van den Driessche, Kirschner, and Yakubu (2001a,2001b), Diekmann, Heesterbeek, and Metz (1990), Hadeler (1989a, 1989b, 1992, 1993), Murray (1989), and Thieme and Castillo-Chavez (1989, 1993) for a few of the directions of current interest. Because both of us are currently interested in the study of disease dynamics we have emphasized epidemiological systems throughout the book. Mathematical and theoretical epidemiology have experienced a great deal of growth over the last two decades due to the emergence and re-emergence of diseases like tuberculosis and AIDS. In our discussion of disease dynamics we have shown our biases. We have focused on the impact at the population (organizational) level and have ignored the outstanding research that is being carried out in immunology. Fortunately, many of the techniques we have illustrated in the context of population dynamics and epidemiology are also of use in population genetics, mathematical physiology, immunology, and other areas of biology.

In this book little has been said about evolution, units of selection, levels of aggregation, and scales (temporal or spatial). Consideration of these factors in the study of population biology and epidemiology is critical. In the study of a disease such as influenza, multiple scales come into play. We have fast (disease dynamics), slow (host demography), and super slow (evolution of cross immunity) time scales. Influenza spread locally (as in schools) and globally. Influenza epidemic waves move across cities, countries, and continents. Locally they are driven by age structure in contact rates and by public transportation, while globally they may be driven by train or airplane flow. We hope that the mathematical techniques and modeling approaches of this book will be useful for the interdisciplinary groups of scientists working on the types of challenges posed by influenza epidemics.

We may summarize our goals by saying that if our book facilitates communication between biologists and mathematicians then we would feel that we have made a contribution to science.

Part IV

Appendix

Appendix A
Answers to Selected Exercises

Chapter 1

Section 1.1

1. 106

3. 435

5. 30

7. (i) x=$10e^{ct}$
   (ii) c=0.0347
   (iii) 60 hours, 66.4 hours

Section 1.2

1. 75

3. 9.758×10^7 kg, 1.547 years

Section 1.4

1. $x = 0$ unstable, $x = K$ asymptotically stable (if $r > 0$, $K > 0$).

3. $x = 0$ unstable, $x = K$ asymptotically stable (if $r > 0$, $K > 0$).

5. $x = 0$ unstable, $x = K\left(1 + \log \frac{r}{d}\right)$ asymptotically stable (if $r > 0$, $K > 0$).

7. $y(0) < 0$ and $y(0)$ greater than solution of $e^{-y} = 2y$.

9. (i) If $e > \beta$, equilibrium 0 is asymptotically stable and equilibrium $1 - \frac{e}{\beta} < 0$ is unstable. If $e < \beta$, equilibrium 0 is unstable and equilibrium $1 - \frac{e}{\beta} > 0$ is asymptotically stable.

   (ii) Only equilibrium is 0, asymptotically stable if $a > 0$ and unstable if $a > 0$.

11. (i) Unstable.

    (ii) Asymptotically stable.

    (iii) Asymptotically stable.

15. For $H = 10$, there is one equilibrium ($V \approx 22$); for $H = 20$, there are two equilibria ($V \approx 16, V \approx 2$); for $H = 30$, there is only one equilibrium ($V \approx 1$). A large herd depletes vegetation discontinuously.

Section 1.5

1. Critical harvest rate 4800, equilibrium with harvest of 3000 per year is 156,900.

3. 7066

5. $\frac{rK}{e}$.

7. Maximum yield is the value of $rxe^{1-\frac{x}{K}} - dx$, with x defined by $re^{1-\frac{x}{K}}\left[1 - \frac{x}{K}\right] = d$; can not be evaluated explicitly.

Section 1.6

1. Lake would be eutrophic.

3. Dump would move lake from current equilibrium past unstable equilibrium to eutrophic equilibrium.

Section 1.7

1. (i) Mean is $\frac{b^2-a^2}{2}$.

   (ii) Cumulative distribution function is

$$f(t) = \begin{cases} 0, & \text{if } t \le a \\ \frac{t-a}{b-a} & \text{if } a < t < b \\ 1, & t \ge b \end{cases}$$

3. Mean is $\frac{\alpha_1+\alpha_2}{\alpha_1\alpha_2}$. Probability density function is $\frac{\alpha_1\alpha_2}{\alpha_1+\alpha_2}(e^{-\alpha_1 t} - e^{-\alpha_2 t})$. If $\alpha_1 = \alpha_2 = \alpha$, probability density function is $\alpha^2 t e^{-\alpha t}$.

Chapter 2

Section 2.1

1. $x_n = 2^{1-n}$.

3. $x_{2n} = \frac{1}{2^n}$, $x_{2n+1} = -\frac{1}{2^n}$.

5. $x_{2n} = r^n$, $x_{2n+1} = -r^n$.

9. 0,1,1,2,3,5,8,13.

Section 2.2

1. Figure A.1.

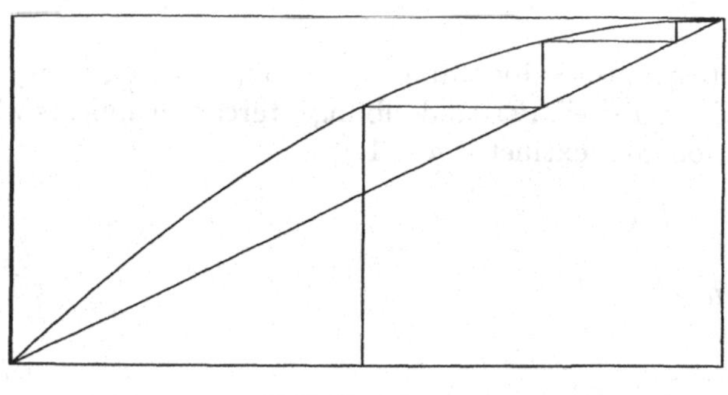

FIGURE A.1.

3. Figure A.2.

5. x_n approaches 1000 if $R = 2$.

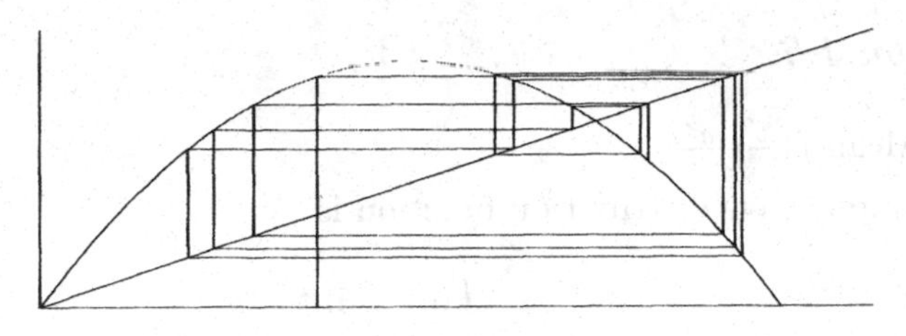

FIGURE A.2.

Section 2.3

1. If $r^2 - 4A < 0$, only equilibrium is 0 (asymptotically stable). If $r^2 - 4A \geq 0$, equilibria are 0, $\frac{r+\sqrt{r^2-4A}}{2}$ (asymptotically stable) and $\frac{r+\sqrt{r^2-4A}}{2}$ (unstable).

3. $x = 0$ (stable if $r < 1$), $x = \frac{1}{\alpha}\left(r^{\frac{1}{\beta}} - 1\right)$, asymptotically stable if $\left(r^{\frac{1}{\beta}} - 1\right)(\beta - 2) < 2$.

5. $x = 0$ (unstable), $x = 1$ (asymptotically stable).

7. (b) $p < 1 - \frac{1}{e} = 0.632$.

9. (i) $x = 0$, $x = 2$ asymptotically stable, $x = 1$ unstable.

   (ii) $\sqrt{\frac{9}{8}} - 1 = 0.06 < a < 1$.

11. $x = \frac{1}{2}$ is asymptotically stable if $0 \leq \alpha < \frac{1}{2\pi}$. Periodic orbit for $\alpha = \frac{1}{4}$ is $\left\{\frac{1}{4}, \frac{3}{4}, \frac{1}{4}, \ldots\right\}$.

13. $x = 0$ (asymptotically stable if $a < 1$). $x = \log a$ (asymptotically stable if $1 < a < e^2$. Period doubling bifurcation appears when $a = e^2$. Population goes extinct if $a < 1$.

Section 2.4

1. $r = e$.

5. Fixed points $0, \frac{5}{2}, \frac{7}{12}, 1$ all unstable. Cycle of period 2 with initial value $\frac{1}{4}$.

9. $r = e$.

Section 2.5

1. $x = 0$ is asymptotically stable for $0 < \alpha < 1$, $x = \frac{\log \alpha}{\beta}$ is asymptotically stable for $1 < \alpha < e^2$.

3. $x = 0$ is asymptotically stable for $1 < \alpha < 1$, $x = \frac{\log \alpha}{\beta}$ is asymptotically stable for $\alpha > 1$.

Section 2.6

1. $x_{n+1} = 6x_n$; (2,3), (6,6), (12,18).

3. (-1,1,1,-1,-1,1,1,...).

5. (0,0) and $\left(\frac{a}{a-1}\frac{\log a}{bc}, \frac{1}{b}\log a\right)$ [if $a > 1$], both unstable.

Section 2.7

1. Equilibria are solutions of $y = \alpha B(y) - D(y)$, with x given by $B(y)$. Asymptotic stability conditions are $|D'(y)| < 1 - \alpha B'(y)$, $-\alpha B'(y) < 1$.

Section 2.8

1. (i) $A = 104.85$, $L = 179.32$, $P = 143.46$.
   (ii) $A = 2.75$, $L = 13.06$, $P = 10.45$.

Chapter 3

Section 3.2

5. $x = \frac{K \pm \sqrt{K^2 - \frac{4KH}{r}}}{2}$.

Section 3.3

1. $x = 0$ asymptotically stable for all T if $r < dA$. $x = \frac{r - Ad}{d}$ asymptotically stable for all T if $r > dA$.

3. (i) $x = 0$ is unstable for all T, $x = 3$ is asymptotically stable if $\sec z < -2$, where $z = -T \tan z$.
   (ii) $x = 0$ is unstable for all T, $x = 3 - \log(1 + p)$ is asymptotically stable if $\log(1 + p) - 2 > \sec z$, where $z = -(1 + p)T \tan z$.

Section 3.4

1. $x = 0$ asymptotically stable for all T if $r < \pi Ad$, $x = \frac{r}{\pi d} - A$ asymptotically stable for all T if $r > \pi Ad$ [for both parts (a) and (b)].

3. $x = 0$ asymptotically stable for all T if $r < \pi d$. $x = \log \frac{r}{\pi d}$, existing only if $r > \pi d$, asymptotically stable for all T.

Section 3.5

1. Positive equilibrium x_∞ is asymptotically stable for all T if it exists, which happens if $H < \max x\left(e^{3-x} - 1\right) = 6.45$.

Chapter 4

Section 4.3

1. $u' = u - v, v' = u + v$ at $(1,1)$

3. $u' = v, v' = 2u + v$ at $(1,-1)$ and $u' = v, v' = -2u + v$ at $(-1,1)$.

5. No equilibrium.

7. $u' = \lambda u, v' = \mu v$ at $(0,0)$, $u' = -ax_\infty u - by_\infty v, v' = -cx_\infty u - dy_\infty v$ at $x_\infty = \frac{d\lambda - b\mu}{ad-bc}, y_\infty = \frac{a\mu - c\lambda}{ad-bc}$.

9. (0,0) with community matrix $\begin{bmatrix} -1 & 1 \\ 0 & -1 \end{bmatrix}$, asymptotically stable.

   (1,1) with community matrix $\begin{bmatrix} -1 & 1 \\ \frac{8}{5} & -1 \end{bmatrix}$, unstable. (4,4) with community matrix $\begin{bmatrix} -1 & 1 \\ \frac{2}{5} & -1 \end{bmatrix}$, asymptotically stable.

11. (i) $S' = -\frac{\beta SI}{N} + \gamma(N - I - S)$ $I' = \frac{\beta SI}{N} - \gamma I$ [N constant].
    (ii) Disease-free equilibrium $S = N$, $I = 0$ is asymptotically stable if $\beta < \nu$. Endemic equilibrium with $\beta S = \nu N$ is asymptotically stable if it exists ($\beta > \nu$).

13. (a) Λ-people/time, β, μ, and γ-1/time.
    (b) $N' = \Lambda - \mu N$.
    (c) $N(t) = \frac{\Lambda}{\mu}\left(1 - e^{-\mu t}\right) + N(0)e^{-\mu t}$.
    (e) $R_0 = \frac{\beta}{\mu + \gamma}$, equilibrium $S = K$, $I = 0$ is asymptotically stable if $R_0 < 1$ and equilibrium $S = \frac{\mu + \gamma}{\beta}$ is asymptotically stable if $R_0 > 1$.

15. (a) $\begin{bmatrix} -\mu & -\beta & 0 \\ 0 & \beta-(\mu+\gamma) & 0 \\ 0 & \gamma & -\mu \end{bmatrix}$, with eigenvalues $-\mu$, $-\mu$, $\beta-(\mu+\gamma)$.

(c) $R_0 = \frac{\beta}{\mu+\gamma}$, same as for (4.9).

(d) Disease-free equilibrium is asymptotically stable if $R_0 < 1$.

Section 4.4

1. (1, 1) unstable.
3. (−1, −1) unstable, (1, 1) unstable.
5. No equilibrium.
7. Orbits depend on values of the parameters; solution is given in Section 5.1.
9. Orbits depend on the values of the parameters; solution is given in Section 5.2.
11. (b) $y \approx \frac{\alpha N}{\gamma-\alpha}\left(e^{-\alpha t} - e^{-\gamma t}\right)$.

(c) $y \approx \frac{N}{1000}$ for $t = 6.915$.

17. Competitive system; all solutions approach $\left(\frac{2}{3}, \frac{2}{3}\right)$.
19. (0,0) has matrix $\begin{bmatrix} 0 & -1 \\ -1 & 0 \end{bmatrix}$ (saddle point). (1,1) has matrix $\begin{bmatrix} 1 & 0 \\ 0 & 1 \end{bmatrix}$ (unstable node).

Section 4.5

1. There are three equilibria (0, 0), (20, 0) and (10, 30). All of them are unstable. There is a stable limit cycle around (10, 30). Two species coexist with oscillation.
3. There are three equilibria (0, 0), (20, 0) and (2, 32.4). All of them are unstable. There is a stable limit cycle around (2, 42.4). Two species coexist with oscillation.
5. There are two equilibria (0, 0), (20, 0). (20, 0) is asymptotically stable. Thus the predator species goes to extinction.

7. The equilibrium $(60, 20)$ is asymptotically stable and all trajectories approach it. Two species coexist.

9. The equilibrium$(0, 45)$ is asymptotically stable and all trajectories approach it. x-species goes to extinction and y-species wins the competition.

11. The equilibrium $(0, 16)$ is asymptotically stable and all trajectories approach it. x-species goes to extinction and y-species wins the competition.

13. If $n < 8$, steady state is asymptotically stable, spiral point if $n < \sqrt{48}$ and node if $\sqrt{48} < n < 8$. If $n > 8$ steady state is a saddle point.

Chapter 5

Section 5.1

1. The equilibrium $(60, 20)$ is asymptotically stable and all trajectories approach it. Two species coexist.

3. The equilibrium$(0, 45)$ is asymptotically stable and all trajectories approach it. x-species goes to extinction and y-species wins the competition.

5. The equilibrium $(0, 16)$ is asymptotically stable and all trajectories approach it. x-species goes to extinction and y-species wins the competition.

Section 5.2

1. There are three equilibria $(0, 0)$, $(20, 0)$ and $(10, 30)$. All of them are unstable. There is a stable limit cycle around $(10, 30)$. Two species coexist with oscillation.

3. There are three equilibria $(0, 0)$, $(20, 0)$ and $(2, 32.4)$. All of them are unstable. There is a stable limit cycle around $(2, 42.4)$. Two species coexist with oscillation.

5. There are two equilibria $(0,0)$, $(20,0)$. $(20,0)$ is asymptotically stable. Thus the predator species goes to extinction.

Section 5.3

1. Figure A.3.

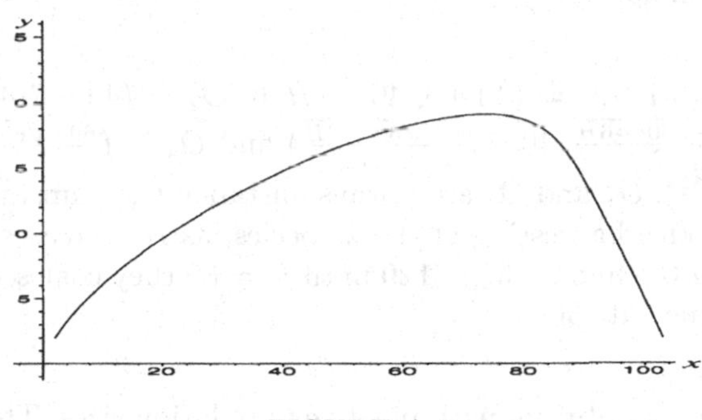

FIGURE A.3.

Section 5.5

1. There are two equilibria $(0,0)$ and $(120,70)$. The first is unstable and the second is asymptotically stable.

3. There are three equilibria $(0,0)$, $(0,10)$ and $(4+\sqrt{34}, 5+\sqrt{34})$. The first two are unstable and the third is asymptotically stable.

5. There are three equilibria $(0,0)$, $(0,M)$ and $(K,0)$. They are all unstable. If $ab \geq 1$, orbits are unbounded. If $ab < 1$, stable equilibrium $(x_\infty, y_\infty) = \left(\frac{K+aM}{1-ab}, \frac{Mb+K}{1-ab}\right)$ appears and all trajectories approach this equilibrium.

Chapter 6

Section 6.1

1. When $E = 0$, the equilibrium $(x_\infty, y_\infty) = (60,20)$ is asymptotically stable and thus the two species coexist. When $E > 0$, the stable

equilibria are on the line $x + y = 80$, and when $E = 120$, it coalesces to $(0, 80)$. Therefore, harvesting decreases the x-species size and increases the y-species, and eventually moves coexistence to x-extinction.

3. Without harvesting, the x-species goes to extinction and the y-species wins the competition. Harvesting the x-species will speed up extinction of the x-species.

5. Four equilibria $O_1 = (40 + \sqrt{40^2 - H}, 0)$,$O_2 = (40 - \sqrt{40^2 - H}, 0)$, $O_3 = (\frac{60+\sqrt{60^2-6H}}{2}, 40 - \frac{60+\sqrt{60^2-6H}}{6})$ and $O_4 = (\frac{60-\sqrt{60^2-6H}}{2}, 40 - \frac{60-\sqrt{60^2-6H}}{6})$. O_1 and O_2 are always unstable; O_4 is unstable and O_3 is stable. With harvesting of the x-species, as H increases $O_3$and O_4 move along the line $x+3y = 120$ until $H = 60$ they coalesce, resulting in coexistence at $(30, 30)$

7. There is no equilibrium with positive population sizes. The x-species becomes extinct in finite time

Section 6.2

1. $H_c = H^* = \frac{5}{6}$

3. $H_c = 5.787, H_1^* = 3.7037, H_2^* = 5.5556$. H must be in the interval$(0, H_1^*)$ or in the interval (H_2^*, H_c)

5. $H_c = 8.8889$. Trace of community matrix is positive, meaning that equilibrium is unstable. There exists a stable limit cycle.

Section 6.3

1. If we repeat the process 5 times, the estimated average instantaneous harvest is 34.5526. If we repeat the process 10 times, the estimated average instantaneous harvest is 34.2373. If we repeat the process 15 times, the estimated average instantaneous harvest is 34.1638. If we repeat the process 100 times, the estimated average instantaneous harvest is 34.0324. If we repeat the process 500 times, the estimated average instantaneous harvest is 34.0318.

Section 6.4

1. $Y(E) = \frac{aE}{E+b}$, always less than a and approaches a as $E \to \infty$. However, maximum fish population size is $\frac{a}{b}$. Thus the yield can not exceed $\frac{a}{b}$.

3. E_{MSY} is the unique solution of the equation $Log(\frac{r}{d+E}) - \frac{E}{d+E} = 0$ for E in the interval $(0, r-d)$. The maximum sustainable yield $Y(E_{MSY}) = \frac{E_{MSY}^2}{d+E_{MSY}}$

Section 6.5

1. $F(y) = rylog\frac{K}{y}$, $h = Ey$, $C(y) = \frac{c}{y}$. y^* is the unique solution to the equation $rplog\frac{K}{y} - (\delta + r)(p - \frac{c}{y}) = 0$, If $\frac{K}{e} < \frac{c}{p}$, $y^* < \frac{K}{e} = y_{MSY}$, implying overfishing.

Chapter 7

Section 7.2

1. 1.4416

3. 2.827%

5. 17.56%

Section 7.3

1. 58.33%

3. 36%

5. $0.9125 foxes/km^2$

7. (a) $C = C_0 e^{-\alpha t}$. (b) $S_0 e^{\frac{\beta C_0}{\alpha}(e^{-\alpha t}-1)}$. (c) $\lim_{t\to\infty} S(t) = S_0 e^{-\frac{\beta}{\alpha}C_0}$.

9. 2 hours.

Section 7.4

1. (a) Check $\frac{dN}{dt} = 0$. (b)$\frac{1}{\mu}$ average life span; $\frac{1}{\gamma}$ average infectious period; $\frac{1}{\xi}$ average quarantine period. Their units are the same, namely time. (c)$\nu = \frac{\mu}{\sigma}, \theta = \frac{\gamma}{\sigma}, \zeta = \frac{\xi}{\sigma}$. (d) $R_0 = \frac{\sigma}{\mu+\gamma}$, $R_0 \leq 1$,$(0,0,0)$ is asymptotically stable. $R_0 > 1$,$(0,0,0)$ is unstable.

Section 7.5

1. 23.1%

3. No

Chapter 8

Section 8.1

1. $\lambda_0 = 0.9491$, $v_0 = [0.8021, 0.4534, 0.1979]$, stable age distribution $[0.5131, 0.3604, 0.1226]$

3. $\lambda_0 = 1.6826$, $v_0 = [0.9552, 0.2838, 0.0843]$, stable age distribution $[0.7281, 0.2145, 0.0637]$

5. Leslie matrix is

$$A = \begin{bmatrix} 1 & 3 \\ 2/3 & 0 \end{bmatrix}$$

$\lambda_0 = 2$, $v_0 = [0.9487, 0.3162]$, the stable age distribution is $[0.75, 0.25]$. The total population $P_n \simeq c2^n$ as n is large.

Section 8.2

1(d). $B(t) = \left(\beta \int_0^\infty \phi(s)ds\right) e^{(\beta-\mu)t}$.

References

[1] Adler, F., L. Smith, and C. Castillo-Chavez (1989) A distributed delay model for the local population dynamics of a parasitoid-host system, *Mathematical Approaches to Ecological and Environmental Problem Solving* (Castillo-Chavez C., S. A. Levin, and C. Shoemaker, eds.), pp. 152-62. *Lecture Notes in Biomathematics 81*, Springer-Verlag, Berlin, Heidelberg, New York, London, Paris, Tokyo, Hong Kong.

[2] Agur,Z.L., G. Mazor, R. Anderson, and Y. Danon (1993) Pulse mass measles vaccination across age cohorts, *Proc. Nat. Acad. Sci.*, **90**:11698-11702.

[3] Allee, W.C. (1931) *The Social Life of Animals*, Heinemann, London.

[4] Anderson, R.M. (1974) Mathematical models of host-helminth parasite interactions. In: *Ecological Stability*, (M. B. Usher and M. H. Williamson,eds.) 43-69. Chapman and Hall, London.

[5] Anderson, R.M., ed.(1982) *Population Dynamics of Infectious Diseases: Theory and Applications,* Chapman and Hall, London-New York.

[6] Anderson,R.M., H.C. Jackson, R.M. May, and A.M. Smith (1981) Population dynamics of fox rabies in Europe, *Nature*, **289**:765-771.

[7] Anderson, R.M. and R.M. May (1982) Co-evolution of host and parasites, *Parasitology*, **85**:411-426.

[8] Anderson, R.M. and R.M. May (1983) Vaccination against rubella and measles: Quantitative investigations of different policies, *J. Hygiene*, **90**:259-325.

[9] Anderson, R.M. and R.M. May (1991) *Infectious Diseases of Humans,* Oxford Science Publications, Oxford.

[10] Aparicio, J.P., A. Capurro, and C. Castillo-Chavez (2000a). Markers of disease evolution: the case of tuberculosis. *Department of Biometrics, Cornell University, Technical Report Series*, BU-1503-M.

[11] Aparicio J. P.,A.F. Capurro, and C. Castillo-Chavez (2000b) On the fall and rise of tuberculosis. *Department of Biometrics, Cornell University, Technical Report Series*, BU-1477-M.

[12] Aparicio J.P., A. Capurro, and C. Castillo-Chavez (2001a). Frequency Dependent Risk of Infection and the Spread of Infectious Diseases. In: *Mathematical Approaches for Emerging and Reemerging Infectious Diseases : Models, Methods and Theory*, Springer-Verlag, Edited by Castillo-Chavez, C. with S. Blower, P. van den Driessche, D. Kirschner, and A.A. Yakubu. (In press).

[13] Aparicio, J.P., A. Capurro, and C. Castillo-Chavez (2001b) On the long-term dynamics and re-emergence of tuberculosis. In: *Mathematical Approaches for Emerging and Reemerging Infectious Diseases: Models, Methods and Theory*, Springer-Verlag, Edited by Castillo-Chavez, C. with S. Blower, P. van den Driessche, D. Kirschner, and A.A. Yakubu. (In press).

[14] Armstrong, R.A. and R. McGehee (1980) Competitive exclusion, *Amer. Naturalist*, **115**:151-170.

[15] Aron, J.L. and R.M. May (1982) The population dynamics of malaria. In: *Population Dynamics of Infectious Disease*, R.M. Anderson, ed., 139-179, Chapman and Hall, London.

[16] Arreola R., A. Crossa, and M.C. Velasco (2000) Discrete-time SEIS models with exogenous re-infection and dispersal between two Patches, *Department of Biometrics, Cornell University, Technical Report Series*, BU-1533-M.

[17] Arriola P., I. Mijares-Bernal, J.A. Ortiz-Navarro, H. Campus, and R.A. Senz (2000) Dynamics of the spruce budworm population under the action of predation and insecticides, *Department of Biometrics, Cornell University, Technical Report Series*, BU-1417-M.

[18] Ayala, F.J., M.E. Gilpin, and J.G. Ehrenfeld (1973) Competition between species: Theorical models and experimental tests, *Theoretical Pop. Biol.*, **4**:331-356.

[19] Bailey, N.T.J. (1975) *The Mathematical Theory of Infectious Diseases and Its Applications,* Griffin, London.

[20] Barbour, A.D. (1978) Macdonald's model and the transmission of bilharzia, *Trans. Roy. Soc. Trop. Med. Hygiene,* **72**:6-15.

[21] Barrera J.H., A. Cintron-Arias, N. Davidenko, L.R. Denogean, and S.R. Franco-Gonzalez (2000) Dynamics of a two-dimensional discrete-time SIS model, *Department of Biometrics, Cornell University, Technical Report Series,* BU-1518-M.

[22] Bélair, J. and L. Glass (1983) Self similarity in periodically forced oscillators, *Phys. Lett.,* **96A**:113-116.

[23] Beck, K. (1984) Co-evolution: Mathematical aspects of host-parasite interactions, *J. Math. Biol.,* **19**:63-77.

[24] Bellman, R.E. and K.L. Cooke (1963) *Differential-Difference Equations,* Academic Press, New York.

[25] Bendixson, T. (1901) Sus les courbes définies par des équations differentielles, *Acta. Math.,* **24**:1-88.

[26] Bernardelli, H. (1941) Population waves, *J. Burma Res. Soc.,* **31**:1-18.

[27] Beverton, R.J. and S.J. Holt (1956) *The theory of fishing,* In *Sea Fisheries; Their Investigation in the United Kingdom,* M. Graham,ed., 372-441, Edward Arnold, London.

[28] Blythe, S.P. and C. Castillo-Chavez (1990) Scaling law of sexual activity, *Nature,* **344**:202.

[29] Blythe, S.P., S. Busenberg, and C. Castillo-Chavez (1995) Affinity and paired-event probability, *Math. Biosci.,* **128**:265-84.

[30] Blythe, S.P., C. Castillo-Chavez, J. Palmer, and M. Cheng. (1991) Towards a unified theory of mixing and pair formation, *Math. Biosci.,* **107**:379-405.

[31] Brauer, F. (1976) Perturbations of the nonlinear renewal equation, Adv. in Math., 22: 32-51.

[32] Brauer, F. (1986) Coexistence and survival of invading species. In *Mathematical Ecology,* T.G. Hallam, S.A. Levin, and L.J. Gross eds., 605-616, World Scientific Press, Singapore.

[33] Brauer, F. (1987a) Harvesting in delayed recruitment population models. In *Oscillation, Bifurcation, and Chaos,* CMS Conference Proceedings **8**, 317-327, AMS, Providence.

[34] Brauer, F. (1987b) A class of Volterra integral equations arising in delayed-recruitment population models, *Nat. Res. Modeling*, **2**:259-278.

[35] Brauer, F. (1989) Multispecies interactions and coexistence. In *Differential equations and Applications*, A.R. Aftabizadeh, ed., 91-96, Ohio University Press, Athens.

[36] Brauer, F. and J.A. Nohel (1969) *The Qualitative Theory of Ordinary Differential Equations,* W. A. Benjamin, New York; reprinted by Dover, New York 1989.

[37] Brauer, F. and D.A. Sánchez (1975) Constant rate population harvesting: equilibrium and stability, *Theor. Pop. Biol.*, **8**:12-30.

[38] Brauer, F. and A.C. Soudack (1979) Stability regions and transition phenomena for harvested predator-prey systems, *J. Math. Biol.*, **7**:319-337.

[39] Brauer, F. and A.C. Soudack (1985) Optimal harvesting in predator-prey systems, *Int. J. Control*, **41**:111-128

[40] Brauer, F., A.C. Soudack, and H.S. Jarosch (1976) Stabilization and destabilization of predator-prey sytems under harvesting and nutrient enrichment, *Int. J. Control*, **23**:553-573.

[41] Bremermann, H.J. and J. Pickering (1983) A game-theoretical model of parasite virulence, *J. Theor. Biol.*, **100**:411-426.

[42] Bremermann, H.J. and H.R. Thieme (1989) A competitive exclusion principle for pathogen virulence, *J. Math. Biol.*, **27**:179-190.

[43] Bulmer, M.G. (1976) The theory of predator-prey oscillations, *Theor. Pop. Biol.*, **9**:137-150.

[44] Busenberg, S. and C. Castillo-Chavez. (1989) Interaction, pair formation and force of infection terms in sexually transmitted diseases. In *Mathematical and Statistical Approaches to AIDS Epidemiology*, Lect. Notes Biomath. **83**, C. Castillo-Chavez, ed., 289-300, Springer-Verlag, Berlin-Heidelberg-New York.

[45] Busenberg, S. and C. Castillo-Chavez (1991) A general solution of the problem of mixing of subpopulations, and, its application to risk- and age-structured epidemic models, *IMA J. Math. Appl. Med. Biol.*, **8**:1-29.

[46] Busenberg, S. and K.L. Cooke (1993) *Vertically Transmitted Diseases: Models and Dynamics,* Biomathematics **23**, Springer-Verlag, Berlin-Heidelberg-New York.

[47] Butler, G.J., H.I. Freedman, and P. Waltman (1986) Uniformly persistent systems, *Proc. Am. Math. Soc.*, **96**:425-430.

[48] Butler, G.J., S.B. Hsu, and P. Waltman (1983) Coexistence of competing predators in a chemostat, *J. Math. Biol.*, **17**:133-151.

[49] Butler, G.J. and P. Waltman (1986) Persistence in dynamical systems, *J. Diff. Equations*, **86**:225-263.

[50] Carpenter, S.R., D. Ludwig, and W.A. Brock (1999) Management of eutrophication for lakes subject to potentially reversible change, *Ecological Applications*, **9**:751-771.

[51] Castillo-Chavez, C. (1987a) Linear character dependent models with constant time delay in population dynamics, *Int. J. Math. Modeling*, **9**:821-836.

[52] Castillo-Chavez, C. (1987b) Non-linear character-dependent models with constant time delay in population dynamics *J. Math. Analysis and Applications,* **128(1)**:1-29.

[53] Castillo-Chavez, C., ed. (1989) *Mathematical and Statistical Approaches to AIDS Epidemiology,* Lect. Notes Biomath. **83**, Springer-Verlag, Berlin-Heidelberg-New York.

[54] Castillo-Chavez, C., S. Blower, P. van den Driessche, D. Kirschner, and A.A. Yakubu, eds. (2001a) *Mathematical Approaches for Emerging and Reemerging Infectious Diseases: Models, Methods and Theory,* Volume I, Springer-Verlag, Berlin-Heidelberg-New York. (In press)

[55] Castillo-Chavez, C. with S. Blower, P. van den Driessche, D. Kirschner, and A.A. Yakubu, eds. (2001b) *Mathematical Approaches for Emerging and Reemerging Infectious Diseases: Models, Methods and Theory,* Volume II, Springer-Verlag, Berlin-Heidelberg-New York. (In press).

[56] Castillo-Chavez, C. and S. Busenberg (1991) On the solution of the two-sex problem. In *Proceedings of the International Conference on Differential Equations and Applications to Biology and Population Dynamics*, S. Busenberg and M. Martelli, eds., Lect. Notes Biomath. **92**, 80-98, Springer-Verlag, Berlin-Heidelberg-New York.

[57] Castillo-Chavez, C., S. Busenberg and K. Gerow (1991) Pair formation in structured populations. In *Differential Equations with Applications in Biology, Physics and Engineering,* J. Goldstein, F. Kappel, and W. Schappacher, eds., 47-65, Marcel Dekker, New York.

[58] Castillo-Chavez, C., K. Cooke, W. Huang, and S.A. Levin (1989a) The role of long incubation periods in the dynamics of HIV/AIDS. Part 1: Single Populations Models, *J. Math. Biol.*, **27**:373-98.

[59] Castillo-Chavez, C., K.L. Cooke, W. Huang, and S. A. Levin (1989b) Results on the dynamics for models for the sexual transmission of the human immunodeficiency virus, *Applied Math. Letters*, **2(4)**:327-331.

[60] Castillo-Chavez, C., and Z. Feng (1998a) Global stability of an age-structure model for TB and its applications to optimal vaccination strategies, *Mathematical Biosciences,* **151**:135-154.

[61] Castillo-Chavez, C. and Z. Feng (1998b) Mathematical models for the disease dynamics of tuberculosis, *Advances in Mathematical Population Dynamics - Molecules, Cells, and Man* (O. Arino, D. Axelrod, M. Kimmel, (eds)), 629-656, World Scientific Press, Singapore.

[62] Castillo-Chavez, C., H.W. Hethcote, V. Andreasen, S.A. Levin & Wei-min Liu (1988) Cross-immunity in the dynamics of homogeneous and heterogeneous populations. In *Mathematical Ecology*, T.G. Hallam, L.G. Gross, and S.A. Levin, eds., 303-316, World Scientific Publishing Co., Singapore.

[63] Castillo-Chavez, C., H.W. Hethcote, V. Andreasen, S.A. Levin and Wei-min Liu (1989) Epidemiological models with age structure, proportionate mixing, and cross-immunity, *J. Math. Biol.*, **27**:233-258.

[64] Castillo-Chavez, C., and W. Huang (1995) The logistic equation revisited: the two-sex case, *Math. Biosci.*, **128**:299-316.

[65] Castillo-Chavez, C., W. Huang, and J. Li (1996a) Competitive exclusion in gonorrhea models and other sexually-transmitted diseases, *SIAM J. Appl. Math,* **56(2)**:494-508.

[66] Castillo-Chavez, C., W. Huang, and J. Li, (1996b) On the existence of stable pairing distributions, *J. Math. Biol.*, **34**:413-441.

[67] Castillo-Chavez, C., W. Huang, and J. Li (1997) The effects of females' susceptibility on the coexistence of multiple pathogen strains of sexually-transmitted diseases, *Journal of Mathematical Biology,* **35**:503-522.

[68] Castillo-Chavez, C. and S-F Hsu Schmitz, (1993) On the evolution of marriage functions: It takes two to tango. In *Structured Population Models in Marine, Terrestrial, and Freshwater Systems*, S. Tuljapurkar and H. Caswell, eds., 533-553, Chapman and Hall, New York.

[69] Castillo-Chavez, C., S.A. Levin, and C. Shoemaker, eds. (1989) *Mathematical Approaches to Ecological and Environmental Problem Solving, Lecture Notes in Biomathematics 81*, Springer-Verlag, Berlin-Heidelberg-New York.

[70] Castillo-Chavez, C., S-F. Shyu, G. Rubin, and D. Umbauch, (1992) On the estimation problem of mixing/pair formation matrices with applications to models for sexually-transmitted diseases. In *AIDS Epidemiology: Methodology Issues*, N.P. Jewell, K. Dietz, and V.T. Farewell, eds., 384-402, Birkhäuser, Boston.

[71] Castillo-Chavez, C., and H.R. Thieme (1993) Asymptotically autonomous epidemic models. In *Mathematical Population Dynamics: Analysis of Heterogeneity, Vol. 1, Theory of Epidemics*, O. Arino, D. Axelrod, M. Kimmel, and M. Langlais, eds., 33-50, Wuerz, Winnipeg.

[72] Castillo-Chavez, C., J.X. Velasco-Hernàndez, and S. Fridman (1994) Modeling Contact Structures in Biology. In *Frontiers of Theoretical Biology*,S.A. Levin, ed., Lect. Notes Biomath. **100**, 454-491, Springer-Verlag, Berlin-Heidelberg-New York.

[73] Castillo-Chavez C., and A.A. Yakubu (2000a) Discrete-time nonlinear pair formation models with geometric solutions, *Department of Biometrics, Cornell University, Technical Report Series*, BU-1496-M.

[74] Castillo-Chavez C., and A.A. Yakubu (2000c) Dispersal,disease and life history evolution, *Department of Biometrics, Cornell University, Technical Report Series*, BU-1538-M.

[75] Castillo-Chavez C., and A.A. Yakubu (2000d) Epidemics models on attractors, *Department of Biometrics, Cornell University, Technical Report Series*, BU-1540-M.

[76] Castillo-Chavez C., and A.A. Yakubu (2000e) Discrete-time nonlinear pair-formation models with geometric solutions, *Department of Biometrics, Cornell University, Technical Report Series*, BU-1496-M.

[77] Castillo-Chavez C., and A.A. Yakubu (2000f) Nonlinear mating models for population with non-overlapping generations, *Department of Biometrics, Cornell University, Technical Report Series*, BU-1497-M.

[78] Castillo-Chavez, C. and A.A. Yakubu (2001a). Discrete-time S-I-S models with complex dynamics. To appear in: *Nonlinear Analysis: Theory Methods and Applications.*

[79] Castillo-Chavez, C. and A.A. Yakubu (2001b). Discrete-time S-I-S models with simple and complex dynamics. In: *Mathematical Approaches for Emerging and Reemerging Infectious Diseases: Models, Methods and Theory*, C. Castillo-Chavez with S. Blower, P. van den Driessche, D. Kirschner, and A.A. Yakubu, eds. Springer-Verlag, Berlin-Heidelberg-New York (In press).

[80] Castillo-Chavez, C. and A.A. Yakubu (2001c). Intra-specific competition, dispersal and disease dynamics in discrete-time patchy environments. In: *Mathematical Approaches for Emerging and Reemerging Infectious Diseases: Models, Methods and Theory*, C. Castillo-Chavez with S. Blower, P. van den Driessche, D. Kirschner, and A.A. Yakubu, eds. Springer-Verlag, Berlin-Heidelberg-New York.(In press).

[81] Castillo-Garsow C., G. Jordan-Salivia, and A. Rodriguez-Herrera (2000) Mathematical models for the dynamics of tobacco use, recovery, and relapse, *Department of Biometrics, Cornell University, Technical Report Series*, BU-1505-M.

[82] Clark, C.W. (1990) *Mathematical Bioeconomics: The Optimal Management of Renewable Resources, 2nd ed.*, Wiley, New York.

[83] Cohen, J.E. (1995) *How Many People Can the Earth Support?* W. W. Norton and Company, New York-London.

[84] Cooke, K.L. and P. van den Driessche (1986) On zeroes of some transcendental functions, *Funkcialaj Ekvacioj*, **29**:77-90.

[85] Cooke, K.L. and J. A. Yorke (1973) Some equations modeling growth processes and gonorrhea epidemics, *Math. Biosci.*, **58**:93-109.

[86] Costantino, R.F., R.A. Desharnais, J.M. Cushing, and B. Dennis (1995) Experimentally induced transitions in the dynamic behavior of insect populations, *Nature*, **375**:227-230.

[87] Costantino, R.F., R.A. Desharnais, J.M. Cushing, and B. Dennis (1995) Chaotic dynamics in an insect population, *Science*, **275**:389-391.

[88] Crane, J. (1991) The epidemic theory of ghettoes and neighborhood effects on dropping out and teenage childbearing,*Amer. J. Sociology*, **96**:1226-1259.

[89] Cushing, J.M. (1998) *An Introduction to Structured Population Dynamics*, CBMS-NSF Regional Conference Series in Applied Mathematics **71**, SIAM, Philadelphia.

[90] Dean, A.M. (1983) A simple model of mutualism, *Amer. Naturalist*, **121**:409-417.

[91] DeRoos, A., O. Diekmann, and J. A. J. Metz (1992) Studying the dynamics of structured population models: A versatile technique and its application to Daphnia, *Amer. Naturalist*, **139**:123-147.

[92] Devaney, R.L. (1992) *A First Course in Chaotic Systems: Theory and Experiment*, Addison-Wesley, Reading, Mass.

[93] Diekmann O.and J.A.P. Heesterbeek (2000). *Mathematical epidemiology of infectious diseases: Model building, analysis and interpretation*, John Wiley and Sons, ltd.

[94] Diekmann, O., J.A.P. Heesterbeek, and J.A.J. Metz (1990) On the definition and the computation of the basic reproductive ratio R_0 in models for infectious diseases in heterogeneous populations, *J. Math. Biol.*, **28**:365-382.

[95] Dietz, K. (1979) Epidemiologic interference of virus populations, *J. Math. Biol.*, **8**:291-300.

[96] Dietz, K. (1988) The first epidemic model: A historical note on P. D. En'ko, *Australian J. Stat.*, **30**:56-65.

[97] Dietz, K. (1988) On the transmission dynamics of HIV, *Math. Biosci.* **90**:397-414.

[98] Dietz, K. and K.P. Hadeler (1988) Epidemiological models for sexually transmitted diseases, *J. Math. Biol.*, **26**:1-25.

[99] Dietz, K. D. Schenzle (1985) Proportionate mixing models for age-dependent infection transmission, *J. Math. Biol.*, **22**:117-120.

[100] Dobson, A.P. (1988) The population biology of parasite-induced changes in host behavior.*The Quarterly Review of Biology*, **63**:139-165.

[101] Dodson, S.I. (1974) Zooplankton competition and predation: An experimental test of the size-efficiency hypothesis, *Ecology*, **55**:605-613.

[102] Dulac, H. (1934) Points Singulieres des Équations Differentielles, *Mém. Sci. Math.*, Fasc. **61**, Gauthier-Villars, Paris.

[103] Durrett,R. and S. A. Levin (1994) Stochastic spatial models: A user's guide to ecological applications, *Phil. Trans. Roy. Soc. B*, **343**:329-350.

[104] Dushoff, J., W. Huang, and C. Castillo-Chavez (1998) Backward bifurcations and catastrophe in simple models of fatal diseases, *Journal of Mathematical Biology*, **36**:227-248.

[105] Dwyer, G., S.A. Levin, and L. Buttel (1990) A simulation model of the population dynamics and evolution of myxomatosis, *Ecological Monographs*, **60**:423-447.

[106] Edelstein-Keshet, L. (1988) *Mathematical Models in Biology*, Random House, New York.

[107] Elaydi, S. (1996) *An Introduction to Difference Equations*, Springer-Verlag, Berlin-Heidelberg-New York.

[108] Ellner, S., R. Gallant, and J. Theiler (1995) Detecting nonlinearity and chaos in epidemic data. In *Epidemic Models: Their Structure and Relation to Data*, D. Mollison,ed., Cambridge University Press, Cambridge, 229-247.

[109] Elton, C.S. and M. Nicholson (1942) The ten year cycle in numbers of lynx in Canada, *J. Animal Ecology* **11**:215-244.

[110] Engbert, R. and F. Drepper (1994) Chance and chaos in population biology-models of recurrent epidemics and food chain dynamics, *Chaos, Solutions & Fractals*, **4(7)**:1147-1169.

[111] En'ko, P.D. (1889) On the course of epidemics of some infectious diseases, *Vrach. St. Petersburg*:1008-1010, 1039-1042, 1061-1063.

[112] Evans, A.S. (1982) *Viral Infections of Humans*, 2nd ed., Plenum Press, New York.

[113] Ewald, W.P. (1994) *The Evolution of Infectious Disease*, Oxford University Press, Oxford.

[114] Feigenbaum, M.J. (1980) The metric universal properties of period doubling bifurcations and the spectrum for a route to turbulence. In *Nonlinear Dynamics*, R.H.G. Helleman, ed., *Ann. N.Y. Acad. Science*, **357**:330-336.

[115] Feller, W. (1941) On the integral equation of renewal theory, *Ann. Math. Stat.*, **12**:243-267.

[116] Feng, Z., C. Castillo-Chavez, and A. Capurro (2000). A model for TB with exogenous re-infection, *Journal of Theoretical Population Biology*, **5**:235-247.

[117] Feng, Z., W. Huang and C. Castillo-Chavez (2001). On the role of variable latent periods in mathematical models for tuberculosis, *Journal of Dynamics and Differential Equations*, **13** (In press).

[118] Fenner, F. and K. Myers (1978) Myxoma virus and myxomatosis in retrospect: the first quarter century of a new disease. In *Viruses and the Environment*, J.I. Cooper and F.O. MacCallum, eds., 539-570, Academic Press, London.

[119] Fenner, F. and F.N. Ratcliffe (1965) *Myxomatosis,* Cambridge University Press, Cambridge.

[120] Fibonacci, L. (1202) *Tipographia, delle Scienze Mathematica,* Roma.

[121] Fitzhugh, R. (1961) Impulses and physiological states in theoretical models of nerve membrane, *Biophys. Journ.*, **1**:445-466.

[122] Franke, J.E. and A.-A. Yakubu (1995) Extinction in systems of bobwhite quail populations, *Can. App. Math. Quarterly,* **3**:173-201.

[123] Fredrickson, A.G. (1971) A mathematical theory of age structure in sexual populations: Random mating and monogamous marriage models, *Math. Biosci.*, **20**:117-143.

[124] Freedman, H.I., G.J. Butler, and P. Waltman P. (1986) Uniformly persistent systems, *Proc. Amer. Math. Soc.*, **96**:425-430.

[125] Freedman, H.I. and P. Waltman (1984) Persistence in models of three interacting predator-prey populations, *Math. Biosc.*, **68**:213-231.

[126] Gause, G.F. (1934a) *The Struggle for Existence,* Williams and Wilkins, Baltimore.

[127] Gause, G.F. (1934b) Experimental demonstration of Volterra's periodic oscillation in the numbers of animals, *J. Experimental Biology,* **11**:44-48.

[128] Gilpin, M.E. (1973) Do hares eat lynx?, *Amer. Naturalist,* **107**:727-730.

[129] Gladwell, M. (2000) *The Tipping Point,* Little, Brown and Co., Boston.

[130] Glass, L. and M.C. Mackey (1979) Pathological conditions resulting from instabilities in physiological control systems, *Ann. N. Y. Acad. Science,* **316**:214-235.

[131] Glass, L. and M.C. Mackey (1988) *From Clocks to Chaos: The Rhythms of Life,* Princeton University Press, Princeton, N. J.

[132] Gompertz, B. (1825) On the nature of the function expressing the law of human mortality, *Phil. Trans.*, **115**:513-585

[133] Gurney, W.S.C., S.P. Blythe, and R.M. Nisbet (1980) Nicholson's blowflies revisited, *Nature*, **287**:17-21.

[134] Gurtin, M.L. and R.C. MacCamy (1974) Nonlinear age dependent population dynamics, *Arch. Rat. Mech. Analysis*, **54** (**3**):281-300.

[135] Hadeler, K.P. (1984). Integral equations with discrete parasites: hosts with a Lotka birth law,In: *Conf. Proc. Autumn Course on Math. Ecology, Trieste, 1982*, (S.Levin and T.Hallam, eds),356-365, *Lecture Notes in Biomathematics*, **54** Springer-Verlag, Berlin-Heidelberg-New York.

[136] Hadeler, K.P. (1984) Population dynamics of killing parasites which reproduce in the host, *J. Math. Biol.*, **21**:5-55.

[137] Hadeler, K.P. (1989a) Pair formation in age-structured populations, *Acta Applicandae Mathematicae*, **14**:91-102.

[138] Hadeler, K.P. (1989b) Modeling AIDS in structured populations, *47th Session of the International Statistical Institute, Paris, August/September. Conf. Proc.*, **C1-2**:83-99.

[139] Hadeler, K.P. (1992) Periodic solutions of homogeneous equations, *J. Diff. Equations*, **95**:183-202.

[140] Hadeler, K.P. (1993), Pair formation with maturation period, *J. Math. Biol*, **32**:1-15.

[141] Hadeler, K.P., and C. Castillo-Chavez (1995) A core group model for disease dransmission, *Math Biosci.*, **128**:41-55.

[142] Hadeler, K.P. and K. Nagoma. (1990) Homogeneous models for sexually transmitted diseases, *Rocky Mountain J. Math.*, **20**:967-986.

[143] Hadeler, K.P., R. Waldstätter, and A. Wörz-Busekros (1988) Models for pair-formation in bisexual populations, *J. Math. Biol.*, **26**:635-649.

[144] Hassell, M.P. (1975) Density dependence in single-spacies populations, *J. Animal. Ecol.*, **44**:283-295.

[145] Hayes, N.D. (1950) Roots of the transcendental equation associated with a certain differential-difference equation, *J. London Math. Soc.*, **25**:226-232.

[146] Heesterbeek, H. (1992) R_0, *Thesis*, CWI, Amsterdam.

[147] Heesterbeek, J.A.P., and M. G. Roberts(1995) Threshold quantities for helminth infections, *J. Math. Biol.*, **33**:425-434.

[148] Hethcote, H.W. (1976) Qualitative analysis for communicable disease models, *Math. Biosc.*, **28**:335-356.

[149] Hethcote, H.W. (1978) An immunization model for a hetereogeneous population, *Theor. Pop. Biol.*, **14**:338-349.

[150] Hethcote, H.W. (1989) Three basic epidemiological models. In *Applied Mathematical Ecology*, S. A. Levin, T.G. Hallam and L.J. Gross, eds., Biomathematics **18**, 119-144, Springer-Verlag, Berlin-Heidelberg-New York.

[151] Hethcote, H.W., and J.A. Yorke (1984) *Gonorrhea Transmission Dynamics and Control,* Lect. Notes Biomath. **56**, Springer-Verlag, Berlin-Heidelberg-New York.

[152] Hethcote, H.W., and J.W. Van Ark. (1992) *Modeling HIV Transmission and AIDS in the United States,* Lect. Notes Biomath. **95**, Springer-Verlag, Berlin-Heidelberg-New York.

[153] Hodgkin, A.L. and A.F. Huxley (1952) A quantitative description of membrane current and its application to conduction and excitation in nerves, *J. Physiology*, **117**:500-544.

[154] Hofbauer, J. (1981) General cooperation theorem for hypercycles, *Monatsheft Math.*, **91**:223-240.

[155] Holling, C.S. (1965) The functional response of predators to prey density and its role in mimicry and population regulation, *Mem. Entomol. Soc. Canada*, **45**:1-60.

[156] Holt, R.D. (1977) Predation, apparent competition, and the structure of prey communities, *Theor. Pop. Biol.*, **12**:197-225.

[157] Hoppensteadt, F.C. (1974) An age dependent epidemic model, *J. Franklin Inst.*, **297**:325-333.

[158] Hoppensteadt, F.C. (1975) *Mathematical Theories of Populations: Demographics, Genetics and Epidemics,* SIAM, Philadelphia.

[159] Huang, W., K. Cooke, and C. Castillo-Chavez (1990) Stability and bifurcation for a multiple group model for the dynamics of HIV/AIDS transmission, *SIAM J. of Applied Math.*, **52(3)**:835-854.

[160] Hudson, P.J., and A. P. Dobson (1995) Macroparasites: Observed patterns in naturally fluctuating animal populations. In: *Ecology of Infectious Diseases in Natural Populations* (B. T. Grenfell and A. P. Dobson, eds), 144-176, Cambridge University Press.

[161] Huffaker, C.B. (1958) Experimental studies on predation: Dispersion factors and predator-prey oscillations, *Hilgardia* **27**:343-383.

[162] Hurewicz, W. (1958) *Lectures in Differential Equations,* MIT Press, Cambridge.

[163] Hsu Schmitz, S-F (1993) Some theories, estimation methods and applications of marriage functions in demography and epidemiology, *Ph. D. dissertation*, Cornell University.

[164] Hsu Schmitz, S-F and C. Castillo-Chavez (1994) Parameter estimation in non-closed social networks related to the dynamics of sexually-transmitted diseases. In *Modeling the AIDS Epidemic: Planning, Policy, and Prediction*, E.H. Kaplan and M.L. Brandeau eds., 533-559, Raven Press, New York.

[165] Hsu Schmitz, S-F and C. Castillo-Chavez, (1996) Completion of mixing matrices for nonclosed social networks. In *Proceedings of the First World Congress of Nonlinear Analysts*, V. Lakshmikantham, ed., 3163-3173, de Gruyter, Berlin.

[166] Huang, W., K.L. Cooke, and C. Castillo-Chavez (1992) Stability and bifurcation for a multiple group model for the dynamics of HIV/AIDS transmission, *SIAM J. App. Math.*, **52**:835-854.

[167] Hutchinson, G.E. (1948) Circular casual systems in ecology, *Ann. N.Y. Acad. Sci*, **50**:221-246.

[168] Hutson, V. and G.T. Vickers (1983) A criterion for permanent coexistence of species with an application to a two-prey one-predator system, *Math. Biosc.*, **63**:253-269.

[169] Ivlev, V.S. (1961) *Experimental Ecology of the Feeding of Fishes,* Yale University Press, New Haven.

[170] Jewell, N.P., K. Dietz, and V.T. Farewell. (1991) *AIDS Epidemiology: Methodological Issues,* Birkhäuser, Boston-Basel-Berlin.

[171] Kaplan, D. and L. Glass (1995) *Understanding Nonlinear Dynamics,* Springer-Verlag, Berlin-Heidelberg-New York.

[172] Kendall, D.G. (1949) Stochastic processes and population growth, *J. Roy. Stat. Soc., Ser. B* **2**:230-264.

[173] Kermack, W.O. and A.G. McKendrick (1927) A contribution to the mathematical theory of epidemics, *Proc. Royal Soc. London,* **115**:700-721.

[174] Kermack, W.O. and A.G. McKendrick (1932) Contributions to the mathematical theory of epidemics, part. II, *Proc. Roy. Soc. London,* **138**:55-83.

[175] Kermack, W.O. and A.G. McKendrick (1933) Contributions to the mathematical theory of epidemics, part. III, *Proc. Roy. Soc. London*, **141**:94-112.

[176] Keyfitz, N. (1949) The mathematics of sex and marriage, *Proceedings of the Sixth Berkeley Symposium on Mathematical Statistics and Probability*, **4**:89-108.

[177] Kierstad, H. and L.B. Slobodkin (1953) The size of water masses containing plankton blooms, *J. Mar. Res.*, **12**:141-147.

[178] Kingsland, S.E. (1985) *Modeling Nature: Episodes in the History of Population Ecology,* University of Chicago Press, Chicago.

[179] Kolmogorov, A.N. (1936) Sulla teoria di Volterra della lotta per l'esisttenza, *Giorn. Inst. Ital. Attuari*, **7**:74-80.

[180] Kurtz, T.G. (1970) Solutions of ordinary differential equations as limits of pure jump Markov processes, *J. Appl. Probab.*, **7**:49-58.

[181] Kurtz, T.G. (1971) Limit theorems for sequences of jump Markov processes approximating differential equations, *J. Appl. Prob.*, **8**:344-356.

[182] Kretzschmar, M. (1989) A renewal equation with birth-death process as a model for parasitic infections, *J. Math. Biol.*, **27**:191-221.

[183] Kretzschmar, M. (1993) Comparison of an infinite dimensional model for parasitic diseases with a related 2-dimensional system, *J. Math. Anal. Appl.*, **173**:235-260.

[184] Lajmanovich, A. and J.A. Yorke (1976) A deterministic model for gonorrhea in a nonhomogeneous population, *Math. Biosci.*, **28**:221-236.

[185] Lee, E.B. and L. Markus (1967) *Foundations of Optimal Control Theory,* Wiley, New York.

[186] Leslie, P.H. (1945) On the use of matrices in certain population mathematics, *Biometrika:* **33**,183-212.

[187] Leslie, P.H. (1948) Some further notes on the use of matrices in population mathematics, *Biometrika*, **35**:213-245.

[188] Levin, S.A. (1970) Community equilibria and stability, and an extension of the competitive exclusion principle, *Am. Naturalist*, **104**:413-423.

[189] Levin, S.A. (1983a) Co-evolution. In *Population Biology*, H.I. Freedman and C. Strobeck, eds., Lect. Notes Biomath. **52**, 328-334, Springer-Verlag, Berlin-Heidelberg-New York.

[190] Levin, S.A. (1983b) Some approaches to the modeling of coevolutionary interactions. In *Co-evolution*, M. Nitecki, ed., 21-66, University of Chicago Press, Chicago.

[191] Levin, S.A. and R. M. May (1976) A note on difference-delay equations, *Theor. Pop. Biol.*, **9**:178-187.

[192] Levin, S.A. and D. Pimentel (1981) Selection of intermediate rates of increase in parasite-host systems, *Amer. Naturalist*, **117**:308-315.

[193] Lewis, E.G. (1942) On the generation and growth of a population, *Sankhyã*, **6**:93-96.

[194] Li, T.Y. and J.A. Yorke (1975) Period three implies chaos, *Amer. Math. Monthly*, **82**:985-992.

[195] Londen, S-O. (1974) On the asymptotic behavior of the bounded solutions of a nonlinear Volterra equation, *S.I.A.M. J. Math. Anal.*, **5**:849-875.

[196] Lorenz, E.N. (1963) Deterministic non-periodic flow, *J. Atmos. Science*, **20**:130-141.

[197] Lotka, A.J. (1922) The stability of the normal age distribution, *Proc. Nat. Acad. Sci.*, **8**:339-345.

[198] Lotka, A.J. (1925) *Elements of Physical Biology,* Williams & Wilkins, Baltimore; Reissued as Elements of Mathematical Biology, Dover, New York (1956).

[199] Lotka, A.J. and F.R. Sharpe (1923) Contributions to the analysis of malaria epidemiology, *Am. J. Hygiene*, **3**, Supplement: 1-21.

[200] Lubkin, S. and C. Castillo-Chavez (1996) A pair formation approach to modeling inheritance of social traits. In *Proceedings of the First World Congress of Nonlinear Analysts, Tampa, Florida*, V. Lakshmikantham, ed., 3227-3234, de Gruyter, Berlin.

[201] Ludwig,D., D.D. Jones, and C.S. Holling (1978) Qualitative analysis of insect outbreak systems: The spruce budworm and forest, *J. Animal Ecol.*, **47**:315-332.

[202] Luo, X. and C. Castillo-Chavez, (1993) Limit behavior of pair-formation for a large dissolution rate, *J. Math. Systems, Estimation, and Control*, **3**:247-264.

[203] McFarland, D. (1972) Comparison of alternative marriage models. In *Population Dynamics*, T.N.E. Greville, ed., 89-106, Academic Press, New York-London.

[204] McKendrick, A.G. (1926) Applications of mathematics to medical problems, *Proc. Edinburgh Math. Soc.*, **40**:98-130.

[205] MacLulich, D.A. (1937) *Numbers of the Varying Hare,* Univ. of Toronto Studies, Biological Series, No. **43**, University of Toronto Press, Toronto.

[206] McNeill, W.H. (1976) *Plagues and Peoples,* Doubleday, New York.

[207] McNeill, W.H. (1992) *The Global Condition,* Princeton University Press, Princeton, N.J.

[208] Mackey, M.C. and L. Glass (1977) Oscillation and chaos in physiological control systems, *Science*, **197**:287-289.

[209] Malthus, T.R. (1798) *An Essay on the Principle of Population*, 1a Ed, J Johnson in St Paul's Churchyard, London.

[210] May, R.M. (1974) *Stability and Complexity in Model Ecosystems,* Princeton University Press, Princeton, N.J.

[211] May, R.M. (1976) Simple mathematical models with very complicated dynamics, *Nature*, **261**:459-467.

[212] May R. (1981) *Theoretical ecology, principles and applications*, Second Edition, Sinauer Associates, Massachusetts.

[213] May, R.M. (1986) Population dynamics of microparasitic infections. In *Mathematical Ecology; An Introduction*, T.G. Hallam and S.A. Levin, eds., Biomathematics **17**, 405-442, Springer-Verlag, Berlin-Heidelberg-New York.

[214] May, R.M., and R.M. Anderson (1978) Regulation and stability of host-parasite population interactions. II. Destabilizing processes, *J. Animal Ecology,* **47**:249-267.

[215] May, R.M. and R.M. Anderson (1983) Epidemiology and genetics in the co-evolution of parasites and hosts, *Phil. Trans. Royal Soc. London, Series B*, **219**:281-313.

[216] May, R.M. and W.J. Leonard (1975) Nonlinear aspects of competition between species, *SIAM J. App. Math.*, **29**:243-253.

[217] Maynard Smith, J. (1974) *Models in Ecology,* Cambridge University Press, Cambridge.

[218] Metz, J.A.J and O. Diekmann (1986) *The Dynamics of Physiologically Structured Populations,* Lect. Notes Biomath., **68**, Springer-Verlag, Berlin-Heidelberg-New York.

[219] Milton, J.G. and J. Bélair (1990) Chaos, noise and extinction in models of population growth, *Theor. Pop. Biol.*, **37**:273-290.

[220] Mollison D.(ed.) (1995) *Epidemic Models: their structure and relation to data,* Cambridge University Press.

[221] Monod, J. (1950) La technique de cultive continue: Théorie et applications, *Ann. Inst. Pasteur*, **79**:390-410.

[222] Murray, J.D. (1989) *Mathematical Biology,* Biomathematics **19**, Springer-Verlag, Berlin-Heidelberg-New York.

[223] Nagumo, J.S., S. Arimoto, and S. Yoshizawa (1962) An active pulse transmission line simulating nerve axon, *Proc. Inst. Radio Engineers*, **50**:2061-2071.

[224] Nicholson, A.J. (1954) An outline of the dynamics of animal populations, *Australian. J. Zoology,* **3**:9-65.

[225] Nicholson, A.J. and V. A. Bailey (1935) The balance of animal populations, *Proc. Zool. Soc. London,* **3**:551-598.

[226] Nisbet, R.M. and W.S.C. Gurney (1982) *Modeling Fluctuating Populations,* Wiley-Interscience, Chichester.

[227] Nisbet, R.M. and W.S.C. Gurney (1983) The systematic formulation of population models for insects with dynamically varying instar duration, *Theor. Pop. Biol.*, **23**:114-135.

[228] Nisbet, R.M. and W.S.C. Gurney (1986) The formulation of age structure models. In *Mathematical Ecology; An Introduction*, T.G. Hallam and S.A. Levin, eds., Biomathematics **17**, 95-115, Springer-Verlag, Berlin-Heidelberg-New York.

[229] Nold, A. (1980) Heterogeneity in diseases-transmission modeling, *Math. Biosc.*, **52**:227-240.

[230] Novick, A. and L. Szilard (1950) Experiments with the chemostat on spontaneous mutation of bacteria, *Proc. Nat. Acad. Sci. USA*, **36**:708-719.

[231] Oaten, A. and W.W. Murdoch (1975) Switching, functional response and stability in predator-prey systems, *Amer. Naturalist,* **109**:299-318.

[232] Odum, E.P. (1953) *Fundamentals of Ecology,* Saunders, Philadelphia.

[233] Oster, G. (1978) The dynamics of nonlinear models with age structure. In *Studies in Mathematical Biology,* S.A. Levin, ed., Vol **16**, 411-438, Math. Assoc. of America.

[234] Paine, R.T. (1966) Food web complexity and species diversity, *Amer. Naturalist*, **100**:65-75.

[235] Parlett, P. (1972) Can there be a marriage function? In *Population Dynamics*, T.N.E. Greville, ed., 107-135, Academic Press, New York-London.

[236] Park, T. (1948) Experimental studies of interspecies competition: I. Competition between populations of the flour beetles, tribolium confusum Duvall and tribolium castaneum Herbst, *Ecological Monographs*, **18**:267-307.

[237] Pearl, R. and L.J. Reed (1920) On the rate of growth of the population of the United States since 1790 and its mathematical representation, *Proc. Nat. Acad. Sci.*, **6**:275-288.

[238] Poincaré, H. (1881) Sur les courbes definies par des equation differentielles, *J. Math. Pures Appl.*, Vol. **1**(1881):375-427, Vol. **8**(1882):251-286, Vol. **11**(1885):162-244, Vol. **12**(1886):151-217.

[239] Pollard, J.H. (1973) *Mathematical Models for the Growth of Human Populations,* Cambridge University Press, Cambridge.

[240] Pontryagin, L.S., V.S. Boltyanskii, R.V. Gamkrelidze, and E.F. Mischenko (1962) *The Mathematical Theory of Optimal Processes,* Wiley-Interscience, New York. (translated from the Russian).

[241] Raggett, G.F. (1982) Modeling the Eyam plague, *IMA Journal*, **18**:221-226.

[242] Real, L.A. and J. H. Brown (1991) *Foundations of Ecology: Clasic Papers with Commentaries,* University of Chicago Press, Chicago.

[243] Renshaw, E. (1991) *Modeling Biological Populations in Space and Time,* Cambridge University Press, Cambridge.

[244] Ricker, W.E. (1954) Stock and recruitment, *J. Fisheries Res. Board Canada,* **11**:559–623.

[245] Ricker, W.E. (1958) *Handbook of Computations for Biological Statistics of Fish Populations,* Bull. Fisheries Res. Board Canada, 119.

[246] Rivera-Salguero, B. (2000) Scramble versus contest competition in a two patch system. *Department of Biometrics, Cornell University, Technical Report Series*, BU-1515-M.

[247] Rosenzweig, M.L. (1969) Why the prey curve has a hump, *Amer. Naturalist*, **103**:81–87.

[248] Rosenzweig, M.L. (1971) Paradox of enrichment: Destabilization of exploitation ecosystems in ecological time, *Science*, **171**:385–387.

[249] Rosenzweig, M.L. and R.H. MacArthur (1963) Graphical representation and stability conditions of predator-prey interactions, *Amer. Naturalist*, **97**:209-223.

[250] Ross, R. (1911) *The Prevention of Malaria*, 2nd ed., (with Addendum), John Murray, London.

[251] Saha, P. and S. Strogatz (1995) The birth of period 3, *Math. Mag.*, **68**:42-47.

[252] Saints, K. (1987) Discrete and continuous models of age-structured population dynamics, *Senior Research Report*, Harvey Mudd College, Claremont, Cal.

[253] Samuelson, P.A. (1941) Conditions that a root of a polynomial be less than unity in absolute value, *Ann. Math. Stat.*, **12**:360-364.

[254] Sánchez B.N., P.A. Gonzalez, R.A. Saenz (2000) The influence of dispersal between two patches on the dynamics of a disease, *Department of Biometrics, Cornell University, Technical Report Series*, BU-1531-M.

[255] Sánchez, D.A. (1978) Linear age-dependent population growth with harvesting, *Bull. Math. Biol.*, **40**:377-385.

[256] Sánchez, D.A. (1968) *Ordinary Differential Equations and Stability Theory: An Introduction,* W. H. Freeman, San Francisco; reprinted by Dover, New York (1979).

[257] Sandefur, J. (1990) *Discrete Dynamical Systems: Theory and Applications,* Oxford University Press, Oxford.

[258] Schoener, T.W. (1973) Population growth regulated by intraspecific competition for energy or time: Some simple representations, *Theor. Pop. Biol.*, **4**:56-84.

[259] Sharpe, F.R. and A.J. Lotka (1911) A problem in age distribution, *Phil. Magazine*, **21**:435-438.

[260] Shulgin, B., L. Stone, and Z. Agur (1998) Pulse vaccination strategy in the SIR epidemic model, *Bull. Math. Biol.*, **60**:1123-1148.

[261] Skellam, J.G. (1951) Random dispersal in theoretical populations, *Biometrika*, **38**:196-218.

[262] Smith, F.E. (1963) Population dynamics in Daphnia magna and a new model for population growth, *Ecology*, **44**:651-663.

[263] Smith, H.L. (1988) Systems of ordinary differential equations which generate an order preserving flow: A survey of results, *SIAM Rev.*, **30**:87-113.

[264] Smith, H.L. and P. Waltman (1996) *The Theory of the Chemostat: Dynamics of Microbial Competition,* Cambridge Studies in Mathematical Biology **13**, Cambridge University Press, Cambridge.

[265] Soberon, J.M. and C.M. Del Rio (1981) The dynamics of a plant-pollinator interaction, *J. Theor. Biol.*, **91**:363-378.

[266] Soper, H.E. (1929) Interpretation of periodicity in disease prevalence, *J. Roy. Stat. Soc., Series B*, **92**:34-73.

[267] Stone, L., B. Shulgin, Z. Agur (2000) Theoretical examination of of the pulse vaccination in the SIR epidemic model, *Math. and Computer Modeling*, **31**:207-215.

[268] Strogatz, S. (1994) *Nonlinear Dynamics and Chaos*, Addison-Wesley, Reading, Mass.

[269] Thieme, H.R. (1994) Asymptotically autonomous differential equations in the plane, *Rocky Mountain J. Math.*, **24**;351-380.

[270] Thieme, H.R. and C. Castillo-Chavez (1989) On the role of variable infectivity in the dynamics of the human immunodeficiency virus. In *Mathematical and statistical approaches to AIDS epidemiology*, C. Castillo-Chavez, ed., Lect. Notes Biomath. **83**, 200-217, Springer-Verlag, Berlin-Heidelberg-New York.

[271] Thieme, H.R. and C. Castillo-Chavez (1993) How may infection-age dependent infectivity affect the dynamics of HIV/AIDS?, *SIAM J. Appl. Math.*, **53**:1447-1479.

[272] U. S. Bureau of the Census (1975) *Historical statistics of the United States: colonial times to 1970*, Washington, D. C. Government Printing Office.

[273] U.S. Bureau of the Census (1980) *Statistical Abstracts of the United States*, 101st edition.

[274] U.S. Bureau of the Census (1991) *Statistical Abstracts of the United States,* 111th edition.

[275] U.S. Bureau of the Census (1999) *Statistical Abstracts of the United States,* 119th edition.

[276] Velazquez, J.P. (2000) SIS nonlinear discrete time models with two competing strains. *Department of Biometrics, Cornell University, Technical Report Series,* BU-1514-M.

[277] Verhulst, P.F. (1838) Notice sur la loi que la population suit dans son accroissement, *Corr. Math. et Phys.*, **10**:113-121.

[278] Verhulst, P.F. (1845) Recherches mathématiques sur la loi d'accroissement de la population, *Mém Acad Roy, Brussels,* **18**:1-38.

[279] Verhulst, P.F. (1847) Recherches mathématiques sur la loi d'accroissement de la population, *Mém Acad Roy, Brussels,* **20**.

[280] Volterra, V. (1926) Variazioni e fluttazioni del numero d'individui in specie animali conviventi, *Mem. Acad. Sci. Lincei,* **2**:31-13.

[281] Volterra, V. (1931) *Leçons sur la Théorie Mathématique de la Lutte pour la Vie,* Gauthier-Villars, Paris.

[282] Waldstätter, R. (1989) Pair formation in sexually transmitted diseases. In *Mathematical and Statistical Approaches to AIDS Epidemiology,* C. Castillo-Chavez, ed., Lect. Notes Biomath. **83**, 260-274, Springer-Verlag, Berlin-Heidelberg-New York.

[283] Waldstätter, R. (1990) Models for pair formation with applications to demography and epidemiology, Ph. D. thesis, University of Tübingen.

[284] Waltman, P. (1986) *A Second Course in Ordinary Differential Equations,* Academic Press, Orlando, Fla.

[285] Webb, G.F. (1985) *Theory of Nonlinear Age-Dependent Population Dynamics,* Marcel Dekker, New York.

[286] Yodzis, P. (1989) *Introduction to Theoretical Ecology,* Harper and Row, New York.

Index

Texts in Applied Mathematics

(continued from page ii)

31. *Brémaud:* Markov Chains: Gibbs Fields, Monte Carlo Simulation, and Queues.
32. *Durran:* Numerical Methods for Wave Equations in Geophysical Fluid Dynamics.
33. *Thomas:* Numerical Partial Differential Equations: Conservation Laws and Elliptic Equations.
34. *Chicone:* Ordinary Differential Equations with Applications.
35. *Kevorkian:* Partial Differential Equations: Analytical Solution Techniques, 2nd ed.
36. *Dullerud/Paganini:* A Course in Robust Control Theory: A Convex Approach.
37. *Quarteroni/Sacco/Saleri:* Numerical Mathematics.
38. *Gallier:* Geometric Methods and Applications: For Computer Science and Engineering.
39. *Atkinson/Han:* Theoretical Numerical Analysis: A Functional Analysis Framework.
40. *Brauer/Castillo-Chávez:* Mathematical Models in Population Biology and Epidemiology.
41. *Davies:* Integral Transforms and Their Applications, 3rd ed.
42. *Deuflhard/Bornemann:* Scientific Computing with Ordinary Differential Equations.
43. *Deuflhard/Hohmann:* Numerical Analysis in Modern Scientific Computing: An Introduction, 2nd ed.
44. *Knabner/Angermann:* Numerical Methods for Elliptic and Parabolic Partial Differential Equations.
45. *Larsson/Thomée:* Partial Differential Equations with Numerical Methods.
46. *Pedregal:* Introduction to Optimization.